The Eponym Dictionary of Amphibians

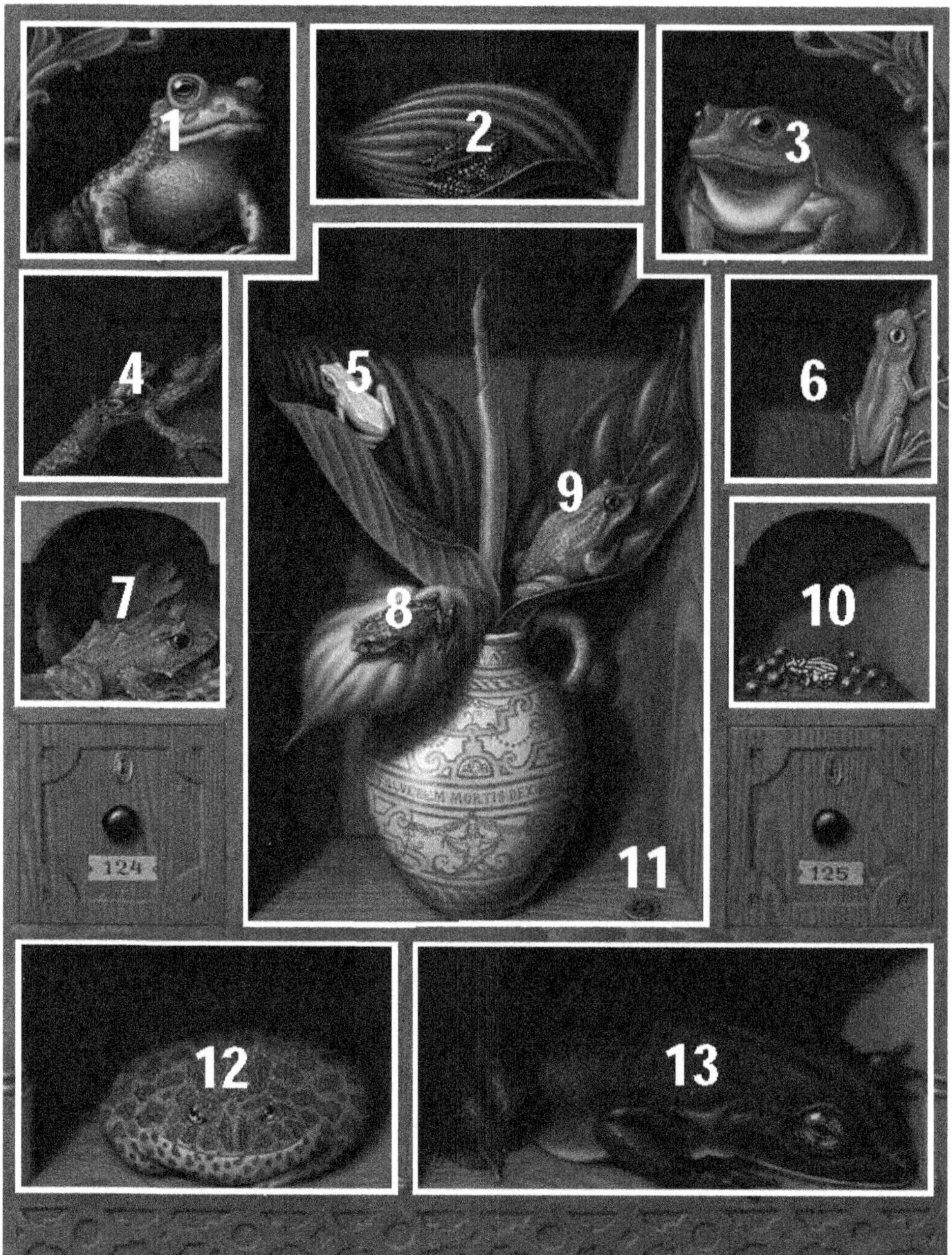

1. Green Toad (*Bufo viridis*). Widespread throughout Continental Europe and Middle East. **2.** Siberut Island Frog (*Rana siberu*). Indonesia. **3.** Tomato Frog (*Dyscophus antongilii*). Madagascar. **4.** Dyeing Poison Arrow Frog (*Dendrobates tinctorius* [formerly *Dendrobates azureus*]). Brazil, French Guiana, Guyana, Suriname. **5.** Panamanian Golden Frog (*Atelopus zeteki*). Panama. **6.** Reinwardt's Flying Frog (*Rhacophorus reinwardtii*). Indonesia, Malaysia. **7.** Eyelash Leaf Frog (*Ceratobatrachus guentheri)* Papua New Guinea, Solomon Islands. **8.** Golden Toad (*Bufo periglenes*). Costa Rica. **9.** Northern Gastric Brooding Frog (*Rheobatrachus vitellinus*). Australia. **10.** The Southern Corroboree Frog (*Pseudophryne corroboree*). Australia. **11.** *Nyctibatrachus minimus*. India. **12.** Ornate Horned Frog (*Ceratophrys ornata*). Argentina, Brazil, Uruguay. **13.** Goliath Frog (*Conraua goliath*). Cameroon, Equatorial Guinea.

The closed drawers in the painting represent all the frog species not yet discovered, and those who might die out before becoming known to us.

The Eponym Dictionary of Amphibians

Bo Beolens | Michael Watkins | Michael Grayson

Published by **Pelagic Publishing**
www.pelagicpublishing.com
PO Box 725, Exeter, EX1 9QU

The Eponym Dictionary of Amphibians

ISBN 978-1-907807-41-1 (Hbk)
ISBN 978-1-907807-42-8 (ePub)
ISBN 978-1-907807-44-2 (Pdf)
ISBN 978-1-907807-43-5 (Mobi)

British Library Cataloguing in Publication Data
A catalogue record for this book is available from the British Library.

Cover image *Frog Cabinet* © Madeline von Foerster
www.madelinevonfoerster.com

For Julia Summers & Andy Bryson

Bo Beolens

For Sebastian, Madeleine and Christie

Mike Watkins

CONTENTS

Acknowledgements *ix*
Introduction *xi*

The Eponym Dictionary of Amphibians

A 1
B 15
C 36
D 50
E 62
F 68
G 75
H 88
I 101
J 103
K 109
L 118
M 132
N 152
O 157
P 161
Q 174
R 175
S 187
T 211
U 219
V 220
W 226
X 237
Y 238
Z 240

Bibliography 243

ACKNOWLEDGEMENTS

Many people have contributed useful bits of information and most of them are people who have an animal named after them and so appear in the book. However a number of other people have been very generous with their time and help and we wish to acknowledge them separately:

Alan Channing, University of the Western Cape, Belville, South Africa

Dr Frank Glaw, Zoologische Staatssammlung, München

Florence Greffe, Conservateur des Archives, Académie des Sciences-Institut de France

Professor Dr Ismail Karaboz, Head of Biology Department, Science Faculty, Ege University, Turkey

Dr Edgar Lehr

Professor Dr Annemarie Ohler, Curator of Reptiles and Amphibians, and Scientific Manager of Vertebrate Collections, Muséum National d'Histoire Naturelle, Paris

Thiago Ribero de Carvalho Tavares, Universidade de São Paulo, Brazil

INTRODUCTION

Who Is It For?

Vernacular names of animals often contain a person's name (such names are called 'eponyms'). Furthermore, many scientific names contain the Latinized name of its discoverer, or some other person thought worthy of the honour, often in the binomial for a species or in the trinomial for a sub-species. Some genera names are also eponyms. So this book is for the amateur herpetologist, the student of zoology or anyone else interested in taxonomy, nomenclature or amphibians.

How to Use This Book

We use a number of abbreviations to describe taxonomic status etc.

Alt	=	Alternative
DNF	=	Dates not found
JrSYN	=	Junior Synonym of
NCR	=	Not currently recognised
NRM	=	Now regarded as monotypic
NPORB	=	Name preoccupied: replaced by
SiN	=	Sometimes included in
Syn	=	Synonymous with
UIi	=	Now usually included in

In the text the following abbreviations are widely used:

AMNH	=	American Museum of Natural History, New York
AOU	=	American Ornithologists' Union
ANWC	=	Australian National Wildlife Collection
BOC	=	British Ornithologists Club
BOU	=	British Ornithological Union
BMNH	=	British Museum (Natural History)
BTO	=	British Trust for Ornithology
CSIRO	=	Commonwealth Scientific and Industrial Research Organisation
MNHN	=	Muséum National d'Histoire Naturelle, Paris
RAOU	=	Royal Australasian Ornithological Union
RSPB	=	Royal Society for the Protection of Birds
Smithsonian	=	Smithsonian Institution
UN	=	United Nations

This book is arranged alphabetically by the names of the people after whom amphibians have been named. Generally, the easiest way to find your animal is to look it up under the name of the person that is apparently embedded in the animal's common or scientific name. We say 'apparently', as things are rarely as simple as they seem: in some names, for example, the apostrophe implying ownership is a transcription error; other names may have been named after places. So, we have included any names where we think confusion might arise, but we do not promise to be completely comprehensive in that respect. You should also beware of spelling. Surf the net and you may well find animals' names spelt in a number of different ways – that greatest resource is also full of inaccuracies and misinformation. We have tried to include entries on those alternatives that we have come across.

Each biography follows a standard format: first, you will find the name of the person honoured; next, there follows a list of amphibians named after that person, arranged in order of the year in which they were described. (This list gives common names, scientific names, names of the people who first described each species and the date of the original descriptions – in that sequence). Alternative English names follow in parentheses and are each preceded by the abbreviation Alt.; different scientific names (where taxonomists disagree) are preceded by the abbreviation Syn. (synonym); finally, there is a brief biography of that individual.

To assist you in your search, we have cross-referenced the entries by highlighting (in bold) the names of those describers who also appear elsewhere in the book. Some amphibians are named in different ways after the same person and we have also tried to marry these up using cross-references.

A person's fame does not get them a correspondingly long entry – in fact, often the opposite. Very famous people, like Charles Darwin, have fairly brief write-ups. He is so well known, and so much has been written about him, that it is sufficient for us merely to indicate that he is commemorated.

Sometimes amphibians are named in the vernacular after the finder, the person who wrote the description or some other person of the latter's choice. When more than one person has thought a species new, the amphibians may get more than one set of names, so it can warrant an entry in several places!

There are a very great number of recent namings of fossil animals. As the rate at which fossil remains are discovered and described seems to be increasing exponentially and disagreement among palaeontologists epidemic, we decided that we would ignore any prehistoric extinctions; in simplistic terms, any occurring before Columbus discovered America.

In the last 100 years many cities and countries have changed their names. We have normally used the name by which the subject would have known it, putting in brackets the name by which it is now known, e.g. Salisbury (Harare) Rhodesia (Zimbabwe).

What's in a Name?

Tracking down the provenance of eponymous amphibian names, and finding out about the individuals responsible for them, proved to be fraught with difficulties. The final count is 2,668 amphibians, including a few where the same animal has been named after two people. The names honour 1,609 individual people, but there are also 128 which sound like people's names but in fact are not, plus 83 miscellaneous, relating to indigenous peoples, fictional characters, conservation groups, guerrilla armies, chartered accountants and biblical and mythological references. Included in the 1,609 are entries for 11 names of people whom we have been completely unable to identify!

Describers and Namers

New species are first brought to the attention of the scientific community in a formal, published description of a type specimen, essentially a dead example of the species, which will eventually be lodged in a scientific collection. The person who describes the species will give it its scientific name, usually in Latin but sometimes in Latinized ancient Greek. Sometimes the 'new' animal is later reclassified and then the scientific name may be changed. This frequently applies to generic names (the first part of a binomial name), but specific scientific names (the second part of a binomial), once proposed, usually cannot be amended or replaced – there are precise and complicated rules governing any such changes.

The scientific names used in this book are largely those used in the AmphibiaWeb Database of Berkeley, California. We may have missed a few recently published taxonomic changes (although we have tried to be comprehensive up until the final proof of the book), but we have put the name of the original describer after every entry. Because alterations to taxonomy have been so radical, and so swiftly changing, we decided never to put brackets round changed entries.

Although we have used current scientific names as far as possible, these are not always as universal as the casual observer might suppose. There is no 'world authority' on such matters.

There are no agreed conventions for English names and indeed the choice of vernacular names is often controversial. Often the person who coined the scientific name will also have given it a vernacular name, which may not be an English name if the describer was not an English-speaker. On the other hand, vernacular names have often been added afterwards, frequently by people other than the describers. In this book, therefore, when we refer to an animal having been NAMED by someone, we mean that that person gave it the ENGLISH name in question. We refer to someone as a DESCRIBER when they were responsible for the original description of the species and hence for its scientific name. As we have said above, it is the describer's name which is given after the scientific name in the biographies.

BIOPAT is a German society that encourages people, organizations and companies to donate money for taxonomic research. They are rewarded by having a species named after themselves or someone they nominate. The effect is that the describers often have little or no knowledge of the people after whom they name species. There are a number of such examples in this book; a very clear description of the process was given to us by Dr Edgar Lehr in his explanation of why a species of Andes Frog was named *Phrynopus horstpauli*.

Amphibians are a class of animal where the rate of discovery of new species does not seem to abate, unlike the discovery rates of mammals and birds. As a result, the reader may find that people associated with more recently named species and sub-species are not included in this book.

Animals Named after More Than One Person

Throughout the text you may come across several different names for the same species. In some cases these names are honorifics; for example, Chari's Bush Frog *Philautus charius* Rao, 1937 is the same as Seshachar's Bush Frog. This peculiarity has sometimes come about through simple mistakes or misunderstandings – such as believing juveniles or females to be a different species from the adult male. In some cases, the same animal was found at about the same time in two different places; only later has it emerged that this is the same animal named twice. Some of these duplications persist even today, with the same amphibian being called something different in different places or by different people. However, sometimes the apparent duplication is due to the fact that a different name may be used for different subspecies to distinguish them from the nominal form.

Male or Female

In some cases we know that an animal is named after a man, even though its scientific name is in the feminine. This mostly occurs when a name ends in the letter 'a'. Presumably, the reason for this is that most singular Latin nouns ending in 'a' are feminine – for example, 'mensa' means 'table' (nothing very feminine about that), and the possessive/genitive case is 'mensae', not 'mensai'. There are a number of masculine Latin nouns, e.g. 'agricola' meaning 'farmer', but they are declined as though they are feminine, the convention being that the feminine form is adopted in such cases. This convention has been falling into disuse in recent years. It is quite striking how many modern namings ignore it.

Red Herrings

Further confusion arises from a number of animals which appear to be named after people, but – upon closer examination – turn out to be named after a place that was itself named after a person. We have included these with an appropriate note, as other sources of reference will not necessarily help the enquirer.

Dubious Names

There are a number of vernacular and scientific names which are dubious and have, because of their origin, proved to be impossible to identify or amplify, or are regarded by most authorities as just plain wrong. These are all omitted from this book but we have fuller particulars of them, should any of them ever become recognized by the scientific community.

The Eponym Dictionary of Amphibians

A

Abbott

Abbott's Robber Frog *Eleutherodactylus abbotti* **Cochran**, 1923

Lowland Litter Frog *Leptobrachium abbotti* Cochran, 1926

William Louis Abbott (1860–1936) was a student, naturalist and collector. Initially qualifying as a physician at the University of Pennsylvania and then working as a surgeon at Guy's Hospital in London, he decided not to pursue medicine but use his private wealth to engage in scientific exploration. As a student (1880) he had collected in Cuba and Hispaniola (1883) in the company of Joseph Krider, son of the taxidermist John Krider. He went to East Africa (1887) spending two years there. He studied the wildlife of the Indo-Malayan region (1891), using his Singapore-based ship 'Terrapin', and made large collections of mammals from Southeast Asia for the Smithsonian. He switched to Siam (1897) and spent 10 years exploring and collecting in and around the China Sea. He provided much of the Kenya material in the Smithsonian, and was the author of *Ethnological Collections in the United States National Museum from Kilima-Njaro, East Africa* in their report (1890/91). He returned to Hispaniola (1917) exploring the interior and discovering yet more new species. He retired to Maryland but continued his life-long study of birds until his death. Eighteen birds, two mammals and two reptiles are named after him.

Abdulali

Abdulali's Wrinkled Frog *Nyctibatrachus humayuni* Bhaduri & Kripalani, 1955

[Alt. Humayun's Wrinkled Frog]

(See **Humayun**)

Abe, A

Beaked Toad sp. *Rhinella abei* Baldissera, **Caramaschi** & **Haddad**, 2004

Dr Augusto Shinya Abe is Professor of Zoology at the Universidade Estadual Paulista, Brazil. He co-wrote *Oxygen uptake in snakes: Is there a reduction in fossorial species?* (1994).

Abe, Y

Abe's Salamander *Hynobius abei* **Sato**, 1934

[Alt. Sato's Oriental Salamander]

Yoshio Abe (1883–1945) was Professor of Zoology, Karahuto Normal University, Japan. He wrote about kinorhynchs (microscopic marine invertebrates) (1930) and at least one was named after him.

Abramalaga

Strabomantid frog sp. *Bryophryne abramalagae* Lehy & Catenazzi, 2010

Abra Málaga is a 4,313-metre-high mountain pass in the Peruvian Andes.

Achaval

Beaked Toad sp. *Rhinella achavali* Maneyro, Arrieta & de Sá, 2004

Federico Achaval (b.1941) is a herpetologist who was (1975–1983) at the Museo Nacional de Historia Natural, Montevideo, Uruguay, where he is now a Research Associate. He taught at the Universidad de la República Oriental del Úruguay (1973–2004), where he was a Professor in the Department of Zoology. He co-wrote *Anfibios y reptiles del Úruguay* (1997).

Achuar

Robber Frog sp. *Pristimantis achuar* Elmer & **Cannatella**, 2008

The Achuar are an indigenous people of the upper Amazon basin, Ecuador.

Adamantina

Nibo Land Frog *Austrochaperina adamantina* **Zweifel**, 2000

Dr Jared Mason Diamond (b.1937) is an American physiologist who is a Professor at the Medical School of University College Los Angeles, California. He has also been a Research Associate in Ornithology and Mammalogy at the Los Angeles County Museum of Natural History (1985). He is interested in nutrition and ornithology. He has published extensively, including detailed studies of the avifauna of Papua New Guinea and the Philippines. Dr Diamond's studies on bird diversity in New Guinea's tropical rainforests have made fundamental contributions to our understanding of species' coexistence, altitudinal segregation of montane species, speciation in rainforest environments and bowerbird evolution. He won the Tyler Prize for Environmental Achievement for pioneering work in conservation biology (2001), has a MacArthur Foundation award, and won the Pulitzer Prize (1998) for his book *Guns, Germs and Steel*. The Latin word *adamantina* meaning 'like a diamond', may seem an unlikely eponym, but Zweifel explained that it 'refers to Jared Diamond, the collector of this and a great many other valuable herpetological specimens from Papua New Guinea.' A mammal and two birds are named after him.

Adelbert

Adelbert Rainforest Frog *Albericus brunhildae* **Menzies**, 1999

The Adelbert Range is a range of mountains in Papua New Guinea. These were themselves named after German botanist Adelbert von Chamisso (1781–1838).

Adelphos

Poison Frog genus *Adelphobates* **T Grant** et al., 2006

Dr Charles William Myers (b.1936) and Dr John W. Daly (1933–2008) are unrelated scientists directly involved with studies of poison frogs, who worked together and co-wrote *Preliminary evaluation of skin toxins and vocalizations in taxonomic and evolutionary studies of poison-dart frogs* (Dendrobatidae) (1976). Daly was a physician and biochemist at the US National Institutes of Health, where he worked (1960–2008). Oregon State College awarded his bachelor's degree (1954) and master's (1954), and Stanford his doctorate (1958). However, the binomial is from the Ancient Greek words *adelphos* meaning 'brother' or 'twin' and *bates* meaning 'walker' or 'climber'. (See **Myers**)

Adler

Adler's Mottled Treefrog *Plectrohyla thorectes* Adler, 1965

Dr Kraig K. Adler (b.1940) is Professor of Biology at Cornell University. He was co-founder and Chairman of the Ohio Herpetological Society, which became the Society for the Study of Amphibians and Reptiles. He co-edited *The New Encyclopedia of Reptiles and Amphibians* (2002). His major research interest is the sensory basis of long-distance orientation and navigation in reptiles and amphibians. Five reptiles are named after him.

Adametz

Bamenda Reed Frog *Hyperolius ademetzi** **Ahl**, 1931

Alexander Friedrich Adametz (b.1882) was a captain in the German Colonial Army in Cameroon where he collected (1909), and was still resident there on the outbreak of the First World War (1914).

* Ahl mis-spelt his name as ademetz, but the original spelling must stand.

Adolf Friedrich

Adolf Friedrich's Squeaker Frog *Arthroleptis adolfifriederici* **Nieden**, 1911

[Alt. Adolf's Squeaker, Rugegewald Screeching Frog]

Duke Adolf Friedrich Albrecht Heinrich von Mecklenburg (1873–1969) was an explorer and colonial politician in Africa and (1949–1951) first President of the German Olympic Committee. He conducted scientific research on the African Rift Valley and crossed Africa from East to West (1907–1908). He led another expedition to Lake Chad and eastwards to the upper reaches of the Congo River (1910–1911). He was the last Governor of the German colony of Togoland (1912–1914). After the First World War he became Vice President, German Colonial Society for Southwest Africa. Eleven birds, a mammal and two reptiles are named after him.

Aecii

Monte Duida Treefrog *Tepuihyla aecii* **Ayarzagüena, Señaris** & **Gorzula,** 1993

Strictly speaking the scientific name is an acronym rather than an eponym. The Agencia Española de Cooperación Internacional is an organisation that, among its activities, offers help and partnership to conservationists in many countries.

Aglave

Anamalozoatra Madagascar Frog *Spinomantis aglavei* Methuen & **Hewitt**, 1913

[Syn. *Mantidactylus aglavei*]

Monsieur Aglave was the administrator of the province of Andevoranto in Madagascar.

Aguilar

Treefrog sp. *Hypsiboas aguilari* **Lehr, Faivovich** & Jungfer, 2010

César Augusto Aguilar Puntriano (b.1971) is a Peruvian herpetologist at the Departamento de Herpetología, Museo de Historia Natural, Universidad Nacional Mayor de San Marcos, Lima, where he is a friend and colleague of the frog's describers.

Aguirre

Linhares Dwarf Frog *Physalaemus aguirrei* **Bokermann**, 1966

Dr Alvaro Coutinho Aguirre (1899–1987) was a Brazilian zoologist, entomologist and ornithologist who worked at the Ministry of Agriculture, Rio de Janeiro. He was a member of the Brazilian Academy of Sciences. He co-wrote *Contribuição para o conhecimento da alimentação das aves brasileiras* (1965). He made a collection of amphibians a decade before Bokermann in the same locality.

Ahl

Ahl's Toad *Duttaphrynus himalayanus* **Günther**, 1864
[*Bufo abatus* (Ahl, 1925) is a synonym]
Ahl's River Frog *Phrynobatrachus latifrons* Ahl, 1924
Ahl's Screeching Frog *Phrynobatrachus brevipalmatus* Ahl, 1925
Reed Frog sp. *Hyperolius ahli* **Loveridge**, 1936 NCR
[JrSYN *Hyperolius argus*]
Ahl's Squeaker *Arthroleptis affinis* Ahl, 1939
Ahl's Painted Reed Frog *Hyperolius pictus* Ahl, 1931
[Alt. Variable Reed Frog]
Ahl's Sedge Frog *Hyperolius marginatus argentovittis* Ahl, 1931
[Alt. Silver-striped Sedgefrog]

Dr Christoph Gustav Ernst Ahl (1898–1945) was an ichthyologist, herpetologist and aquarist. He served in the artillery in the First World War (1916). He studied natural science at Humboldt University, Berlin (1919–1921), where he was awarded his doctorate. He was at the Department of Ichthyology and Herpetology, Zoological Museum (1921–1941), becoming Curator of Herpetology (1923) and later Director. He was Editor-in-Chief of the magazine *Das Aquarium* (1927–1934). Having joined the Nazi party (1930s) to keep his job, he was expelled for indiscipline (1939). He was sacked by the Museum (1941), probably as his scientific work was 'superficial and careless and his knowledge of the literature poor', rather than because he had been recalled to the Wehrmacht (1939). He fought in Poland and North Africa and was reported as missing in action in Herzegovina (1945). He wrote 170 papers on fishes and amphibians, but many of the names he coined are no longer considered valid. Two reptiles are named after him.

Ahuitzotl

Imperial salamander *Pseudoeurycea ahuitzotl* **Adler**, 1996

Ahuitzotl (d.1502) was King (1486–1502) of Tenochtitlan, the eighth Aztec ruler. He was perhaps the greatest known military leader of Pre-Columbian Mesoamerica having, during his reign, doubled the size of land under Aztec dominance. He was in power at the time when the Spanish conquered Mexico.

Ainsworth

Ainsworth's Salamander *Plethodon ainsworthi* Lazell, 1998
[Alt. Bay Springs Salamander, Catahoula Salamander]

Jackson Harold Ainsworth (died before 1998) was an American naturalist who collected (for the Southern Mississippi University) (1964) the only two known specimens from a single site in Jasper County, Mississippi. Lazell made several subsequent collecting attempts to locate more. He had identified the salamanders as a new species when he came across them in the museum collection. All fifteen attempts have proved unsuccessful.

Akha

Torrent Frog sp. *Amolops akhaorum* **Stuart** et al., 2010

The Akha are an indigenous group in Laos.

Alberch

Robber Frog sp. *Eleutherodactylus alberchi* **Flores**, 1988 NCR
[JrSYN *Pristimantis surdus*]
Alberch's (Mushroom-tongue) Salamander *Bolitoglossa alberchi* García-París et al., 2002

Pere Alberch (1954–1998) was a Catalan-born theoretical biologist who was educated in the USA. His bachelor's degree was awarded (1976) by Kansas University and his doctorate (1980) by the University of California, Berkeley. He was at Harvard (1980–1989) as a lecturer in biology, and Curator of Herpetology at the Museum of Comparative Zoology. He returned to Spain as Research Professor at the Museo Nacional de Ciencias Naturales, Madrid, and was Director of the Museum (1989–1995). He resigned his directorship on health grounds and was due to move to Valencia but died suddenly of heart disease.

Alberich

Microhylid frog genus *Albericus* **Burton** & **Zweifel**, 1995

In the epic German poem the *Nibelungenlied*, Alberich was the dwarf who guarded the Rheingold. Menzies named eight new species of *Albericus* (1999), one of which is *Albericus rhenaurum – rhenaurum* being Latin for Rheingold. The other seven all have Wagnerian connotations and appear in this book.

Alberico

Robber Frog sp. *Pristimantis albericoi* **Lynch** & **Ruiz-Carranza**, 1996

Professor Dr Michael Alberico (1937–2005) was an American biologist and zoologist who was Professor at Del Valle University, Colombia. He moved to Colombia (1980) after graduating in biology from the University of Illinois and taking his master's degree and doctorate in zoology at the University of New Mexico (1979). He was murdered; shot dead after withdrawing money from an ATM. In the citation he is described as one 'who devoted his scientific career to the study of Colombian mammals.' A mammal is also named after him.

Alcorn

Leaf Frog sp. *Agalychnis alcorni* **Taylor**, 1952
[JrSYN *Agalychnis dacnicolor*]

J. Ray Alcorn (DNF) was a collector and wildlife biologist. Much of his career was spent working for the U.S. Fish and Wildlife Service. He wrote an article entitled 'The Introduced Fishes of Nevada with a History of Their Introduction' (1943) and, as an Assistant District Agent of the Fish and Wildlife Service, he wrote an article for the Journal of Mammalogy entitled 'On the Decoying of Coyotes'. (1946). He also spent over 50 years gathering data for *The Birds of Nevada* (1988). He collected small mammals along the Alaska Highway in British Columbia, southern Yukon, and southern Alaska (1947–1948) and was collecting in Mexico (1949 and 1954–1956). He and his family sent specimens from Nicaragua to the Mammalogy Division, Museum of Natural History, University of Kansas (1956). There is a JR Alcorn Collection held at the University of Nevada, made up of specimens he collected and donated. A mammal is named after him.

Alemán

Cagua Treefrog *Hypsiboas alemani* **Rivero**, 1964

G. César Alemán was a herpetologist, Curator at the Museo de Historia Natural La Salle, Venezuela. He wrote on herpetology, such as *Contribución al estudio de los reptiles y batracios de la Sierra de Perijá* (1953). A reptile is named after him.

Alessandro

Rocket Frog sp. *Allobates alessandroi* **Grant** & **Rodriguez**, 2001

Dr Alessandro Catenazzi is a Swiss-Peruvian herpetologist. He is Research Fellow, Department of Biology, San Francisco State University and was previously a Research Associate at the Smithsonian and a post-doctoral researcher at the University of California, Berkeley. Florida International University awarded his PhD (2006). He was Field Assistant, Cocha Cashu Biological Station, Manu National Park, Peru (1996–1998) and previously studied Salamanders in Switzerland (1992–1996). He has written or co-written c.50 scientific papers, particularly on Peruvian frogs. The etymology reads: 'The specific epithet is a patronym for Alessandro Catenazzi in recognition of his field studies of the Peruvian herpetofauna.'

Alfaro

Limon Worm Salamander *Oedipina alfaroi* **Dunn**, 1921

Dr Don Anastasio Alfaro (1865–1951) was a Costa Rican archaeologist, geologist, ethnologist, zoologist and writer. From childhood he collected birds, insects, minerals and plants. He graduated at the University of Santo Tomás (1883). He asked the President (1885) to create a National Museum and then dedicated much of his life to it, becoming Director soon after it was established (1887). He spent his life teaching and exploring as well as continuing to collect. He wrote poetry and a number of books, including one on Costa Rican mammals. Three mammals and two birds are named after him.

Alfred

Alfred's Striped Caecilian *Ichthyophis alfredi* Mathew & Sen, 2009

Dr J. R. B. Alfred is an independent researcher who was Director, Zoological Survey of India (1997–2006). He is now Vice-President of Nature Environment & Wildlife Society (NEWS) and was formerly (2000–2002) Vice-President of the National Institute of Ecology. His MSc was awarded by Karnatak University. Among other papers and books he wrote *Waterbirds of Northern India* (2001).

Alfredo

Alfredo's Rain Frog *Craugastor alfredi* **Boulenger**, 1898
[Alt. Alfred's Rainfrog]

Professor Alfredo Augusto Dugès (1826–1910) was Professor of Natural History, Guanajuato University, Mexico. He is regarded as the father of Mexican herpetology, and was the first to define Mexican herpetofauna in Linnaean terms.

Alice

Alice's Night Frog Nyctibatrachus aliciae **Inger** *et al., 1984*
[Alt. Grandison's Wrinkled Frog]

Alice Grandison (see **Grandison**).

Alipio

Bahia Forest Frog *Macrogenioglottus alipioi* **Carvalho**, 1946
Saddleback Toad sp. *Brachycephalus alipioi* **Pombal** & Gasperini, 2006
Redbelly Toad sp. *Melanophryniscus alipioi* Langone et al., 2008

Dr Alipio de Miranda-Ribeiro (1874–1939) was a Brazilian zoologist, herpetologist and ichthyologist who did much, with Adolpho and Bertha Lutz, to enhance the collections at Brazilian National Natural History Museum, Rio de Janeiro. He initially studied medicine but joined the National Museum (1894) before completing the course. He was Assistant Naturalist for the Museum (1897) and Secretary, Department of Zoology (1899), Deputy Head & Professor, Zoology Department (1910–1929). He created the Inspectorate of Fisheries (1911), the first South American oceanographic service and was its first Director (1911–1912). He wrote the 5-volume *Fauna Brasiliensis* (1907–1915). Carvalho was his successor as Curator of Herpetology at the Museum. (See **Miranda**)

Alisan

Alisan's Salamander *Hynobius arisanensis* **Maki**, 1922
[Alt. Arisan Hynobiid]

The holotype was obtained at Arisan, Taiwan. It seems that 'Alisan' is a corruption of Arisan, caused by the difficulty experienced by many by Chinese speakers in pronouncing the letters R and L.

Allen, GM

Allen's Slippery Frog *Conraua alleni* **Barbour & Loveridge**, *1927*
Allen's River Frog *Phrynobatrachus alleni* **Parker**, *1936*

Dr Glover Morrill Allen (1879–1942) was a collector, curator, editor, librarian, mammalogist, ornithologist, scientist, taxonomist, teacher and writer. He was Librarian at the Boston Society of Natural History (1901–1927). He was employed to oversee the mammal collection at the Museum of Comparative Zoology, Harvard (1907), having taken his PhD (1904). He was Curator of Mammals (1925–1938) and then Professor of Zoology (1938–1942). He was keen on all vertebrates, particularly birds (editing *Auk* 1939–1942) and mammals (President of the American Society of Mammalogists 1927–1929). He made many collecting trips (1903–1931), variously to Africa, including the Harvard African Expedition to Liberia (1926), Australia, the Bahamas, Brazil, Labrador and the West Indies. He wrote a great many scientific papers and articles, and a number of books. Early works include *The Birds of Massachusetts* (1901) where he notes taking a specimen of passenger pigeon (1904), which might well have been the last one recorded in the wild. A mammal, two birds and two reptiles are named after him.

Allen, PH

Allen's Worm Salamander *Oedipina alleni* **Taylor**, 1954

Paul Hamilton Allen (1911–1963) was an American botanist whose early training was at the Missouri Botanical Garden. He went to Panama (1934) to collect and returned to manage a field station in the Canal Zone (1936–1939). During the Second World War he worked collecting rubber in the Colombian Amazon and (1946) joined the United Fruit Company in Costa Rica. He was Director, Fairchild Tropical Botanic Garden, Florida (1953–1954), then returned to Central America to teach. He worked there until his early death from cancer. He wrote *The Rain Forests of Golfo Dulce* (1956). He was with Taylor when the holotype was collected.

Allen (family)

Cuzco Reserve Treefrog *Dendropsophus allenorum* **Duellman & Trueb**, 1989

Arly, Constance & Harold Allen. The etymology reads: 'The specific epithet is a patronym for Arly, Constance, and the late Harold Allen of the Allen Press of Lawrence, Kansas. We associate this name with them in recognition of their patient and dedicated efforts to provide scientists with an improved understanding of English usage and the realities of the publication process. Through the years, the Allens and their employees have improved the quality of scientific journals and contributions to these journals in significant, but all-too-often unrecognized, measures.'

Allison

Morobe Land Frog *Liophryne allisoni* **Zweifel,** 2000
Allison's Mountain Frog *Choerophryne allisoni* **Richards & Burton**, 2003

Dr Allen Allison (b.1950) is a herpetologist who joined the staff of the Bishop Museum, Hawaii (1973) and is currently Assistant Director of Research. He gained his doctorate from the University of California (1979). A reptile is also named after him.

Alluaud

Fort Dauphin Digging Frog *Rhombophryne alluaudi* **Mocquard**, 1901

Charles Alluaud (1861–1949) was an entomologist, botanist and naturalist who came from a wealthy family, living in a chateau where the painter Corot was a frequent visitor. His father was President of the Royal Porcelain Factory, Limoges. He travelled extensively, including on scientific expeditions to the Seychelles and Madagascar (1892–1893). Four reptiles are named after him.

Aloysius

Aloysius Skittering Frog *Euphlyctis aloysii* Joshy et al., 2009

St Aloysius College, Mangalore, India

Altig

Microhylid frog genus *Altigius* Wild, 1995

Dr Ronald Altig is a zoologist who is Professor Emeritus at Mississippi State University. The University of Illinois awarded his bachelor's degree, Southern Illinois University his master's and Oregon State University his doctorate. The etymology mentions his extensive contribution to tadpole biology.

Alvarado

Alvarado's Salamander *Bolitoglossa alvaradoi* **Taylor**, 1954

[Alt. Moravia de Chirripo Salamander]

Don Fernando Alvarado Chacón was owner of Moravia de Chirripo, Limón Province, Costa Rica, where the salamander holotype was collected.

Alvarenga

Alvarenga's Treefrog *Bokermannohyla alvarengai* **Bokermann**, 1956

[Alt. Santa Barbara Treefrog]

Colonel Moacyr Alvarenga (1915–2010) was a Brazilian Air Force officer as well as being a zoologist and entomologist associated with the Museu Nacional, Rio de Janeiro. One of his collecting companions and writing collaborators was Bokermann, whose obituary he wrote (1995). Among the many taxa named after him is a dung beetle.

Alvarez del Toro

Alvarez del Toro's Salamander *Cryptotriton alvarezdeltoroi* **Papenfuss** & **Wake**, 1987

Miguel Alvarez del Toro (1917–1996) was a life-long naturalist working mostly in Chiapas State, Mexico, having always collected insects and small vertebrates. He collected birds for the Academy of Natural Sciences, Philadelphia (1938–1939). He became a taxidermist at the Museum of Natural History in Tuxtla Gutierrez, Chiapas (1942) and collected vertebrates near the city and surveyed more remote areas. He had no formal education yet was a great teacher, scientist and conservationist. He taught at the Colegio de Ciencias y Artes de Chiapas and at the Universidad Nacional Autónoma de México. Among many honours the Universidad de Chiapas awarded him an honorary doctorate. He was the principal force behind the creation of six protected areas in Chiapas. He published many papers, including over 40 on birds including *Birds of Chiapas*. Four reptiles and a bird are named after him.

Amadeus

Haitian Robber Frog *Eleutherodactylus amadeus* **Hedges**, **Thomas** & Franz, 1987

[Alt. Mozart's Frog]

Wolfgang Amadeus Mozart (1756–1791) is too well known to require a biography here! The frog was named after the composer 'for the remarkable resemblance of the wide-band audiospectrogram of this species to musical notes.'

Ambrosi

Ambrosi's Cave Salamander *Speleomantes ambrosii* **Lanza**, 1954

[Alt. Spezia Cave Salamander; Syn. *Hydromantes ambrosii*]

Carlo d'Ambrosi (1898–1998) was an Italian geologist and speleologist. He theorized (1920s) on underground connections in the cave and river systems of the Trieste area. He wrote *Note illustrative della carta geologica delle tre Venezie: Foglio Trieste* (1955).

Amélie

Glass Frog sp. *Teratohyla amelie* Cisneros-Heredia & Meza-Ramos, 2007

[Syn. *Cochranella amelie*]

Amélie is the protagonist of the film *Le Fabuleux Destin d'Amélie Poulain*. The describers state that this is 'a movie where little details play an important role in the achievement of joie de vivre; like the important role that Glassfrogs and all amphibians and reptiles play in the health of our planet.'

Amicorum

Treefrog sp. *Dendropsophus amicorum* Mijares-Urrutia, 1998

The binomial *amicorum* means 'of the friends', and is to collectively thank all of the author's numerous friends who helped him in his work. Their number is apparently so substantial that he felt only a genitive plural would be appropriate.

Amiet

Pyxicephalid frog genus *Amietia* **Dubois**, 1987

Toad genus *Amietophrynus* **Frost** et al. 2006

Amiet's Long-fingered Frog *Cardioglossa melanogaster* Amiet, 1972

Pangi Territory Frog *Amietia amieti* **Laurent**, 1976

Volcano Clawed Frog *Xenopus amieti* Kobel et al., 1980

Professor Jean-Louis Amiet is a French herpetologist, entomologist, ecologist and ichthyologist, formerly at the Université de Yaoundé, Cameroon, until retiring to Europe. He wrote *Faune du Cameroun* (1987). A reptile is also named after him.

Amoman

Amoman Choerophryne *Choerophryne amomani* **Günther**, 2008

Amoman Mountain, Yapen Island, Papua, Indonesia.

Anamiato

Cross Frog sp. *Oreophryne anamiatoi* Kraus & **Allison**, 2009

Jim Anamiato of the Papua New Guinea National Museum went on several of the authors' expeditions, including the one on which the frog holotype was collected.

Ananjeva

Ananjeva Asian Treefrog *Kurixalus ananjevae* **Matsui** & **Orlov**, 2004

Professor Dr Natalia Borisovna Ananjeva (b.1946) is Head of the Department of Herpetology and Vice-Director at the Zoological Institute, Russian Academy of Sciences, St Petersburg, being the first female zoologist member. She graduated in biology at Leningrad State University.

Anchieta

Anchieta's Ridged Frog *Ptychadena anchietae* **Bocage**, 1868

[Alt. Benguella Grassland Frog, Plain Grass Frog]

Anchieta's Tree Frog *Leptopelis anchietae* Bocage, 1873

José [Alberto] de [Oliveira] Anchieta (1832–1897) was an independent naturalist and collector who travelled widely in Africa. He left Lisbon (1857) to join a close friend who had settled in Cabo Verde off the West African coast. Here, although self-taught, he practised medicine. A cholera outbreak nearly killed him, so he returned to Portugal (1859). He studied medicine in Lisbon but left for Angola before completing his studies. There he collected all manner of natural history specimens, which he donated to museums on his return to Portugal, before leaving for the last time (1866). Little is known about the next period of Anchieta's life because most of the museum specimens, as well as his letters to Bocage, disappeared in a fire (1978). We know that he is recorded at various locations (1866–1897) in Angola and Mozambique and died, probably from chronic malaria, when returning from an expedition to Caconda. He was responsible for identifying at least 25 new mammals, 46 birds and as many amphibians and reptiles. Six birds, a mammal, seven reptiles and many other taxa are named after him.

Anderson, J

Anderson's Khasi Hills Frog *Limnonectes khasianus* Anderson, 1871

Anderson's Frog *Odorrana andersonii* **Boulenger**, 1882

[Alt. Yunnan Odorous Frog]

Anderson's Crocodile Newt *Echinotriton andersoni* Boulenger, 1892

[Alt. Ryukyu Spiny Newt]

Anderson's Bush Frog *Theloderma andersoni* **Ahl**, 1927

[Syn. *Philautus andersoni*]

Dr John Anderson (1833–1900) was a naturalist who was Professor of Natural History at Free Church College, Edinburgh. He became Curator of the Indian Museum, Calcutta (1865) and collected for the trustees. He went on scientific expeditions to Yunnan (1867), Burma (1875–1876) and the Mergui Archipelago (1881–1882). He became Professor of Comparative Anatomy at the Calcutta Medical School (1885), then returned to London (1886). He wrote *Guide to the Calcutta Zoological Gardens* (1883). Four birds, three mammals and eight reptiles are named after him.

Anderson, JD

Anderson's Salamander *Ambystoma andersoni* Krebs & Brandon, 1984

Dr James Donald Anderson (1928–1977) was an American herpetologist at Rutgers, The State University of New Jersey (1972–1977), where he was Professor of Zoology and where a James D Anderson Memorial Research Fund is now administered. He was Curatorial Assistant (1958–1959) at the Museum of Vertebrate Zoology, University of California, Berkeley, which awarded his doctorate (1960). He took part in a number of expeditions to Mexico, organized by the AMNH, New York. He wrote *The life history and systematics of Ambystoma rosaceum* (1961).

Anderson (South Carolina)

Anderson's Tree Frog *Hyla andersonii* **Baird**, 1854

[Alt. Pine Barrens Treefrog]

This frog is named after the holotype locality, Anderson, South Carolina.

Andersson

Andersson's Toad *Rhaebo anderssoni* Melin, 1941

[Probably a junior synonym of *R. guttatus* (Schneider, 1799)]

Andersson's Stubfoot Toad *Atelopus palmatus* Andersson, 1946

Professor Dr Lars Gabriel Andersson (1868–1951) was a zoologist and herpetologist at the Natural History Museum, Stockholm. He studied, from undergraduate to doctorate, at Uppsala University (1887–1909). He was an Assistant at the museum (1894–1895 & 1897–1902). He became Professor in Stockholm, taking part in the Swedish Australia Expedition (1910–1913) and collected many reptiles. He co-wrote *Results of Dr. Mjobergs Swedish scientific expeditions to Australia 1910–13* (1913). A reptile is also named after him.

Andi

Atlantic Robber Frog *Craugastor andi* **Savage**, 1974

Dr Andrew Starrett (1930–2008) was Emeritus Professor, University of Southern California, where he taught biology and zoology. Savage wrote: 'The name *andi* is an arbitrary combination of letters and is dedicated to my good friend and Costa Rican field companion Dr Andrew Starrett.' Starrett co-wrote *Records of bats from Costa Rica* (1968).

Andre

Andre's Clawed Frog *Xenopus andrei* Loumont, 1983

As the original description says nothing, the etymology is unknown. A likely contender is Dr André Villiers (1915–1983) who was a French entomologist and herpetologist. Interested in natural history from an early age, after military service he joined the Paris Metropolitan Museum (1937) as an Assistant. He took part in the Cameroon expeditions led by Bernard Gèze. He was awarded his doctorate (1943). After the war he was in charge of entomology at the Institut Français d'Afrique Noire. He joined the Muséum National d'Histoire Naturelle, Paris (1956) as Professor, taking over the entomology laboratory until retiring (1980). He published 761 papers, primarily on entomology but also on herpetology and ornithology. He had a life long interest in the people and natural history of Cameroon, where the clawed frog holotype was collected.

Andre (Göldi)

Lowland Tropical Bullfrog *Leptodactylus andreae* **Müller**, 1923

André Göldi, who was primarily a botanist, was Director of the experimental agricultural establishment at Peixeboi, Pará, Brazil, part of the museum in Pará associated with his brother, Emil August **Goeldi**. The holotype was collected at the Peixeboi establishment.

Andreone

Andreone's Treefrog *Boophis andreonei* **Glaw** & Vences, 1994

[Alt. Benavony Bright-eyed Frog]

Dr Franco Andreone (b.1961) is an Italian zoologist whose master's degree (1985) was awarded by the Università degli Studi di Torino and his doctorate (1990) in animal biology by the Università di Bologna. He is Curator, Zoology Department, Museo Regionale di Scienze Naturali, Torino (1991).

Andrews, EA

Andrews' Robber Frog *Eleutherodactylus andrewsi* **Lynn**, 1937

Professor Dr Ethan Allen Andrews (1859–1956) was a biologist at Johns Hopkins University, Baltimore, having studied there for his bachelor's degree (1881) and doctorate (1887) before becoming Assistant Professor of Biology. He was at the Johns Hopkins research station at Montego Bay in Jamaica for many years, becoming Associate Professor (1892) and full Professor of Zoology (1908). He was a leading expert on Jamaica's flora and fauna. He was President of the Society of American Zoologists (1904).

Andrews, RC

Andrews' Toad *Bufo gargarizans andrewsi* **Schmidt**, 1925

Roy Chapman Andrews (1884–1960) was a larger-than-life American who became an explorer, collector and curator whom many believe to have been the real-life Indiana Jones. He always maintained that from his earliest childhood he had a desire for travel and adventure. 'I was born to be an explorer', he wrote (1935) in *The Business of Exploring*. 'There was never any decision to make. I couldn't do anything else and be happy.' He said too that his only ambition was to work at the AMNH. He first worked as a taxidermist, and after graduating (1906) went to New York City and applied for a job at the Museum. The Director told him there were no jobs but Andrews persisted saying, 'You have to have somebody to scrub floors, don't you?' The director took him on and from this humble beginning he went on to become the Museum's most famous explorer. Initially a taxidermist, he developed an interest in whales and travelled to Alaska, Japan, Korea and China to collect various marine mammals. He was Naturalist on the USS Albatross voyage to the Dutch East Indies and Borneo (1909–1910). He led an expedition to China and Outer Mongolia (1921–1923) where he collected both live specimens, including the toad, and fossils, including the first eggs to be positively identified as those of a dinosaur. He continued to make further expeditions over a number of years until returning to the USA (1930). He later became Director of the Museum (1934). He retired (1942), moved to California and spent the rest of his life writing about his exploits, including his autobiography *Under a Lucky Star*. Three mammals and two birds are named after him.

Angel

Angel's Madagascar Frog *Boehmantis microtympanum* Angel, 1935
Angel's Caecilian *Geotrypetes angeli* **Parker**, 1936
False Angel's Caecilian *Geotrypetes pseudoangeli* **Taylor**, 1968

Fernand Angel (1881–1950) was a zoologist and herpetologist. He joined the Muséum National d'Histoire Naturelle, Paris (1905) as an Assistant Taxidermist for Leon Vaillant and then François Mocquard, working there until he died. The only break in his career was his French Army service (1914–1918). Nine reptiles are named after him.

Anil

Anil's Bush Frog *Raorchestes anili* **Biju** & Bossuyt, 2006
[Syn. *Philautus anili*]

Dr A. H. Anil Zachariah is a wildlife veterinary surgeon and naturalist who teaches at Kerala Agricultural University, India. He has specialized in the amphibians of Kerala and has been collecting them ever since (1994). He wrote *The Story of Frog Princess* (2008).

Anita

Anita's False Brook Salamander *Pseudoeurycea anitae* **Bogert**, 1967

Anita Jones. The etymology reads: 'While my wife, Martha R. Bogert, and I were carrying out investigations in the mountains of Oaxaca, the residents of the state were unfailingly friendly and courteous. Among those who facilitated our investigations, however, no one was more hospitable or tolerant than our gracious and knowledgeable hosts in Oaxaca de Juarez, Roy and Anita Jones. The enthusiasm, fortitude, and perseverance displayed by Anita when she accompanied us on trips into the rugged backcountry of Oaxaca won our gratitude and admiration. As an acknowledgment of her assistance, therefore, I am associating Anita's name with the species discovered near San Vicente Lachixio.'

Ann

Blue-sided Leaf Frog *Agalychnis annae* **Duellman**, 1963

Ann S. Duellman was the describer's wife and collector of the leaf frog holotype in Costa Rica.

Annandale

Annandale's Frog *Clinotarsus alticola* **Boulenger**, 1882
[Alt. Point-nosed Frog, Pointed-headed Frog]
Annandale's Bush Frog *Raorchestes annandalii* Boulenger, 1906
Annandale's High Altitude Frog *Kurixalus naso* Annandale, 1912
[Alt. Long-snouted Treefrog]
Annandale's Shrub Frog *Pseudophilautus semiruber* Annandale, 1913
Annandale's Tree Frog *Chirixalus simus* Annandale, 1915
[Syn. *Chiromantis simus*]
Annandale's Paa Frog *Nanorana annandalii* Boulenger, 1920
[Alt. Boulenger's Hill Frog; Syn. *Paa annandalii*]

Dr (Thomas) Nelson Annandale (1876–1924) was a zoologist and Director of the Indian Museum, Calcutta. He wrote a number of scientific papers (1900–1930) including *Fauna of the Chilka Lake: mammals, reptiles and batrachians* (1915). He was instrumental in establishing a purely zoological survey, not combined with anthropology, undertaking several expeditions, most notably the Annandale-Robinson Expedition that collected in Malaya (1901–1902). A mammal and four reptiles are named after him.

Anson

Stream Toad genus *Ansonia* **Stoliczka**, 1870

Major-General Sir Archibald Edward Harbord Anson (1826–1925) was in the British Army (1844–1867) serving in the United Kingdom (1847–1855), Mauritius (1857–1862), Madagascar (1862–1865) and India (1865–1867). He then became the last Lieutenant Governor of Penang (1867–1882), where the type species of the genus was collected.

Ansorge

Ansorge's Ridged Frog *Ptychadena ansorgii* **Boulenger**, 1905

Dr William John Ansorge (1850–1913) was an explorer and collector in Africa (late nineteenth century.) He crossed the African continent from East to West (1890), and later became District Medical Officer in Nigeria. He wrote *Under the African Sun* (1899). Nineteen birds, four mammals, three reptiles and several African fishes are named after him. His son, Sir Eric Cecil Ansorge (1887–1977), was a lepidopterist.

Antenor

Ecuador Silent Frog *Syncope antenori* **Walker**, 1973

Dr Antenor Leitao de Carvalho (see **Carvalho**).

Anthony, AS

Anthony's Cross Frog *Oreophryne anthonyi* **Boulenger**, 1897

Alearce Savery Anthony (fl.1906) was an immigrant in New Guinea from Mauritius. He was employed as a collector by Rothschild, and led an expedition (1898) from Port Moresby to the Owen Stanley Range. He later managed a coffee plantation at Sogeri until 1905 and was at Redscar Bay (1906), a gold-mining area.

Anthony, HE

Anthony's Poison-arrow Frog *Epipedobates anthonyi* **Noble**, 1921

[Syn. *Ameerega anthonyi*]

Dr Harold Elmer Anthony (1890–1970) was Curator of Mammals, AMNH, New York, and a noted collector of neotropical animals. His prime interest was mammals. He was President, American Society of Mammalogists (1935–1937). He was on the Vernay-Cutting expedition to Burma (1938), where 'he was delighted with the native Lisu who brought him many species of voles and mice. Delighted, that is, until he realised they were selling him animals that they had stolen from his own traps.' He wrote *A new rabbit and a new bat from Neotropical regions* (1917). Three mammals and a bird are named after him.

Antongil

Tomato Frog *Dyscophus antongilii* **Grandidier**, 1877

Bay of Antongil or Antongila Bay is an area of northeast Madagascar, and the type locality.

Api

Api Dwarf Toad *Pelophryne api* **Dring**, 1983

Gunung Api is a mountain in Sarawak, the type locality for this toad.

Appleby

Appleby's Asian Toad *Leptolalax applebyi* **Rowley** & **Cao**, 2009

Robert Appleby (b.1962) of ADM Capital Foundation, a Hong Kong-based hedge fund, is a generous supporter of biodiversity conservation and scientific capacity building in Asia. Oxford awarded his bachelor's degree in zoology.

Aracamun

Robber Frog sp. *Ceuthomantis aracamuni* Barrio-Amoros & Molina, 2006

[Syn. *Pristimantis aracamuni*]

Cerro Aracamuni is the type locality in Venezuela.

Arathoon

Djikoro Wart Frog *Limnonectes arathooni* **Smith**, 1927

Lazarus Sarkies Arathoon MBE (1883–1932) was British Vice-Consul at Macassar, Sulawesi, for 18 years, during which time (1925) he was awarded the MBE. At the time of his death he was Managing Director of the merchants Michael Stephens & Co.. Earlier he had worked at Raffles Hotel in Singapore, as did his father. Smith says of Arathoon in his etymology '...to whom I was indebted for much assistance in collecting during my stay on the island.'

Araya

Asian Toad sp. *Leptolalax arayai* **Matsui**, 1997

Dr Kunio Araya is an entomologist who is Associate Professor, Department of Environmental Changes, University of Kyushu, Japan. He wrote, with Matsui and others, *Acoustic characteristics and systematics relationships of arboreal microhylid frogs of the genus* Metaphrynella *from Malaysia* (1996). He collected the toad holotype in Sabah.

Archbold

Archbold's Land Frog *Austrochaperina archboldi* **Zweifel**, 2000

Richard A. Archbold (1907–1976) was an American patron of science who financed and led expeditions, particularly to New Guinea (1933–1964). He had been a member of the (1929–1931) Madagascar Expedition (funded by his father). He also set up a permanent research station at Lake Placid in Florida and became a zoologist at the AMNH. Many taxa, including eight birds and a mammal genus, are named after him.

Archey

Archey's Frog *Leiopelma archeyi* Turbott, 1942

Sir Gilbert Edward Archey (1890–1974) was a zoologist, ethnologist and museum director who was born in England and taken to New Zealand as a child. He took both his bachelor's degree and doctorate at Canterbury University College, Christchurch. He was Assistant Curator, Canterbury Museum (1914–1923), though First World War service in the New Zealand Army interrupted his career, as did similar Second World War service. He was Director, Auckland Institute (1924–1964), noted for his work on moas and on Maori Art. He wrote *The Moa, a Study of the Dinornithiformes* (1941).

Ardila

Stubfoot Toad sp. *Atelopus ardila* **Coloma** et al., 2010

Maria Cristina Ardila-Robayo (see **Cristina**).

Arfak

Large River-Frog *Hylarana arfaki* **Meyer**, 1875
[Alt. Arfak Mountains Frog]

As one of its vernacular names suggests, this frog is named after the Arfak Mountains, West Papua (Indonesian New Guinea).

Argus

Argus Reed Frog *Hyperoilus argus* **Peters**, 1854

In Greek mythology, Argus was a giant with supernumerary eyes: anything from four to a hundred, depending on the version. His name is often used in taxonomy for species that are decorated with *ocelli* (eye-like spots).

Ari

Ari's Caecilian *Chthonerpeton arii* Cascon & Lima-Verde, 1994

Dr Ari Santiago Lima-Verde is a physician who is a Professor at Fundação Universidade Estadual do Ceará, Brazil, and a relative of the junior author.

Ariadne

Serra da Bocaina Snouted Treefrog *Scinax ariadne* **Bokermann**, 1967

In Greek mythology, Ariadne was the daughter of King Minos of Crete. She aided the hero Theseus in overcoming the monstrous Minotaur (actually her half-brother), and was later the bride of the god Dionysus.

Arikan

Lycian Salamander sp. *Lyciasalamandra arikani* Göçmen & Akman, 2012

Dr Huseyin Arikan was a Turkish herpetologist and Professor at the biology department of Ege University, Izmir, where he is a colleague of the senior author.

Arilda

Teresopolis Treefrog *Aplastodiscus arildae* **Cruz** & **Peixoto**, 1987

Arilda M. Gonçalves da Cruz is the wife of the senior describer, Carlos Alberto G. da Cruz.

Armstrong

Armstrong's Robber Frog *Eleutherodactylus armstrongi* **Noble** & Hassler, 1933

John C. Armstrong was an Assistant Curator at the AMNH, New York (1930s–1950s). He wrote a number of articles including *New Caridea from the Dominican Republic* (1949). The original etymology says the frog was 'named in honor of Mr. John C. Armstrong who planned the second expedition and assisted in much of the herpetological collecting.' [Armstrong Expedition to the Dominican Republic (1932–1933)]. A reptile is named after him.

Arndt

Arndt's Choerophryne *Choerophryne arndtorum* **Günther**, 2008
Fanged Frog sp. *Xenorhina arndti* Günther, 2010

Dr Rudolf G. Arndt of Pomona, New Jersey, and his German-based family have given Günther considerable support for his research. Dr Arndt, a marine scientist and ichthyologist, is Professor Emeritus at Richard Stockton College, Pomona, where he taught (1974). He co-wrote *Fresh Water Fishes of South Carolina* (2009). The (plural) binomial for the *Choerophryne* honours Dr Arndt's family as well as himself. A reptile is named after him.

Arne

Kassewe Grassland Frog *Ptychadena arnei* **Perret**, 1997

Dr Arne Schiøtz (see **Schiøtz**)

Arnold

Arnold's Paa Frog *Nanorana arnoldi* **Dubois**, 1975
[Alt. Arnold's Spiny Frog]

Arnold Arboretum was founded (1872) in memory of James Arnold (1781–1868) whose bequest of the parkland to Harvard made its creation possible. The frog holotype was collected in Burma during the (1937) Vernay-Cutting Expedition, and the Arnold Arboretum was one of the organisations for which collecting was undertaken.

Arnold, EN

Montseny Brook Newt *Calotriton arnoldi* Carranza & Amat, 2005

Dr Edwin Nicholas 'Nick' Arnold (b.1940) is Director, Herpetology Research at the Natural History Museum, London, where he acted as the senior author's mentor and guide during his six years of herpetological research there. His publications include *Reptiles and Amphibians of Europe* (2003). Five reptiles are named after him.

Arntzen

Balkan Crested Newt *Triturus arntzeni* Litvinchuk et al., 1999
[Syn. *Triturus karelinii arntzeni*]

Dr J. W. 'Pim' Arntzen, who works at Naturalis-Nationaal Natuurhistorisch Museum, Leiden, has long studied newts. He wrote *Seasonally changing sex ratio and asynchronous pond presence of male and female* Triturus *newts* (2002).

Arrabal

Arrabal's Surinam Toad *Pipa arrabali* **Izecksohn**, 1976

Jailton Aguiar Arrabal collected the first known examples of this toad, in Amazonas state, northern Brazil.

Arthur

Arthur's Stubfoot Toad *Atelopus arthuri* **Peters**, 1973

Arthur J Peters collected the holotype (1959). The describer, James Arthur Peters, wrote: 'The species name is derived from the first name of the collector of the holotype, who provided me with pleasant company on the long trip to Papallacta and back, and who collected the first specimens of the new taxon while I was busy fixing our supper.' From the similarity of their names, we conjecture that they were related. James Arthur Peters died the year before his description was published.

Asanka

Asanka's Shrub Frog *Pseudophilautus asankai* Manamendra-Arachchi & Pethiyagoda, 2005

Asanka Goonewardena is a Sri Lankan businessman who founded a private medical equipment company (1986), and was co-founder (1991) of the Wildlife Heritage Trust. The trust runs the Agra Arboretum, Sri Lanka, where the frog holotype was collected, and Goonewardena personally provides support for research into herpetology.

Asmat

Bangladeshi Cricket Frog *Fejervarya asmati* Howlander, 2011

Professor Dr Ghazi S. M. Asmat was the describer's teacher and mentor. He works at Department of Zoology, University of Chittagong, Bangladesh, where he is a Professor of Conservation Biology.

Astarte

Paranapiacaba Treefrog *Bokermannohyla astartea* **Bokermann**, 1967

The description has no etymology. Perhaps, given that Bokermann has been fond of names from mythology, it is named after Astarte, whom the ancient Syrians regarded as a goddess of fertility, sexuality and war.

Atahualpa

Amazonas Water Frog *Telmatobius atahualpai* **Wiens**, 1993

Atahualpa (d.1533) was the last ruler of the Incas of Peru. He became emperor (1532) after defeating his half-brother Huáscar in a civil war. When the Spanish arrived, Francisco Pizarro captured Atahualpa and eventually executed him. The asteroid 4721 Atahualpa is also named after him.

Atif

Atif's Lycian Salamander *Lyciasalamandra atifi* Basoglu, 1967

Professor Dr Atif Sengün (1915–2002) was a Turkish biologist who graduated (1939) from Universität Berlin and was awarded a doctorate by Istanbul University (1943) where he worked throughout his career, being a Professor of Zoology and Dean of the University (1960–1962), later retiring from his chair (1975). He was one of the original members of the Scientific and Technological Research Council of Turkey (1964) and wrote *Evolution* (1984). He was a friend of the author and gave him the salamander holotype (1966).

Atkins

Atkins' Robber Frog *Eleutherodactylus atkinsi* **Dunn**, 1925
[Alt. Cuban Groin-spot Frog]

Edwin Farnsworth Atkins (1850–1926) was the leading sugar plantation owner in Cuba. His company owned enormous estates, acquired before the Spanish-American War. He had been sent to Cuba (1869) to learn more about the sugar business as his family were brokers and shippers of sugar, and among the properties he acquired were a number in the region of Cienfuegos, where he established the Atkins Institute for Tropical Plant Research, later (1899) the Harvard Botanic Station for Tropical Research and Sugar Cane Investigation, of which Robert M Grey (q.v.) was in charge. Atkins provided Harvard with a large endowment (1920), in exchange for which institute and botanical gardens became part of the university. Dunn wrote (1925): 'I was so fortunate last year as to be able to spend nearly two months at the Harvard Biological Station on the Atkins estate at Soledad, near Cienfuegos, on the southern coast of Cuba... ...in dry hill-forest, a large and beautiful frog with brilliant markings of scarlet and black was found… The brilliantly colored form lacks a name and it seems appropriate to christen it in honor of Mr. Edwin F. Atkins, who, by the establishment of the Harvard Biological Station, has given a great impetus to the study of Cuban, and of tropical, biology.'

Atukorale

Atukorale's Dwarf Toad *Duttaphrynus atukoralei* **Bogert** & Senanayake, 1966
[Alt. Yala Toad; Syn. *Bufo atukoralei*]

Mr Vicky Atukorale was a Sri Lankan naturalist who collected the holotype (1965). He wrote *Notes on Birds Collected at Kumana Sanctuary on 17th May, 1949* for the Dehiwela Zoo. (1949).

Aua

Tree Frog sp. *Litoria auae* **Menzies** & **Tyler**, 2004

Aua was the beautiful daughter of Pam, a legendary New Guinea chieftain.

Aubry

Pyxicephalid frog genus *Aubria* **Boulenger**, 1917
Gaboon Forest Treefrog *Leptopelis aubryi* **Duméril**, 1856
Kala Forest Treefrog **Leptopelis aubryioides* **Andersson**, 1907

Charles Eugène Aubry-Lecomte (1821–1879) was a French civil servant and colonial administrator in New Caledonia and West Africa. An amateur naturalist, he collected wherever he was posted and discovered many new reptiles, also making a collection of fish in Gabon (1850s). He was also amongst the first to describe the hallucinogenic iboga root. A bird and two reptiles are named after him.

* *Leptopelis aubryioides* was named owing to its resemblance to *L. aubryi* rather than specifically to honour Aubry.

Aucoin

Toad sp. *Incilius aucoinae* O'Neill & **Mendelson**, 2004
[Syn. *Bufo aucoinae*]

Lisa Louise Aucoin (1975–2001) was a herpetologist who was killed in a car crash shortly after returning from a trip to Costa Rica. Southeastern Louisiana University, Hammond awarded her master's degree in biology posthumously.

Audant

Colonia Robber Frog *Eleutherodactylus audanti* **Cochran**, 1934

Dr André Audant (b.1909) was a Haitian entomologist who studied at Kansas State University (1931) and was (1934) Government entomologist at Haiti's Department of Agriculture, Port au Prince. He and Barbour collected the frog holotype (1934). A number of insects are named after him.

Auguste

Barking Frog *Craugastor augusti* Duges, 1879

Auguste Henri Duméril (see **Duméril**).

Aureol

Freetown Long-fingered Frog *Arthroleptis aureoli* **Schiøtz**, 1964

Mount Aureol is in Sierra Leone.

Avicupi

Robber Frog sp. *Pristimantis avicuporum* **Duellman** & Pramuk, 1999

Avicuporum is Latin for 'of the bird-catchers' and commemorates the efforts of the ornithologists of the Museum of Zoology, Louisiana State University, in collecting amphibians!

Avila Pires

Humming Frog sp. *Chiasmocleis avilapiresae* Peloso & Sturaro, 2008

Professor Dr Teresa Cristina Sauer de Ávila Pires is a Brazilian zoologist and herpetologist at Museum Paraense Emilio Goeldi, where she researches and teaches. She wrote *Lizards of Brazilian Amazonia (Reptilia: Squamata)* (2005). To her friends she is known as 'TC'.

Ayanganna

Ayanganna Stefania Treefrog *Stefania ayangannae* MacCulloch & Lathrop, 2002

Mount Ayanganna in Guyana is the type locality.

Ayarzagüena

Ayarzagüena's Frog *Dischidodactylus colonnelloi* Ayarzagüena, 1985
Ayarzagüena's Treefrog *Tepuihyla edelcae* Ayarzagüena, **Señaris** & **Gorzula**, 1993
Rocket Frog sp. *Colostethus ayarzaguenai* **La Marca**, 1997
[Syn. *Anomaloglossus ayarzaguenai*]

Dr José Ayarzagüena Sanz (1952–2011) was a Spanish zoologist and herpetologist at the Museo de Historia Natural La Salle, Caracas, and Director of the Biological Station of El Frio, where endangered Orinoco Crocodiles are bred for release to the wild. He co-wrote *Fauna of the Venezuelan llanos: Notes on their morphology and ecology* (1985). He died of lung cancer that was detected too late for treatment. A reptile is also named after him.

Azara

Azara's Granulated Toad *Rhinella azarai* **Gallardo**, 1965

[Syn. *Rhinella granulosa azarai*]

Brigadier-General Féliz Manuel de Azara (1746–1811) was a soldier, engineer and naturalist who distinguished himself in various expeditions. He was on the Spanish commission in South America (1781–1801) to try to settle the boundaries between Portuguese and Spanish colonies. There he began to study animals, particularly observing their behaviour. His notes, sent to his brother, the Spanish Ambassador in Paris, are generally acknowledged to be meticulous, but they also contained reports from others that were not so accurate. He wrote *Voyage dans l'Amérique méridionale depuis 1781 jusqu'en 1801* (1801). Fourteen birds, five mammals and a reptile are named after him.

Azzurra

Madagascar Frog sp. *Gephyromantis azzurrae* Mercurio & **Andreone**, 2007

Kintana Azzurra Andreone (b. 2006) is the junior author's younger daughter.

B

Bacchus

Wine Robber Frog *Pristimantis bacchus* **Lynch**, 1984

Bacchus was the Roman god of wine and general jollification, hence Bacchanalian orgies. Lynch wrote that the name was 'in loose allusion to the blood-red eyes' of this frog.

Baetke

Baetke's Bright-eyed Frog *Boophis baetkei* **Köhler, Glaw** & Vences, 2008

Dr Claus Bätke has, since 2007, looked after the East African operations of GTZ (Gesellschaft für Technische Zusammenarbeit) through its Tropical Ecology Support Programme. His support has been very important for BIOPAT, which he helped to found (1999) and of which he was President (2007).

Baggins

Bilbo's Rain Frog *Breviceps bagginsi* Minter, 2003

Bilbo Baggins, the hero of J. R. R. Tolkien's *The Hobbit*, who also features in *The Lord of the Rings*.

Bagrecito

Bagrecito Andes Frog *Psychrophrynella bagrecito* **Lynch**, 1986

[Syn. *Phrynopus bagrecito*]

Bagrecito means 'small catfish' in Spanish and is Lynch's nickname for his colleague David Cannatella. (See **Cannatella**)

Bailey

Bailey's Treefrog *Hyla goughi baileyi* **Cochran**, 1953 NCR

[JrSYN *Dendropsophus werneri*]

Professor Dr Joseph R. Bailey (1913–1998) ended his career as Emeritus Professor of Zoology at Duke University. He was in Brazil (1940–1942) and was working at the Zoological Museum of the University of Michigan (1946). He co-wrote *Amphibians and Reptiles of the Carolinas and Virginia* (1989).

Baird

Baird's Cricket Frog *Pseudacris feriarum* **Baird**, 1854

[Alt. Upland Chorus Frog]

Professor Spencer Fullerton Baird (1823–1887) was a zoologist who became a giant of ornithology and sufficient has been written elsewhere to obviate a long entry. The young Baird became a friend of John James Audubon (1838), sending him specimens. After studying medicine for a time, Baird became Professor of Natural History, Dickinson College, Pennsylvania (1845). He was Assistant Secretary of the Smithsonian (1850–1878), later becoming Secretary. Among his publications is *Catalogue of North American Reptiles* (1853). Fourteen birds, two reptiles and five mammals are named after him.

Baker

Baker's Robber Frog *Eleutherodactylus bakeri* **Cochran**, 1935

Professor Dr Horace Burrington Baker (1889–1971) was an American zoologist who was an expert on terrestrial gastropods. He was awarded his doctorate at the University of Michigan (1920). After army service (1917–1918) he became first an instructor (1920) at the University of Pennsylvania, then Assistant Professor (1926), Associate Professor (1928) and full Professor (1939–1959). He was also Business Manager of 'Nautilus' (1932) and Editor (1957–1970). He collected in Mexico (1910 & 1926), Venezuela (1919) and he collected the frog holotype (1922) in Haiti. He wrote many articles (1920s–1960s). A reptile is named after him.

Balfour

Balfour's Reed Frog *Hyperolius balfouri* **Werner**, 1908

J. W. Balfour was a missionary in Uganda.

Balletto

Balletto's Toad *Bufo tihamicus* Balletto & Cherchi, 1973

Professor Emilio Balletto (b.1943) is an Italian herpetologist and zoologist at the Dipartimento di Biologia Animale e dell'Uomo, Università degli Studi di Torino. He is President of the Italian Zoological Union (2001) and is a member, Executive Committee, World Congress of Herpetology.

Balzan

Yungas Treefrog *Hypsiboas balzani* **Boulenger**, 1898

Dr Luigi Balzan (1865–1893) was an Italian naturalist who set out on a grand tour of South America (1890), travelling alone by whatever means he could find through Brazil, Paraguay, Peru and Bolivia. He wrote *Voyage de M E Simon au Venezuela 1887–1888* (1892). A reptile and a bird are named after him.

Ba Na

Bana Leaf-litter Frog *Leptobrachium banae* Lathrop et al., 1998

Tram Lap Frog *Odorrana banaorum* Bain et al., 2003

Ba Na is the name of the ethnic minority people in the area of Vietnam where these species were collected.

Barber

Barber's Sheep Frog *Hypopachus barberi* **Schmidt**, 1939

Charles M. Barber was a biologist, palaeontologist and collector with the Field Museum, Chicago. He collected the frog holotype on their Guatemalan Expedition (1906). He was in Guatemala again for them (1930s) and made a survey of manatees in Lake Izabel. He lived in Hot Springs, Arkansas (1930s–1940s), where he bred Nubian goats and prospected for fossils. He discovered a number of Cretaceous fossils, including turtles, one of which Schmidt named after him.

Barbour, R

Streamside Salamander *Ambystoma barbouri* Kraus & Petranka, 1989

Professor Dr Roger W. Barbour (1919–1993) taught and researched at the University of Kentucky. Morehead State Teachers College awarded his bachelor's degree in biology (1938), Cornell University his master's (1939) and his doctorate (1949). His studies were interrupted by army service (WW2). Most of his career was with the Department of Zoology, University of Kentucky, Lexington as Instructor (1950–1952), Assistant Professor (1952–1956), Associate Professor (1956–1968) and then Professor (1968–1984). He was in Indonesia (1957–1959) as a member of the Kentucky Contract Team at the Institute Teknologi di Bandung. He wrote *The Amphibians & Reptiles of Kentucky* (1971).

Barbour, T

Jungle Toad genus *Barbourula* **Taylor** & **Noble**, 1924

Barbour's Forest Treefrog *Leptopelis barbouri* **Ahl**, 1927

Yoro Salamander *Nototriton barbouri* **Schmidt**, 1936

Barbour's Giant Treefrog *Platypelis barbouri* Noble, 1940

Dr Thomas Barbour (1884–1946) was an American zoologist who graduated from Harvard (1906) and obtained his doctorate there (1910). He worked at the Harvard Museum (1911–1946), firstly as an Associate Curator of Reptiles and Amphibians, and rising to become Director and Custodian of Harvard Biological Station and Botanical Garden, Soledad, Cuba (1927). He was Executive Officer in charge of Barro Colorado Island Laboratory, Panama (1923–1945). During his time at the museum he explored in the East Indies, the West Indies, India, Burma, China Japan and South & Central America. He was famously jovial good company and would invite all and sundry to eat and converse. His special area of interest was Central American herpetology. He co-wrote *Checklist of North American Amphibians and Reptiles* (1933). Two mammals, three birds and twenty-four reptiles are named after him.

Barisan

Barisan's Flying Frog *Rhacophorus barisani* Harvey, Pemberton & Smith, 2002

Bukit Barisan mountain range, Sumatra.

Barlagne

Matouba Robber Frog *Eleutherodactylus barlagnei* **Lynch**, 1965

Patrice Barlagne collected the holotype at Matouba, Guadeloupe, with James Lazell, Jr (q.v.)

Baron

Baron's Mantella *Mantella baroni* **Boulenger**, 1888

[Alt. Harlequin Mantella, Variegated Mantella]

The Revd Richard Baron (1847–1907) was an evangelistic missionary, geologist and botanist on Madagascar (1872–1907). He was co-founder and editor (1875) of the *Antananarivo Annual*. He sent about 1,000 new species of plants to Kew. He became a Fellow of the Linnean Society (1882) and was (1902) nominated as a founder member of the Académie Malgache, which the French authorities had started. He was widowed three times and survived by his fourth wife after he died of malaria whilst on leave in England. A scorpion, also from Madagascar, is named after him.

Barrio

Bocaina Dwarf Frog *Physalaemus barrioi* **Bokermann**, 1967

Barrio's Frog *Insuetophrynus acarpicus* Barrio, 1970

Cabreria Spiny-chested Frog *Alsodes barrioi* Veloso et al., 1981

Barrio's Escuerzo *Odontophrynus barrioi* **Cei**, **Ruiz** & Beçak, 1982

Dr Avelino Barrio (1920–1979) was a Spanish botanist and herpetologist who lived in Argentina. A graduate of the Colegio Nacional de Buenos Aires, he undertook his doctorate at Buenos Aires University (1948–1954). He spent most of his career there and at the Instituto Nacional de Microbiología 'Gustav Malbrán', concentrating on the genetics of snakes and anurans. He founded (1969) the Centro Nacional de Investigaciones Biológicas and dedicated much time to the study of venoms. A reptile is named after him.

Barthlen

Andes Frog sp. *Phrynopus barthlenae* **Lehr** & **Aguilar,** 2002

Dr Marianne Barthlen is a German patron of BIOPAT.

Barton

Barton's Robber Frog *Eleutherodactylus bartonsmithi* **Schwartz**, 1960

Barton L Smith was an American herpetologist who was a student of Schwartz and helped him collect specimens on his field trips to the Caribbean.

Bartram

Southern Spring Peeper *Pseudacris crucifer bartramiana* Harper, 1939

William Bartram (1739–1823) was an American naturalist, explorer and botanist who has been called the grandfather of American ornithology, perhaps because he was the protector of Alexander Wilson, who was known as the father of American ornithology in his time. He accompanied his father, John Bartram (America's 'first' botanist), on expeditions; to the Catskill Mountains (1753–1754), New York & Connecticut (1755), and Florida (1765). William remained in the South drawing flora, collecting botanical specimens, becoming an accomplished ornithologist and befriending both colonial planters and Native American tribes. When the American War of Independence started, he served in the Georgia Militia (1776), then returned to Pennsylvania (1777). He was offered the position of Professor of Botany, University of Pennsylvania (1782), but never lectured there. As well as befriending Alexander Wilson, he took his nephew Thomas Say under his wing. Bartram put together *Catalogue of Birds of North America*. Three birds are named after him.

Bassler

Pleasing Poison Frog *Ameerega bassleri* Melin, 1941

Pastaza River Caecilian *Oscaecilia bassleri* **Dunn**, 1942

Bassler's Humming Frog *Chiasmocleis bassleri* Dunn, 1949

Dr Harvey Bassler (1882–1950) was a geologist, palaeontologist and expert on the Amazon River Basin. He was a Research Associate, Herpetology, AMNH, New York. He spent many years in Peru, including one visit lasting over 4 years. He made extensive herpetological collections, including Peruvian caecilians. He and Dunn were friends and Bassler sent Dunn many specimens, often with details of who had collected them and where, making suggestions on how they should be named. His collection of Peruvian fossils at Lehigh University, Pennsylvania remains regarded as one of the finest of its kind.

Batanta

Batanta Wrinkled Ground Frog *Platymantis batantae* **Zweifel**, 1969

Batanta Island is in West Papua (Indonesian New Guinea).

Bates

Benito River Night Frog *Astylosternus batesi* **Boulenger**, 1900

Bates' River Frog *Phrynobatrachus batesii* Boulenger, 1906

Bates' Tree Toad *Nectophryne batesii* Boulenger, 1913

George Latimer Bates (1863–1940) was an American naturalist who travelled in West Africa (1895–1931), residing in Cameroon for some years. He sent specimens to, inter alia, the BMNH and the Philadelphia Academy. He wrote *Handbook of the Birds of West Africa* (1930). Twenty birds, four mammals and a reptile are named after him.

Battersby

Battersby's Caecilian *Indotyphlus battersbyi* **Taylor**, 1960

Battersby's Treefrog *Dendropsophus battersbyi* **Rivero**, 1961

James Clarence Battersby (1901–1993) was a British herpetologist at the BMNH, London (1916–1961), joining that institution aged 15 as a boy attendant for George Albert Boulenger, and retiring as Curator with his wife working alongside him. Five reptiles are named after him.

Baudin

Baudin's Treefrog *Smilisca baudinii* **Duméril** & **Bibron**, 1841

[Alt. Common Mexican Treefrog]

Nicolas Thomas Baudin (1754–1803) was a French naval officer and explorer. Captain Baudin set out (1800) from Le Havre with two corvettes, the

'Geographe' and the 'Naturaliste', with nine zoologists and botanists aboard. The expedition's main task was to map the coast of Australia. 'Geographe' crawled back to Le Havre (1804) with a living cargo of 72 birds and other animals, but Baudin had died of tuberculosis in Mauritius. The expedition was a great scientific and artistic success. Two birds and a reptile are named after him.

Baumann, E

Baumann's Reed Frog *Hyperolius baumanni* **Ahl**, 1931

Ernst Baumann (1863–1895) was a German zoologist who initially trained for a career in commerce, but switched to natural history and studied privately at the Zoologisches Museum, Berlin. He was sent to the German research station at Misa-Höhe, Togo (the type locality of the reed frog) and explored extensively in Togo (1892–1895). He was on leave in Germany when he died from recurrent malaria. A bird is named after him.

Baumann, F

Baumann's Tree Toad *Hylodes lateristrigatus* Baumann, 1912

Dr Franz Baumann (1885–1961), who was a long-serving Director of the Bern Museum, described the tree toad in *Brazilianische batrachier des Berner naturhistorischen Museums* (1912). He was awarded his doctorate (1910) and became a Professor (1921).

Baumgartner

Baumgardner's Snouted Treefrog *Scinax* **baumgardneri* **Rivero**, 1961

Dr Hans Baumgartner of Puerto Ayacucho, Venezuela, collected specimens of this treefrog for Rivero, who described him as '….a great friend of naturalists of Territorio Amazonas and a man with a great desire to help others.'

*Rivero spelt the binomial *baumgardneri* rather than *baumgartneri* - presumably in error.

Baxter

Baxter's Toad *Anaxyrus baxteri* **Porter**, 1968
[Alt. Wyoming Toad; Syn. *Bufo baxteri*]

Dr George Theodore Baxter (1919–2005) was an American herpetologist and ichthyologist. His bachelor's and master's (1946) degrees were awarded by the University of Wyoming, where he spent his working life (1947–1984), starting as Instructor and retiring as Emeritus Professor with a break to complete his doctorate at the University of Michigan (1951). He was an expert in the wildlife of Wyoming and the Rocky Mountain region, and wrote *Amphibians and Reptiles of Wyoming* (1980).

Bayon

Bayon's (Common) Reed Frog *Hyperolius viridiflavus* **bayoni* **Boulenger**, 1911

Francisco Antonio Pinheiro Bayão was a Portuguese planter and administrator whose family had settled in Duque de Bragança district of Angola in the seventeenth century. He was a keen naturalist and collector and sent many specimens to Lisbon, often using Anchieta (q.v.) as an intermediary. Three reptiles are named after him.

* His surname Bayão seems always to have been rendered as *bayoni*.

Bearse

Bearse's Robber Frog *Pristimantis bearsei* **Duellman**, 1992

Dr Robert Carleton Bearse (b.1938) is a nuclear physicist whose doctorate was awarded (1964) by Rice University, Texas. He was Associate Vice Chancellor for Research, Graduate Studies and Public Service, University of Kansas. His administration '…continuously enhanced the programs of the Museum of Natural History.' He is now Professor Emeritus, Experimental Nuclear Physics, University of Kansas.

Beccari

Beccari's Giant Frog *Conraua beccarii* **Boulenger**, 1911
[Alt. Filfil Slippery Frog]

Dr Odoardo Beccari (1843–1920) was an Italian botanist. He met (c.1864) Darwin and Hooker at Kew and James Brooke, first Rajah of Sarawak, who enabled him to stay (1865–1868) in Sarawak, Malaya and Brunei – recorded in *Wanderings in the Great Forests of Borneo*. After this he visited Ethiopia before going to New Guinea with d'Albertis (1872–1873). He also collected in the Celebes, New Guinea, and Sumatra, where he found (1878) the Titan Arum (Corpse Flower), the world's largest flower. Five birds, six reptiles and four mammals are named after him.

Beck

Beck's Treefrog *Litoria becki* **Loveridge**, 1945

Sergeant William M Beck Jr supplied the describer with a collection of New Guinea herpetological material.

Becker

Treefrog sp. *Hypsiboas beckeri* **Caramaschi** & **Cruz**, 2004

Professor Johann Becker (1932–2004) was a Brazilian zoologist working first as a researcher then Curator, Museu Nacional, Rio de Janeiro. As a student (1951), he was admitted as apprentice in the Division of Insects at the Museu Nacional. His initial degree was BSc. (1954) and he became zoologist researcher, and later on a full Professor of that Museum, being a specialist in genetics, evolution, and invertebrate palaeontology. He had collected insects, especially beetles, ever since he was a student and, after his death, his entomological collection of over 14,000 specimens was presented to the Zoology Department of the Universidade Estadual de Feira de Santana. A bird is named after him.

Beddome

Beddome's Bush Frog *Raorchestes beddomii* **Günther**, 1876

[Alt. Beddome's Bubble-nest Frog; Syn. *Philautus beddomii*]

Beddome's Leaping Frog *Indirana beddomii* Günther, 1876

Beddome's Toad *Duttaphrynus beddomii* Günther, 1876

Beddome's Caecilian *Ichthyophis beddomei* **Peters**, 1880

[Alt. Nilgiri Yellow-striped Caecilian]

Beddome's Dwarf Wrinkled Frog *Nyctibatrachus beddomii* **Boulenger**, 1882

[Alt. Beddome's Night Frog, Pygmy Wrinkled Frog]

Colonel Richard Henry Beddome (1830–1911) was an officer in India and a keen naturalist and botanist. He joined the army there (1848) and became Assistant Conservator of Forests in the Madras Presidency (1857), eventually becoming Chief Conservator (1860–1882) and a member of the University of Madras (1880). He was first to recognize and describe the great diversity of Indian herpetofauna. After returning to England he lived near, and frequently visited, the botanical gardens at Kew. He wrote *Handbook of the ferns of British India, Ceylon and Malaya Peninsula* (1892). A mammal and fifteen reptiles are named after him.

Bedriaga

Bedriaga's Frog *Pelophylax bedriagae* Camerano, 1882

[Alt. Levant Green Frog; Syn. *Rana bedriagae*]

Dr Jacques Vladimir von Bedriaga (aka Jacob Vladimirovich Bedriaga or Johann von [Jean de] Bedriaga) (1854–1906) was a herpetologist. He studied at the University of Moscow but bad health forced him to leave. He went to the University of Jena, Germany, studied comparative anatomy, and after receiving his doctorate (1875) undertook a series of expeditions to Mediterranean countries to study reptiles, periodically visiting Russia to study specimens that Prjevalsky was sending back from his expeditions. His health deteriorated further so he settled in Nice (1881), later moving to Florence, where he died. He wrote *Die Amphibien und Reptilien Griechenlands*, (1880), the first monograph on Greek herpetology. Five reptiles are named after him.

Beebe

Beebe's Rocket Frog *Anomaloglossus beebei* **Noble**, 1923

Beebe's Toad *Rhinella beebei* **Gallardo**, 1965

Dr Charles William Beebe (1877–1962) was a zoologist, marine biologist, conservationist, explorer and writer. He began his working life looking after the birds at the Bronx Zoo, New York, later becoming Curator of Ornithology, New York Zoological Society (1899–1952), and Director, Department of Tropical Research (1919). He was greatly interested in deep-sea exploration and made a number of descents in the bathysphere, including (1934) a record descent of 923 metres (3,028 feet) off Nonsuch Island, Bermuda. He set up a camp (1942) at Caripito in Venezuela for jungle studies and (1950) bought 92 hectares (228 acres) of land in Trinidad and Tobago, which became the New York Zoological Society's Tropical Research Station (Asa Wright Nature Centre). He married Helen Elswyth Thane Ricker (1900–1981), who wrote romantic novels as 'Elswyth Thane'. Much of his writings were popular books on his expeditions and he made enough money from them to finance his later expeditions. He retired to Trinidad. A number of other taxa including several fishes and a bird are named after him.

Beira

Beira's Toad *Poyntonophrynus beiranus* **Loveridge**, 1932

[Alt. Beira Pygmy Toad; Syn. *Bufo beiranus*]

Beira in Mozambique was the location where the type was taken.

Beireis

Beireis' Treefrog *Dendropsophus leucophyllatus* Beireis, 1783

Professor Gottfried Christoph Beireis (1730–1809) was a German chemist and physician. He studied at Jena, but his medical degree was awarded later for other work and after he became Professor of Medicine and Chemistry (1759). He taught anatomy, medicine, surgery, chemistry, botany, natural history, pharmacy, mineralogy, metallurgy, agriculture, forestry, music, painting and numismatics. He made a fortune as he discovered a way of converting ammonium sulphide to cinnabar and selling the cinnabar as a red dye. He also collected curiosities, especially automata.

Bejarano

Bolivian Cochran Frog *Nymphargus bejaranoi* **Cannatella**, 1980

Dr Gastón Bejarano is a zoologist who is Director of Forestry and National Parks, Ministry of Agriculture, La Paz, Bolivia. Cannatella wrote: 'Prof Bejarano, often at tremendous personal sacrifice, has furthered the study of Bolivian fauna and flora through his encouragement and aid to visiting scientists.' A bird is named after him.

Bell

Bell's Toad *Ascaphus truei* **Stejneger**, 1899
[Alt. American Bell Toad, Coastal/Pacific Tailed Frog]

This is a good example of a mistaken transcription, with a spurious apostrophe.

Bell, H

Robber Frog sp. *Pristimantis bellae* Reyes-Puig & Yánez-Muñoz, 2012

Hilary Bell works for the accountancy firm PricewaterhouseCoopers (PwC), which company is a corporate sponsor of the World Land Trust and helped to fund the reserve 'Bosque Protector Cerro La Candelaria' in Ecuador where the frog is found. PwC decided to hold a competition for their staff, with the prize of having the frog named after them, and Hilary was the lucky winner.

Bell, LN

Everglades Dwarf Siren *Pseudobranchus axanthus belli* **Schwartz**, 1952

Dr L. Neil Bell (b.1933) is a biologist and herpetologist who also had a career in data processing for Kaiser Hospitals, Oakland, California (1960–1991). The University of Florida awarded his bachelor's degree in zoology (1954). He was a wildlife biologist for the Florida Fish and Game Department (1955–1957) and served in the US Air Force (1958–1960). He attended (1977–1981) the University of California, Berkeley, as a mature student (presumably part-time as he was working for Kaiser Hospitals) and took a bachelor's degree in entomology and a doctorate. He collected the siren holotype (1952).

Bell, T

Bell's False Brook Salamander *Pseudoeurycea bellii* **Gray**, 1850

Professor Thomas Bell (1792–1880) was a naturalist and dental surgeon who worked at Guy's Hospital, London (1816–1860). He became Professor of Zoology at King's College (1834) and described many of the reptile species Darwin collected. He was an assistant to Gray, who simply says that the salamander holotype was 'Presented by T. Bell, Esq.' He was President of the Linnean Society of London and wrote *History of British Reptiles* (1839). Ten reptiles are named after him.

Bellon

Snouted Treefrog sp. *Scinax belloni* **Faivovich**, Gasparini & **Haddad**, 2010

José Bellon discovered and collected the treefrog holotype, and has helped others with their fieldwork in southeast Brazil.

Bellona

Murri Robber Frog *Pristimantis bellona* **Lynch**, 1992

In Roman mythology, Bellona was the companion of Mars and goddess of war. The name was apparently inspired by the fact that in this species '… only the females have co-ossifed helmets.'

Belly

Belly's Madagascar Frog *Mantidactylus bellyi* **Mocquard**, 1895

Monsieur Belly was a collector and collected the frog with Charles Alluaud. There is a reptile also named after him (1894) by Mocquard, who gave no details of Belly beyond his family name.

Benítez

Benítez's Treefrog *Hypsiboas benitezi* **Rivero**, 1961

Professor Dr Jaime Benítez (1908–2001) was the longest serving Chancellor of the University of Puerto Rico. He qualified in law at Georgetown University, Washington DC taking bachelor's and master's degrees (1930 & 1931) and an MA from the University of Chicago (1938). He worked at the University of Puerto Rico (1931–1971), starting by teaching social and political sciences and rising to be Chancellor (1942–1966) and President (1966–1971). He entered politics (1971) and was one of Puerto Rico's representatives in the US

Congress (1972–1976). He failed to get re-elected and returned to teaching in Puerto Rico, being Professor of Government at the American College in Bayamón, Puerto Rico. He was an enthusiastic supporter of the expedition during which the treefrog holotype was collected.

Bennett

Holy Cross Toad *Notaden bennettii* **Günther**, 1873
[Alt. Crucifix Toad]

Dr George Bennett (1804–1893) was a surgeon, botanist and zoologist. He was surgeon-naturalist on board the 'Sophia's' passage to the South Seas and Australia (1834). He returned to Britain but eventually settled permanently in Australia (1836). He was the first Curator and Secretary of the Australian Museum (1835) and an early conservationist, writing (1860): 'Many of the Australian quadrupeds and birds are not only peculiar to that country, but are, even there, of comparatively rare occurrence: and such has been the war of extermination recklessly waged against, that they are in a fair way of becoming extinct.... The Author hopes that what he has been induced to say with reference to this important subject will not be without weight to every thoughtful colonist.' He spent 50 years unsuccessfully trying fully to understand monotreme and marsupial biology. He wrote *Gatherings of a Naturalist in Australia – Being Observations Principally on the Animal and Vegetable Productions of New South Wales, New Zealand, and some of the Austral Islands* (1860). Three birds, two reptiles and a mammal are named after him.

Bequaert

Vissoke River Frog *Phrynobatrachus bequaerti* **Barbour** & **Loveridge**, 1929
Gbanga Forest Treefrog *Leptopelis bequaerti* Loveridge, 1941

Professor Dr Joseph Charles Bequaert (1886–1982) was a Belgian botanist, entomologist and malacologist. He graduated with a doctorate in botany from the University of Ghent (1906) and worked for the colonial government in the Belgian Congo (1910–1915). He went to the US (1916), later taking citizenship citizen (1921). He was a Research Assistant at the AMNH, New York (1917–1922), then worked at Harvard (1923–1956); initially teaching entomology at Harvard Medical School, finally becoming Professor of Zoology, Museum of Comparative Zoology. In retirement he became Professor of Biology, University of Houston (1956–1960) and Visiting Entomologist, University of Arizona. Among other works he co-wrote *The Mollusks of the Arid Southwest* (1973). A reptile is named after him.

Berdmore

Berdmore's Narrow-mouthed Frog *Microhyla berdmorei* **Blyth**, 1856
[Alt. Berdmore's Chorus Frog]

Major Thomas Mathew Berdmore (1811–1859) was Assistant Commissioner in Northern Tenasserim (Myanmar). He amassed a large natural history collection, which he presented to the Asiatic Society of Bengal (1856). Two reptiles and two mammals are named after him.

Berg

Lesser Granulated Toad *Rhinella bergi* Céspedez, 2000

Dr Frederico Guillermo Carlos (Friedrich Wilhelm Carl) Berg (1843–1902) was a Latvian entomologist and naturalist. After a number of years working in commerce, he became a conservator of entomological specimens at the Riga Museum. He joined the Museo Nacional, Buenos Aires (1873) and went on expeditions to Patagonia (1874) and Chile (1879). He worked at the Museo Nacional de Historia Natural, Montevideo (1890–1892) and was Director, Museo Nacional, Buenos Aires (1892–1901). A mammal and three birds are named after him.

Berger, L

Italian Pool Frog *Pelophylax bergeri* **Günther**, 1986

Leszek Berger (d.2012) was a Polish herpetologist who was the first to realize that the Edible Frog (formerly *Rana esculenta*, now *Pelophylax kl. esculentus*) was a natural hybrid of *Pelophylax ridibundus* and *P. lessonae*.

Berger, TJ

Berger's Glass Frog *Hyalinobatrachium bergeri* **Cannatella**, 1980

Thomas J. Berger collected with Cannatella and Duellman in Peru (1979) and Ecuador (1991). He appears to have been the expedition's comedian as the etymology says he '....provided comic relief while securing part of the type series.'

Bergmann

Bergmann's Treefrog *Litoria christianbergmanni* **Günther**, 2008

Professor Dr Christian Bergmann of Berlin was honoured in this treefrog's binomial due to the support he gave to Günther's work on Papuan herpetofauna.

Berkenbusch

Berkenbusch's Robber Frog *Craugastor berkenbuschii* **Peters**, 1870

Georg Heinrich August Berkenbusch (1821–1901) was the Prussian Vice-Consul at Puebla in Mexico (1869), where he made a collection that was sent to the Museum für Naturkunde in Berlin.

Berlandier

Rio Grande Leopard Frog *Rana berlandieri* **Baird**, 1859

[Syn. *Lithobates berlandieri*]

Jean Louis Berlandier (1805–1851) was a Belgian botanist, who went to Mexico (1826) to work as a collector. The Mexican Government employed him (1827–1828) and he stayed on, marrying a local woman, dividing his time between a pharmaceutical business and collecting botanical specimens. The Mexican Government employed him again (1834), as an interpreter to General Arista, and in charge of the hospitals at Matamoros during the Mexican War. He drowned whilst trying to cross the San Fernando River. A reptile and a bird are named after him.

Bernal

Argelia Robber Frog *Pristimantis bernali* **Lynch**, 1986

Dr Rodrigo Bernal (b.1959) is a Colombian botanist who is an expert on palms. The University of Aarhus, Denmark awarded his doctorate (1996). He is Associate Professor, Instituto de Ciencias Naturales, Universidad Nacional de Colombia and has published more than 60 papers and longer works and co-wrote *Field guide to the palms of the Americas* (1995). He and Lynch were two of the group that collected the frog holotype.

Bernardo

Toad sp. *Rhinella bernardoi* Sanabria et al., 2010

Dr Joseph Bernardo is a specialist in salamanders at the Department of Zoology, Duke University, and a senior research associate at Cornell University. He is editor-in-chief of the magazine *Contemporary Herpetology* (2002).

Bernhard

Bernhard's Mantella *Mantella bernhardi* Vences et al., 1994

Dr Bernhard Meier is a German biologist and primatologist who was at Ruhr-Universität, Bochum. The describers say he is their friend and honour him for his work on lemurs.

Bernis

Indonesian Caecilian *Ichthyophis bernisi* Salvador, 1975

Dr Francisco Bernis Madrazo (1916–2003) was an ornithologist, pioneer conservationist and ecologist, who was one of the leading lights in creating the Doñana Biological Reserve. The Spanish Civil War (1936–1939) interrupted his studies, but they were finally completed (1941) and his doctorate in botany was awarded (1950). He was a founder member of the Spanish Ornithological Society (1954). He was appointed Professor of Vertebrate Zoology, Madrid University (1956), retiring as Emeritus Professor (1985). He wrote *Migration en Aves, Tratado Teórico y Práctico* (1966).

Bertha

Bertha's Treefrog *Dendropsophus berthalutzae* **Bokerman**, 1962

Dwarf Snouted Treefrog *Scinax berthae* **Barrio**, 1962

Bertha Lutz (see **Lutz, BMJ**)

Bertin

Isaka-Ivondro Madagascar Frog *Spinomantis bertini* **Guibé**, 1947

[Syn. *Mantidactylus bertini*]

Léon Bertin (1896–1956) was a French ichthyologist.

Betancur

Ituango Rocket Frog *Hyloxalus betancuri* **Rivero** & **Serna**, 1991

John J. Betancur was one of the collectors of the species (1989), as was Serna. The etymology says nothing more about him.

Betsileo

Betsileo Reed Frog *Heterixalus betsileo* **Grandidier**, 1872

Bronze Mantella *Mantella betsileo* Grandidier, 1872

Betsileo Bright-eyed Frog *Boophis rhodoscelis* **Boulenger**, 1882

Betsileo Giant Treefrog *Platypelis cowanii* Boulenger, 1882

The Betsileo are a Madagascan ethnic group of Malayo-Indonesian origin.

Beyer

Beyer's Waterdog *Necturus beyeri* **Viosca**, 1937

[Alt. Gulf Coast Waterdog]

George Eugene Beyer (1861–1926) was a German-born naturalist who immigrated to the USA (1880). He travelled in Central America (1881–1883), collecting for the Dresden Zoological Museum. He heard that all his specimens had been lost in a shipwreck so he went to the USA and settled in New Orleans. He became Curator of Vertebrates, Tulane Museum (1893), where he taught (1895–1918), leaving as full Professor to work for Louisiana State Board of Health (1918–1926) (where Viosca also worked). Among his many activities, he was special inspector for Biological Survey of the US Department of Agriculture. He was considered to be the first resident herpetologist for the state of Louisiana, enhancing his reputation by allowing poisonous snakes to bite him, often as a display for journalists.

Bibron

Four-eyed Frog *Pleurodema bibroni* **Tschudi**, 1838
Broad-banded Grass Frog *Ptychadena bibronii* **Hallowell**, 1845
Bibron's Toadlet *Pseudophryne bibronii* **Günther**, 1859

Gabriel Bibron (1806–1848) was a zoologist and herpetologist. He worked closely with André Marie Constant Duméril at the Paris Museum. Duméril was most interested in dissection and left the naming of species largely to Bibron, technically only his assistant. Bibron resigned (1845), later dying of tuberculosis. Sixteen reptiles are named after him.

Biju

Indian Tree Frog sp. *Polypedates bijui* Zachariah et al., 2011

Dr Sathyabhama Das Biju is an Indian systematic biologist teaching (2006) at the University of Delhi where he is Associate Professor. He was initially a plant taxonomist but gradually developed a keen interest in amphibian biology. The University of Kerala awarded his MSc and the University of Calicut his PhD in plant taxonomy. Later, Vrije Universiteit, Brussels awarded his PhD in frog systematics (2007). He has discovered more than 100 amphibians including five new genera. He has written or co-written many papers including *A synopsis to the frog fauna of the Western Ghats, India* (2001). The etymology read: 'This elegant species of tree frog is named after Dr S. D. Biju, University of Delhi, India for his outstanding contribution to the modern amphibian systematics in India.'

Bilbo

Bilbo's Rain Frog *Breviceps bagginsi* Minter, 2003

(See **Baggins**)

Bingham

Robber Frog sp. *Eleutherodactylus binghami* **Stejneger**, 1913 NCR
[JrSYN *Gastrotheca marsupiata*]

Dr Hiram Bingham (1875–1956) was an academic, explorer and politician. He was on the Yale Peruvian Expedition (1911) during which the frog holotype was collected and he rediscovered the Incan city of Machu Pichu. Yale awarded his bachelors degree (1898) and Harvard his doctorate (1905). He entered politics (1923) as Lieutenant Governor of Connecticut and became a Senator for Connecticut in the US Congress (1924–1933). A bird is also named after him.

Biró

New Guinea Cross Frog *Oreophryne biroi* **Méhely**, 1897
Biró's Frog *Cophixalus biroi* Méhely, 1901
[Alt. Sattelberg Rainforest Frog]

Dr Lajos Biró (1856–1931) was a Hungarian ethnographer, ornithologist, entomologist, collector and photographer. He was in German New Guinea (1896–1901), and later worked at the Budapest Natural History Museum (1903–1931). His honorary doctorate was awarded (1926) by University of Sciences, Szeged, Hungary. He also travelled in India, Sri Lanka, Southeast Asia, Anatolia and North Africa. A bird is also named after him.

Bischoff

Bischoff's Treefrog *Hypsiboas bischoffi* **Boulenger**, 1887

Theodor Bischoff was a schoolmaster in Rio Grande do Sul, Brazil, where he collected the treefrog holotype. He is recorded as 'Senhor Bischoff' in a description of a marsupial that he presented to his friend the zoologist Hermann Ihering. We infer he was just one of the many German immigrants in that area of Brazil. He wrote *Sobre os Sambaquis do Estado do Rio Grande do Sul* (1898).

Bishop

Reticulated Flatwoods Salamander *Ambystoma bishopi* **Goin**, 1950
Ozark Hellbender *Cryptobranchus alleganiensis bishopi* **Grobman**, 1943

Professor Dr Sherman Chauncey Bishop (1887–1951) was an American herpetologist and arachnologist. He quarrelled with his local school authorities and so left without graduating, but still managed to get a place at Cornell where he worked as Assistant Entomologist (1909–1913), graduating with a bachelor's degree (1913). He worked at Cornell (1914–1917) as an entomologist and plant pathologist and later as a biologist. Cornell awarded his doctorate (1925). He was on the Cornell

Okefenokee Expedition (1912). He was Zoologist at New York State Museum in Albany (1916–1928), though his career was disrupted as he worked in Naval Intelligence (WW1). He was at the University of Rochester, New York (1928–1951), progressing to become Professor of Zoology. He wrote *Handbook of Salamanders* (1943).

Blair

Blair's Snouted Treefrog *Scinax blairi* **Fouquette** & **Pyburn**, 1972

Plains Leopard Frog *Rana blairi* Mecham et al., 1973 [Syn. *Lithobates blairi*]

Professor Dr William Franklin Blair (1912–1984) was a zoologist who specialized in the hybrid zones of mammals, found that this led him to the study of frogs and, thereafter, he taught herpetology at the University of Texas, becoming Professor of Biology (1955). The University of Tulsa awarded his bachelor's degree (1934), his master's was from the University of Florida (1935), and his doctorate from the University of Michigan (1938). He remained at the latter as a Research Associate before service in the US Army Air Corps (1942–1946). After the Second World War he joined the faculty of the University of Texas. He started the Texas Natural History Collections (1946), concentrating on mammals, amphibians and reptiles. He retired as Professor Emeritus (1982). The annual W. Frank Blair Eminent Naturalist Award was established in his honour. A reptile is named after him.

Blanc

Blanc's Madagascar Frog *Gephyromantis blanci* **Guibé**, 1974

Lope Egg Frog *Leptodactylodon blanci* Ohler, 1999

Emeritus Professor Charles Pierre Blanc (b.1933) is a French herpetologist who worked at the Laboratoire de Zoogéographie, Université Paul Valéry, Montpellier. He has made many trips to Madagascar. He co-wrote *Inventaire systématique des Amphibiens Anoures du centre du Gabon* (2001). A reptile is named after him.

Blanchard

Blanchard's Cricket Frog *Acris blanchardi* Harper, 1947

Dr Frank Nelson Blanchard (1888–1937) was an American herpetologist. He graduated from Tufts University, Massachusetts (1913) and then taught zoology (1913–1916) at an agricultural college. After a zoology fellowship and a doctorate from the University of Michigan (1919), and working briefly at the Smithsonian under Stejneger (q.v.), he returned to the University of Michigan (1920) to teach zoology, becoming an Assistant Professor (1926), then Associate Professor (1934). He was in New Zealand, Australia and Tasmania (1927–1928). He was Vice-President of the American Society of Ichthyologists and Herpetologists (1936–1937). He wrote *Amphibians and Reptiles of Western Tennessee* (1922). Four reptiles are named after him.

Blanford

Blanford's Whipping Frog *Rhacophorus dennysi* Blanford, 1881

Blanford's Frog *Nanorana blanfordii* **Boulenger**, 1882

Blanford's Toad *Amietophrynus blanfordii* Boulenger, 1882

William Thomas Blanford (1832–1905) was a geologist and zoologist. He studied at the Royal School of Mines (1852–1854) and at Freiberg, Saxony, before joining the Indian Geological Survey (1854) where he investigated coalmines at Talchir (1854–1857). He undertook a geological survey of Burma (1860) and was appointed Deputy Superintendent. He surveyed in Bombay (1862–1866) and was then attached to the Abyssinian expedition (1867). He wrote the mammals section of *The Scientific Results of the Second Yarkand Mission: Mammalia* (1879). Ill health forced early retirement and he returned to England (1881). In retirement he edited works for the government on Indian fauna and was President of the Royal Geographical Society (1888–1890). Seventeen reptiles, twelve birds and seven mammals are named after him.

Blomberg

Blomberg's Toad *Rhaebo blombergi* **Myers** & Funkhouser, 1951 [Alt. Colombian Giant Toad]

Leaf Frog sp. *Phyllomedusa blombergi* Funkhouser, 1957 NCR [JrSYN *Phyllomedusa vaillantii*]

Rolf Blomberg (1912–1996) was a Swedish explorer, filmmaker, photographer and author who first went to Ecuador (1934). He was a neutral war correspondent (WW2) and was based in Indonesia where he helped people held in Japanese concentration camps. He returned to Ecuador after the Second World War and made a number of expeditions to contact indigenous tribes, also exploring in Colombia, Brazil and Peru, and made films in Indonesia and Australia. He settled permanently in Ecuador (1968). He wrote *Såna djur finns* (1951).

Blommers

Mantellid frog genus *Blommersia* **Dubois**, 1992
Moramanga Madagascar Frog *Blommersia blommersae* **Guibé**, 1975
Blommers' Madagascar Frog *Guibemantis flavobrunneus* Blommers-Schlösser, 1979
Bright-eyed Frog sp. *Boophis blommersae* **Glaw** & Vences, 1994

Rose Marie Antoinette Blommers-Schlösser (b.1944) is a Dutch entomologist and herpetologist at the Zoölogisch Museum, Amsterdam. She co-wrote *Karyotypes of eight species of phytoseiid mites* (Acarina: Mesostigmata) *from Madagascar* (1975).

Blum

Kosarek Land Frog *Austrochaperina blumi* **Zweifel**, 2000

Dr J. Paul Blum is a herpetologist who made a number of important collections of frogs from New Guinea. He co-wrote *Notes on Xenobatrachus and Xenorhina* (Amphibia: Microhylidae) *from New Guinea with description of nine new species* (1988).

Blyth

Blyth's Frog *Limnonectes blythii* **Boulenger**, 1920
[Alt. Asian Giant Stream Frog, Malayan Giant Frog]

Edward Blyth (1810–1873) was an English zoologist and author. He was Curator of the Museum, Asiatic Society of Bengal (1842–1864). Arthur Grote said of him: 'Had he been a less imaginative and more practical man, he must have been a prosperous one... All that he knew was at the service of everybody. No one asking him for information asked in vain.' His *Catalogue of the mammals and birds of Burma* was published posthumously (1875). Twenty six birds, three reptiles and five mammals are named after him.

Bob

Bob's Robber Frog *Craugastor punctariolus* **Peters**, 1863
[Alt. Streamside Rain Frog]

We have failed to find which 'Bob' has been honoured in the common name.

Bocage

Bocage's Tree Frog *Leptopelis bocagii* **Günther**, 1865
[Alt. Bocage's Burrowing Tree Frog]
Bocage's Grass Frog *Ptychadena subpunctata* Bocage, 1866
[Alt. Spotted Ridged Frog, Speckled-bellied Grass Frog]
Bocage's Reed Frog *Hyperolius bocagei* **Steindachner**, 1867
Bocage's Rocket Frog *Hyloxalus bocagei* **Jiménez de la Espada**, 1870

José Vicente Barboza du Bocage (1823–1907) was Director, National Zoological Museum, Lisbon, Portugal, which is now named in his honour. He became known as the father of Angolan ornithology and wrote *Ornithologie d'Angola*. Twelve birds, ten reptiles and six mammals are named after him.

Bocourt

Bocourt's Robber Frog *Craugastor bocourti* **Brocchi**, 1877
Bocourt's Toad *Incilius bocourti* Brocchi, 1877
Bocourt's Treefrog *Hyla bocourti* **Mocquard**, 1899

Marie Firmin Bocourt (1819–1904) was a zoologist and artist. He followed his father who engraved copperplates for the Paris Natural History Museum. He became (1834) a preparator for Bibron (q.v.). He was officially designated 'museum painter' (1854). He was sent to Siam (1861) (Thailand), where he made an important collection of specimens that he took back to Paris. He published, with Duméril & Mocquard, *Études sur les reptiles et les batraciens* (1870). Eighteen reptiles are named after him.

Boeseman

Boeseman's Snouted Treefrog *Scinax boesemani* **Goin**, 1966

Dr Marinus Boeseman (1916–2006) was a Dutch ichthyologist, Department of Zoology, Leiden Museum, becoming Curator of Fishes (1947–1981). His master's degree was awarded by Leiden (1941). He aided the Dutch resistance (WW2), was arrested (1943), but survived imprisonment at Dachau - though for years his health was so badly affected he could not work. He collected fishes in El Salvador (1953) and was in New Guinea (1954–1955). He caught polio (1957) and suffered from permanent disability in his right arm, but still took part in collecting expeditions. A reptile and at least one fish are named after him.

Boettger, J

Andes Frog sp. *Psychrophrynella boettgeri* **Lehr**, 2006

José Boettger of Oxapampa, Peru, who comes from a family of naturalists, discovered the holotype.

Boettger, O

Boettger's Grainy Frog *Gephyromantis horridus* Boettger, 1880
Boettger's Dainty Frog *Cacosternum boettgeri* **Boulenger**, 1882
Boettger's Robber Frog *Leptodactylus andicola* Boettger, 1891 NCR

[JrSYN *Gastrotheca marsupiata*]
Boettger's Caecilian *Siphonops paulensis* Boettger, 1892
Boettger's Colombian Treefrog *Dendropsophus columbianus* Boettger, 1892
Zaire Dwarf Clawed Frog *Hymenochirus boettgeri* **Tornier**, 1896
Boettger's Spadefoot Toad *Xenophrys boettgeri* Boulenger, 1899
[Alt. Pale-shouldered Horned Toad]
Boettger's Callulops Frog *Callulops boettgeri* **Méhely**, 1901
Boettger's Reed Frog *Heterixalus boettgeri* **Mocquard**, 1902

Professor Dr Oskar Boettger (or Böttger) (1844–1910) was a German zoologist who specialized in herpetology and malacology. He went to university in Frankfurt-am-Main and then to the School of Mines, Freiberg, intending to become a mining engineer but political unrest in Germany prevented him from finding work. He returned to studying, taking a doctorate in palaeontology, University of Würzburg (1869). He joined the Senckenberg Museum, Frankfurt (1870) as a palaeontologist, becoming Curator of Herpetology (1875). Unpaid, he supported himself by taking teaching posts in Offenbach and Frankfurt. He was severely agoraphobic, hardly leaving the Museum (1876–1894), except, as an avid philatelist, for the prospect of a rare stamp. He always asked his assistants to bring him back postage stamps from the countries they visited. Twenty-five reptiles are named after him.

Bogert

Treefrog sp. *Dendropsophus bogerti* **Cochran** & **Goin**, 1970

Charles Mitchill Bogert (1908–1992) was an American herpetologist. He worked as a technician with the Los Angeles City Schools' Division of Nature Study (1928). He was a guide, Rocky Mountain National Park (1930–1932) then a Ranger Naturalist, Grand Canyon (1932–1934). The University of California, where he worked as a teaching assistant (1934–1936), awarded his bachelor's (1934) and master's (1936) degrees. He was Assistant Curator, Herpetology Department, AMNH (1936–1940), becoming Associate Curator (1941) and Curator (1943), retiring eventually as Emeritus Curator. He did much work on the fauna of Mexico, where he felt at home, and made recordings of indigenous folk music that were later commercially released. He researched (1948–1959) in Central America and Bimini Island (Bahamas). He travelled widely in later years, to Sri Lanka (1965) and, after 1979, most of the Americas, Australia and Africa. He was President of the American Society of Ichthyologists and Herpetologists (1952–1954). Among other works he co-wrote *The Gila Monster and its Allies* (1956). Sixteen reptiles are named after him.

Böhme

Mantellid frog genus *Boehmantis* **Glaw** & Vences, 2006
Böhme's Bright-eyed Frog *Boophis boehmei* Glaw & Vences, 1992
Poison Frog sp. *Ameerega boehmei* Lötters et al., 2009

Professor Dr Wolfgang Böhme (b.1944) is a German zoologist and herpetologist. He studied zoology, botany and palaeontology for his doctorate at the University of Kiel (1971). He became (1971) Curator of Herpetology & Deputy Director, Zoological Department, Alexander König Research Institute, Bonn, and (1988) Professor of Zoology, Friedrich William University, Bonn. Nine reptiles are named after him.

Boie

Boie's (Horned) Frog *Proceratophrys boiei* **Wied-Neuwied**, 1824
Boie's Wart Frog **Fejervarya limnocharis* Gravenhorst, 1829
[Alt. Paddyfield Frog, Indian Cricket Frog; Syn. *Rana limnocharis*]

There were two German naturalists called Boie who were brothers. Heinrich Boie (1794–1827) was an explorer and zoologist. He studied under Blumenbach at Göttingen and worked as an assistant to Temminck at Leiden. After Kuhl (q.v.) died in Java, Boie replaced him, went to the Dutch East Indies and was at Buitenzorg (Bogor), Java when he died. He wrote some papers with his brother Friedrich (1798–1870) who was a lawyer, herpetologist, ornithologist and entomologist. It was Friedrich who described reptiles, whilst Heinrich was more interested in exploring. Between them, ten reptiles are named after them.

* Boie's name seems to have become attached to *Fejervarya limnocharis* in error, due to 'Boie 1835' sometimes being given as the species' citation.

Bokermann

Treefrog genus *Bokermannohyla* **Faivovich** et al., 2005
Bokermann's Humming Frog *Chiasmocleis urbanae* Bokermann, 1952 NCR
[JrSYN *Chiasmocleis leucosticta*]
Bokermann's Tarauaca Treefrog *Dendropsophus bokermanni* **Goin**, 1960
Bokermann's Casque-headed Treefrog *Trachycephalus atlas* Bokermann, 1966

Bokermann's Caecilian *Caecilia bokermanni* **Taylor**, 1968
Bokermann's Treefrog *Dendropsophus novaisi* Bokermann, 1968
Bokermann's Lime Treefrog *Sphaenorhynchus prasinus* Bokermann, 1973
Bokermann's Tropical Bullfrog *Leptodactylus bokermanni* **Heyer**, 1973
[Alt. Bokermann's Potter Frog; Syn. *Adenomera bokermanni*]
Bokermann's Bromeliad Frog *Crossodactylodes bokermanni* **Peixoto**, 1983
[Alt. Bokermann's Stream Froglet]
Bokermann's Dwarf Frog *Physalaemus bokermanni* **Cardoso** & **Haddad**, 1985
Bokermann's Spinythumb Frog *Crossodactylus bokermanni* **Caramaschi** & **Sazima**, 1985
Bokermann's Leaf Frog *Phrynomedusa bokermanni* **Cruz**, 1991
Bokermann's Casque-headed Frog *Aparasphenodon bokermanni* **Pombal**, 1993

Dr Werner Carlos Augusto Bokermann (1929–1995) was a Brazilian herpetologist and ornithologist. He received his doctorate in zoology at the Bioscience Institute, São Paulo University. He became Head, Bird Section, Fundação Parque Zoológico de São Paulo, and stayed there throughout his working life. He described a great number of new species, including at least 60 frogs and toads, and published a great many scientific papers. Many other taxa, including a mammal and a bird, are named after him.

Bonaespe

Banded Stream Frog *Strongylopus bonaespei* **Dubois**, 1981
[Alt. Banded Sand Frog, Jonkersberg Frog]

The binomial is actually a toponym rather than an eponym. It is Latin for 'of good hope' and is a reference to the area of South Africa where the frog is found.

Boneberg

Boneberg's Frog *Natalobatrachus bonebergi* **Hewitt** & Methuen, 1912

Father Pascalis Boneberg of Mariannhill Monastery, Natal, collected the holotype.

Bonet

Oaxacan False Brook Salamander *Pseudoeurycea boneti* **Alvarez** & **Martin**, 1967

Dr Federico Bonet Marco (1906–1980) was a Spanish zoologist, entomologist, palaeontologist, geologist and speleologist, who graduated as a physician and became Assistant Professor of Entomology at Madrid University (1930). He joined the Communist Party (1934) and fought in the artillery for the Republicans in the Spanish Civil War, after which he went into exile in Mexico. He wrote extensively on Collembola (Springtails) and has a species named after him.

Bopelet

Dizangue Reed Frog *Hyperolius bopeleti* **Amiet**, 1980

Dr M. Bopelet is a biologist who worked at MESIRES (Ministère de l'Enseignement Superieur de l'Informatique et de la Recherche Scientifique), Yaounde, Cameroon.

Bordoloi

Bordoloi's Tree Frog *Rhacophorus suffry* Bordoloi, Bortamuli & Ohler, 2007
[Alt. Suffry Red-webbed Treefrog]

Dr Sabitry Choudhury Bordoloi is a herpetologist at Department of Zoology, Cotton College, Guwahati, Assam, India, and is head of the Resource Management and Environment Division, Institute of Advanced Study in Science and Technology, Assam. She co-wrote *The Amphibian Fauna of Nagaland, India: Species and Habitats* (2002).

Borelli

Rufous Four-eyed Frog *Pleurodema borellii* **Peracca**, 1895

Dr Alfredo Borelli (1858–1943) was an ornithologist who worked at the Turin Museum (1900–1913). He explored and collected in Argentina and Paraguay (1893–1896). Four birds are named after him.

Boring

Emei Moustache Toad *Leptobrachium boringii* **Liu**, 1945
Warty Newt ssp. *Triturus sinensis boringi* Herre, 1939 NCR
[JrSYN *Paramesotriton chinensis*]

Dr Alice Middleton Boring (1883–1955) was a zoologist, biologist, geneticist and cytologist. Her bachelor's degree (1904) and her doctorate (1910) were both awarded by Bryn Mawr College, Pennsylvania. She taught zoology at the University of Maine (1910–1918). She worked at Yenching University, China (1923–1950), though she spent the Second World War in an internment camp before being repatriated to America. She returned to America permanently (1950) and taught at Smith College (1950–1955). She co-wrote *Handbook of North China Amphibia and Reptiles* (1932). The binomial should really be *boringiae* (feminine) and it is sometimes amended to that.

Borja

Amalfi Rocket Frog *Hyloxalus borjai* **Rivero** & **Serna**, 2000
[Syn. *Colostethus borjai*]

Professor Rafael Antonio Borja Acuña is a Colombian biologist at the Universidad del Atlántico and Director of Zoology at Santa Fé de Bogotá, Colombia. He collected the holotype.

Bornmüller

Bornmüller's Caecilian *Crotaphatrema bornmuelleri* **Werner**, 1899

Dr Joseph Friedrich Nicolaus Bornmüller (1862–1948) was a botanist. He studied horticulture at Potsdam (early 1880s). He was Director of the Herbarium, Weimar (1903–1938) and undertook many trips, including one east of Turkestan (1913) with Fedchenko, the Russian botanist. He worked in Macedonia, then occupied by German forces (1917–1918). His doctorate was honorary, given by the University of Jena. He published *Repertorium specierum novarum regni vegetabilis* (1938). His personal collection was sold after his death to the Museum für Natuurkunde, Berlin. A reptile is named after him.

Bosca

Bosca's Newt *Lissotriton boscai* **Lataste**, 1879
Midwife Toad ssp. *Alytes obstetricans boscai* Lataste, 1879

Eduardo Boscá y Casanoves (1844–1924) was a Spanish herpetologist. He catalogued and described much herpetological fauna of the Iberian Peninsula (1870s–1880s). He described and named the Spanish Cylindrical Skink *Chalcides bedriagai* (1880), so Bedriaga returned the compliment. He wrote *Catalogue des Reptiles et Amphibiens de la Péninsule Ibérique et des Iles Baléares* (1880). A reptile is named after him.

Bott

Andasibe Bright-eyed Frog *Boophis bottae* Vences & **Glaw,** 2002

Mrs Ursula Bott is the long-time assistant of Wolfgang Böhme at the Zoologisches Forschungsmuseum Alexander Koenig, Bonn.

Boulenger

Caecilian genus *Boulengerula* **Tornier**, 1896
North African Green Toad *Pseudepidalea boulengeri* **Lataste**, 1879
[Syn. *Bufo boulengeri*]
Boulenger's Bubble-nest Frog *Raorchestes signatus* Boulenger, 1882
Boulenger's Digging Frog *Plethodontohyla inguinalis* Boulenger, 1882
Boulenger's Indian Frog *Indirana leptodactyla* Boulenger, 1882
[Alt. Long-toed Frog]
Boulenger's Malabar Frog *Micrixalus silvaticus* Boulenger, 1882
Boulenger's Narrow-eyed Frog *Nyctibatrachus major* Boulenger, 1882
[Alt. Large Wrinkled Frog]
Boulenger's Tree Frog *Philautus dubius* Boulenger, 1882*
Boulenger's Tree Frog *Rhacophorus lateralis* Boulenger, 1883*
[Alt. Small Gliding Frog]
Boulenger's Snouted Treefrog *Scinax boulengeri* **Cope**, 1887
Boulenger's Brown Frog *Indirana leithii* Boulenger, 1888
[Alt. Leith's (Leaping) Frog]
Boulenger's Giant Treefrog *Platypelis grandis* Boulenger, 1889
Boulenger's Spiny Frog *Quasipaa boulengeri* **Günther,** 1889
[Alt. Boulenger's Paa Frog]
Boulenger's Asian Tree Toad *Pedostibes hosii* Boulenger, 1892
[Alt. Hose's Tree Toad, Brown Tree Toad]
Boulenger's Burrowing Toad *Bufo fergusonii* Boulenger, 1892 NCR
[JrSYN *Duttaphrynus scaber*]
Boulenger's Climbing Frog *Anodonthyla boulengerii* **Müller**, 1892
Boulenger's Wrinkled Ground Frog *Platymantis boulengeri* **Boettger**, 1892
[Alt. Boulenger's Platymantis]
Boulenger's Bushfrog *Kurixalus verrucosus* Boulenger, 1893
Boulenger's Tree Frog *Chiromantis doriae* Boulenger, 1893*
[Alt. Doria's Bush Frog, Doria's Tree Frog]
Boulenger's Flathead Toad *Pelophryne macrotis* Boulenger, 1895 NCR
[JrSYN *Pelophryne guentheri*]
Boulenger's Caecilian *Boulengerula boulengeri* Tornier, 1896
Boulenger's Callulops Frog *Callulops robustus* Boulenger, 1898
Boulenger's High Altitude Toad *Scutiger boulengeri* **Bedriaga**, 1898
[Alt. Boulenger's Lazy Toad, Xizang Alpine Toad]
Victoria Forest Treefrog *Leptopelis boulengeri* Werner, 1898

Boulenger's Rocket Frog *Hyloxalus vertebralis* Boulenger, 1899
Boulenger's Arrow-poison Frog *Atelopus boulengeri* **Peracca**, 1904
[Alt. Boulenger's Stubfoot Toad]
Boulenger's Short-headed Frog *Breviceps macrops* Boulenger, 1907
[Alt. Desert Rain Frog]
Boulenger's Earless Toad *Mertensophryne anotis* Boulenger, 1907
[Alt. Chirinda Toad, Chirinda Forest Toad]
Marbled Poison Frog *Epipedobates boulengeri* **Barbour**, 1909
Boulenger's Egg Frog *Leptodactylodon boulengeri* **Nieden**, 1910
Boulenger's Oriental Salamander *Hynobius boulengeri* **Thompson**, 1912
[Alt. Odaigahara Salamander]
Boulenger's Backpack Frog *Cryptobatrachus boulengeri* **Ruthven**, 1916
Boulenger's Flying Frog *Rhacophorus modestus* Boulenger, 1920
Boulenger's Garo Hills Frog *Hylarana garoensis* Boulenger, 1920
Boulenger's Hill Frog *Nanorana annandalii* Boulenger, 1920
[Alt. Annandale's Paa Frog; Syn. *Paa annandalii*, *Rana annandalii*]
Boulenger's Madagascar Frog *Gephyromantis boulengeri* Methuen, 1920
Eastern Dwarf Clawed Frog *Hymenochirus boulengeri* **de Witte**, 1930
Boulenger's Robber Frog *Pristimantis boulengeri* **Lynch**, 1981
Boulenger's Striped Frog *Phlyctimantis boulengeri* **Perret**, 1986
Beaked Toad *Rhinella boulengeri* Chaparro et al., 2007
[Replacement name for *Phryniscus proboscideus* Boulenger, 1882]

George Albert Boulenger (1858–1937) was a Belgian-British herpetologist at the British Museum, London and one of the giants of herpetology. He graduated from university in Brussels (1876) and worked at its Museum of Natural History until moving to London (1880). He took British nationality (1882). His output was prodigious: nearly 2,600 species described, including 872 reptiles, and 877 scientific papers. He was also a violinist and polyglot. He retired (1920) to grow roses. No less than 74 reptiles are named after him. (See **Georg**)

* NB. Three taxa have at times been given the vernacular name of Boulenger's Tree Frog

Bourret

Bourret's Asian Toad *Leptolalax bourreti* **Dubois**, 1983
Bourret's Frog *Nanorana bourreti* Dubois, 1987

René Leon Bourret (1884–1957) was a French zoologist. He undertook a comprehensive herpetological survey of Vietnam and studied Indochinese fauna (1922–1942). He published *Les Tortues de l'Indochine* (1941), the first detailed monograph to deal with all the chelonians of Southeast Asia. Among other taxa, seven reptiles, a bird and a mammal are named after him.

Bouwens

Ocellated Snouted Frog *Xenorhina bouwensi* **De Witte**, 1930

M. J. Bouwens was the Civil Governor at Manokwari, Dutch New Guinea.

Boyle

Foothill Yellow-legged Frog *Rana boylii* **Baird**, 1854

Dr Charles Elisha Boyle (1821–1870) was a physician, amateur naturalist, and a '49er'. After working for a printer, and teaching, he graduated from medical college (1847). He got 'goldrush fever' (1848–1849), joining the Columbus and California Industrial Association party as its physician, and travelling the Oregon Trail to California. He practiced as a physician in California, and collected (particularly herpetological) specimens for the Smithsonian. During the American Civil War (1860–1865) he was a surgeon-captain in the army, and then returned to private practice. He was fluent in 32 languages and '…gave much of his time and practice to the poor of the city and as a result never amassed much money and died poor himself.' A mammal and a reptile are named after him.

Brack

Brack's Andes Frog *Phrynopus bracki* **Hedges**, 1990

Antonio José Brack Egg (b.1940) is an agronomist and ecologist who was instrumental in the creation of Parque Nacional Yanachaga-Chemillen (Peru). He wrote *Las Ecoregiones del Perú* (1986) and is now the Peruvian Minister of the Environment.

Brade

Itatiaia Highland Frog *Holoaden bradei* **Lutz**, 1958

Dr Alexandre Curt Brade (1881–1971) was a German architect whose hobby was botany. He eventually went to Brazil, collected plants in Central and South America, and became employed at the Botanical Gardens, Rio de Janeiro, where he was regarded as an expert on ferns and begonias. He collected the frog holotype (1951).

Braestrup

Braestrup's Caecilian *Chthonerpeton braestrupi* **Taylor**, 1968

Dr Frits Wimpffen Bræstrup (1906–1999) was Curator, Zoological Museum, University of Copenhagen, and an expert on earwigs. He travelled in East and southern Africa. He wrote *Remarks on faunal exchange through the Sahara* (1947). A reptile is named after him.

Brame

Brame's Web-footed Salamander *Bolitoglossa bramei* **Wake, Savage** & **Hanken**, 2007

Dr Arden H. Brame Jr II (1934–2004) was a herpetologist who specialized in salamanders and was so devoted to them that he kept a four-foot long Japanese Giant Salamander *Andrias japonicus* in a spare bathtub at his home. He was at the University of Southern California and the Los Angeles Museum (1965). He wrote *New Species of Large Black Salamander, Genus* Bolitoglossa *(Plethodontidae) from Western Panama* (2001). He was also a keen genealogist. His family appear to have been among the very early settlers from the Old World and he published *Early History of the Brame-Brim Family, 1674–1725, in Middlesex County, VA.* He called himself 'Arden H. Brame Junior the Second' as he discovered that he had a half-brother with exactly the same names as himself.

Branch

Branch's Rain Frog *Breviceps branchi* **Channing**, 2012

Dr William Roy Branch (b.1946) is an English-born South African herpetologist. The University of Southampton awarded his bachelor's degree (1968) and his doctorate (1971). He was Curator of Herpetology, Port Elizabeth Museum (1979–2011). He is now retired, but his 'retirement' appears to be hardly recognisable as such!

Branner

Treefrog sp. *Dendropsophus branneri* **Cochran**, 1948

Dr John Casper Branner (1850–1922) was a geologist who graduated (1882) from Cornell and was awarded his doctorate (1885) from the University of Indiana. He was at Stanford University as Professor of Geology (1891–1913), and as President of the University (1913–1915). He was an expert on Brazilian geology and visited the country several times, notably as leader of the Branner-Agassiz Expedition (1899). He wrote *Notes on the fauna of the islands of Fernando de Noronha.* (1888). A mineral, Brannerite, is also named after him. Branner collected the holotype.

Bransford

Bransford's Robber Frog *Craugastor bransfordii* **Cope**, 1886

Dr John F. Bransford (1846–1911) was an American naval surgeon. He was commissioned as Assistant Surgeon (1872) and posted to the 'USS Hartford'. He made herpetological collections in Nicaragua (1873, 1885) and Panama (1875). He resigned from the US Navy (1890) but was recalled to duty for the Spanish-American War, being appointed Acting Assistant Surgeon (1898) and promoted to Surgeon on his retirement (1901).

Brauer

White-lipped Treefrog *Polypedates braueri* **Vogt**, 1911

Professor Dr August Bernhard Brauer (1863–1917) was a German zoologist, herpetologist, and ichthyologist. He graduated from Humboldt-Universität, Berlin, in natural sciences (1885) and took his doctorate there (1892). He collected in the Seychelles (1897) and was on the *Valdivia* expedition (1898), describing (1908) the fishes they collected. He became Professor at Berlin University (1905) and Director of the university's Zoological Museum (1906). He was appointed Professor of the Zoological University, Berlin (1914). A mammal and a reptile are named after him.

Braun, K

Braun's Toad *Amietophrynus brauni* **Nieden**, 1911
[Alt. Dead-leaf Toad; Syn. *Bufo brauni*]

Professor Dr Karl Braun (1870–1935) was a German botanist who studied in Munich and Basel before working in Stuttgart, where he taught chemistry in a high school (1902–1904). He joined the Amani Institute of Botanical Research, Tanganyika, German East Africa (Tanzania) (1904–1914). Still there when the First World War started, he became a POW. After the war's end he took up his duties again at Amami, but left (1920) to return to Europe and later became Professor at Leiden. Loveridge reported that he visited Amani, where the holotype was collected, and was amused to collect a specimen of *Bufo brauni* on the path to what had been Braun's house there. He wrote *Der Reis in Deutsch-Ostafrika* (1907).

Braun, PC & CAS

Brauns' Dwarf Frog *Physalaemus lisei* Braun & Braun, 1977

Smooth horned frog sp. *Proceratophrys brauni* Kwet & **Faivovich**, 2001

Pedro Canisio Braun (1938–1992) and Cristina Assunção Sirangelo Braun. Cristina was a Professor at University, Porto Alegre, Rio Grande do Sul, Brazil. They were a herpetological husband and wife team. Together they wrote *Lista prévia dos anfibios do Estado do Rio Grande do Sul, Brasil* (1980). The Smooth horned frog is named only after him.

Bressler

Rio Yumuri Robber Frog *Eleutherodactylus bresslerae* **Schwartz**, 1960

Miss Sandra L Bressler was an artist and ornithologist. She co-wrote *Observations of Joint Feeding Activities of Certain Fish-eating Birds* (1963). She drew figures of Cuban frogs for Schwartz.

Brewer

Brewer's Carrying Frog *Stefania breweri* Barrio-Amorós & Fuentes-Ramos, 2003

Rocket Frog sp. *Anomaloglossus breweri* Barrio-Amorós, 2006

Charles Brewer-Carıás (b.1939) is a Venezuelan dentist who, after practising for 20 years, became an explorer and speleologist. He is currently out of favour with the Chávez regime as his brother, a lawyer, had to flee into exile for being open in his dislike of the regime. He is well known to the native tribes of the Venezuelan highlands, as he fixed their teeth as well as explored their territory. He was shot by thieves (2003), but despite some loss of flexibility in his shoulder, still explores. The world's largest quartzite cave is named after him, as is a bird.

Briceño

Merida Robber Frog *Pristimantis briceni* **Boulenger**, 1903

Salomón Briceño Gabaldón (1826–1912) was a Venezuelan collector. Among his clients was Lord Lionel Walter Rothschild, for whom he supplied many specimens, particularly birds. A mammal and a bird are named after him.

Brien

Brien's Snouted Treefrog *Scinax brieni* **De Witte**, 1930

Paul Brien (1894–1975) was Professor, Free University Brussels (1928–1964). He was a zoologist and ichthyologist who collected fishes (1937) in the Belgian Congo. He was violently anti-Fascist, being President of the Popular Front (1935–1936), and was elected local senator for the Communist Party in Brabant (1936) but resigned after a few months. During the Second World War his political opinions did not find favour with the occupiers of Belgium and he was arrested, interrogated and imprisoned by the Gestapo (1942).

Brimley, CS & HH

Brimley's (Dusky) Salamander *Desmognathus brimleyorum* **Stejneger**, 1895

[Alt. Ouachita Dusky Salamander]

Brimley's Chorus Frog *Pseudacris brimleyi* Brandt & **Walker**, 1933

Dr Clement Samuel Brimley (1863–1946) and his brother, Herbert Hutchinson Brimley (1861–1946) emigrated from England to North Carolina (1880). They started collecting, preparing and selling natural history specimens. Clement worked for the Entomology Division, North Carolina Department of Agriculture (1919). His doctorate was an honorary Doctorate of Laws, conferred by the University of North Carolina. He wrote *Artificial key to the species of snakes and lizards which are found in North Carolina* (1907). The brothers co-wrote *Birds of North Carolina* (1919). Herbert prepared exhibits (1884) for the North Carolina Board of Agriculture and Immigration and, from this work, the North Carolina State Museum developed and he took charge (1895). The North Carolina Board of Agriculture honoured him (1941) by naming the library 'Brimley Library of Natural History'. Whereas the salamander is named after both of them, only Clement is mentioned in the name of the frog.

Britton

Forester's Cabin Robber Frog *Eleutherodactylus brittoni* **Schmidt**, 1920

Professor Dr Nathaniel Lord Britton (1857–1934) was a botanist and taxonomist who was founding Director, New York Botanical Garden (1896). He studied geology, mineralogy and palaeobotany at the School of Mines, Columbia University (1875–1879), and became a Field Assistant with the Geological Survey of New Jersey (1880–1884). He returned to Columbia University as Instructor in botany and geology (1886) and became Professor of Botany (1890).
His wife, Elizabeth Gertrude Britton née Knight (d.1934) was an eminent bryologist and they both visited Puerto Rico many times during the Scientific Survey of Puerto Rico. He wrote *An Illustrated Flora of the Northern United States and Canada* (1896–1898). Mount Britton in Puerto Rico is named after him.

Broadley

Broadley's Mountain Frog *Nothophryne broadleyi* **Poynton**, 1963
[Alt. Mongrel Frog]
Broadley's Ridged Frog *Ptychadena broadleyi* Stevens, 1972
Broadley's Tree Frog *Leptopelis broadleyi* Poynton, 1985

Dr Donald George Broadley (b.1932) is a specialist in East African herpetology and Curator of Herpetology, Natural History Museum of Zimbabwe, Bulawayo, where his wife Sheila is also a herpetologist. He gained his doctorate from the University of Natal (1966). Among his publications is *On the status of* Simocephalus riggenbachi *Sternfeld 1910* (2007). Eight reptiles are named after him.

Brocchi

Brocchi's Rainfrog *Craugastor brocchi* **Boulenger**, 1882
[Alt. Guatemala Robber Frog]

Paul Louis Antoine Brocchi (1838–1898) was a herpetologist who worked at the Muséum National d'Histoire Naturelle, Paris, with Milne-Edwards. He wrote *Sur quelques batraciens raniformes et bufoniformes de l'Amérique* (1877). A reptile is also named after him.

Brodie

Cerro Pozo de Agua Moss Salamander *Nototriton brodiei* **Campbell** & Smith, 1998

Dr Edmund D. Brodie Jr (b.1941) is a professor of biology at Utah State University. He originally trained to be a secondary school teacher but returned to do a doctorate at Oregon State University. He was Assistant Professor, and later full Professor, at the Biology Department, Adelphi University, New York (1974–1984). He co-wrote *Amphibians and Reptiles of the Pacific Northwest* (1983). He has worked extensively with Campbell on Guatemalan herpetology. His son Edmund D. 'Butch' Brodie III is also a herpetologist and Professor of Biology at the University of Virginia.

Brongersma

Puddle Frog sp. *Phrynobatrachus brongersmai* **Parker**, 1936
Brongersma's Treefrog *Litoria brongersmai* **Loveridge**, 1945
Brongersma's Toad *Pseudepidalea brongersmai* Hoogmoed, 1972
[Alt. Tiznit Toad; Syn. *Bufo brongersmai*]

Dr Leo Daniel Brongersma (1907–1994) was an author and zoologist. His doctorate was from the University of Amsterdam (1934). He began studying herpetology as an Assistant, Amsterdam Zoological Museum. He lectured at the University of Leiden and was Curator of Reptiles & Amphibians, then Director of the Leiden Museum until retiring (1972). He explored in New Guinea and Surinam. His works include *European Atlantic Turtles* (1972). Nine reptiles are named after him.

Brooks

Brooks' Burrowing Frog *Calluella brooksii* **Boulenger**, 1904

Cecil Joslin Brooks (1875–1953) was a collector, mainly botanical, in Borneo and Sumatra early in the twentieth century. He was a metallurgical chemist employed by the Borneo Company for gold exploitation in Sarawak (1900–1910), where he collected, helped by Hewitt, Curator of the Sarawak Museum. He was employed at a goldmine in Sumatra (1912–1923), travelled subsequently in the Dutch East Indies, and then sailed, via Australia and New Zealand, to Europe (1924). In England, he studied his collections and the butterflies in the British Museum. He wrote mainly on ferns and several are named after him, as are a mammal and two reptiles.

Brown, WC

Brown's Wrinkled Ground Frog *Platymantis browni* **Allison** & Kraus, 2001

Dr Walter Creighton Brown (1913–2002) was a herpetologist who specialized in the herpetofauna of the South Pacific and the Philippines. He served in New Guinea during the Second World War. He received his doctorate from Stanford (1955) and went to Silliman University, Philippines, as a Fulbright Professor of Sciences. Seven reptiles are named after him.

Brown (Family)

Browns' Leopard Frog *Lithobates brownorum* Sanders, 1973

The Brown Family comprises Dr Bryce Cardigan Brown (1920–2008), Director Emeritus, Strecker Museum, and Professor Emeritus in Biology, Baylor University in Texas; his wife, Lilian, and their 5 children: Alton, Brent, Carol, Leo and Roy. Bryce graduated in zoology at the University of Texas (1942), joined the Army Air Corps and served in India. After the Second World War he returned to Texas University and finished his master's and doctorate there. He was President, Texas Herpetological Society (1946) and hired as Curator of the Strecker Museum, becoming Director (1966–1981). Among his publications is *An Annotated Checklist of the Reptiles and Amphibians of*

Texas (1950). The etymology reads: 'It is my pleasure to name this new subspecies in honor of the Bryce Brown family who by their yearly excursions into Mexico have added richly to the collections of the Strecker Museum of Baylor University and made this study possible.' The frog was originally described as a subspecies of *Rana berlandieri*; the holotype was collected by Brent Brown (1964).

Bruce

Patch-nosed Salamander *Urspelerpes brucei* Camp et al., 2009

Dr Richard (Dick) C. Bruce is a salamander ecologist who was Director of the Highlands Biological Station, North Carolina, assisted by his wife Elizabeth (1972–1999). Since then he has been at the Department of Biology, Western Carolina University, where he is Emeritus Professor of Biology. He was lead editor of *The Biology of Plethodontid Salamanders* (2000) and has written numerous scientific papers.

Bruna

Bruna's Stream Frog *Spinomantis brunae* **Andreone** et al., 1998

Bruna Cugnetto is the senior author's mother.

Brunhilda

Adelbert Rainforest Frog *Albericus brunhildae* **Menzies**, 1999.

Brunhilda (or Brynhildr) was a shieldmaiden and Valkyrie in Norse mythology and is a central character in Wagner's *Der Ring des Nibelungen*. She is the Valkyrie of the title of the second opera in the Ring cycle - *Die Walküre* - and appears in the third and fourth operas, *Siegfried* and *Die Götterdämmerung*. Menzies named several members of this genus after Wagnerian characters.

Bruno

Bruno's Casque-headed Frog *Aparasphenodon brunoi* **Miranda-Ribeiro**, 1920

Dr Bruno Lobo was Professor and Director of the National Museum in Rio de Janeiro, Brazil (1915–1923).

Buchard

Moustache Toad sp. *Leptobrachium buchardi* Ohler, Teynié & David, 2004

Michel Buchard is a businessman in Clermont-Ferrand, France. Since 1998 he has given very generous support to the study of natural history in general and to herpetology and entomology in particular. A reptile is named after him.

Buchner

Buchner's Toad *Amietophrynus buchneri* **Peters**, 1882

Dr Max Buchner (1846–1921) was Curator, Staatliches Museum für Volkerkunde, Munich. He undertook a number of collecting expeditions to Western Africa (1878–1882) and later to the Pacific and Asia. He collected many anthropological objects and natural history specimens, particularly grasses.

Buckley

Buckley's Giant Glass Frog *Centrolene buckleyi* **Boulenger**, 1882

Buckley's Robber Frog *Pristimantis buckleyi* Boulenger, 1882

Buckley's Slender-legged Treefrog *Osteocephalus buckleyi* Boulenger, 1882

Warty Leaf Frog *Agalychnis buckleyi* Boulenger, 1882

Clarence Buckley (fl.1839, d.1889) was a collector who made several expeditions to Ecuador in the 1880s. He collected over 10,000 specimens of many plant and animal taxa in a small area of Ecuador, sending them to various institutions and scientists but mainly to the British Museum (Natural History), London. A reptile and six birds are named after him.

Budgett

Budgett's Burrowing Frog *Hildebrandtia ornata* **Peters**, 1878

[Alt. Ornate Burrowing Frog; *Rana budgetti* is a junior synonym]

Budgett's Frog *Lepidobatrachus laevis* Budgett, 1899

John Samuel Budgett (1872–1904) was an embryologist, ichthyologist, naturalist, artist and explorer, employed by Trinity College, Cambridge. He made a minimum of four West African expeditions, making a very extensive collection of African fishes. He suffered so badly from malaria [eventually fatally] that he was not passed as fit for service in the Boer War, so he went exploring again. He succeeded in documenting the embryonic development of *Polypterus*, an African freshwater fish genus. A mammal is named after him.

Bullock

Bullock's False Toad *Telmatobufo bullocki* **Schmidt**, 1952

Professor Dr Dillman Samuel Bullock (1878–1971) was an agriculturalist, naturalist and Methodist missionary. He graduated from Michigan State University (1902) and worked for American Missionary Society, Bunster Agricultural School in Angol, Chile (until 1912). He then returned and taught in the USA (1912–1921), before returning to South America. For a year he worked as Agricultural Commissioner,

US Department of Agriculture at the US embassy, Buenos Aires. He returned to teach at the School in Chile (1923–1958). He collected the holotype (1931). Schmidt wrote: '...We are indebted to Dr D. S. Bullock of El Vergel, Angol, Chile, for the two specimens of this frog, collected in the Nahuelbuta forest... ...it is pleasant to name the new form for Dr Bullock, in recognition of his services to Chilean natural history and of his long-continued friendly cooperation with Chicago Natural History Museum.' A bird is named after him.

Bulmer

Bulmer's Treefrog *Litoria bulmeri* **Tyler**, 1968

Ralph N. H. Bulmer (1928–1988) was a social anthropologist who was a member, University of Auckland. He worked and collected in the Papua New Guinea highlands (1955–1976). He co-wrote *Karam classification of frogs* (1971).

Bürger, H

Rhacophorid frog genus *Buergeria* **Tschudi**, 1838

Buerger's Frog *Buergeria buergeri* Temminck & **Schlegel**, 1838

[Alt. Kajika Frog]

Heinrich Bürger (1804–1858) was a German physicist and biologist employed by the Dutch government (he may have been two years younger than he claimed, to appear older than he was). He studied mathematics and astronomy at Göttingen (1821–1822) and called himself 'doctor' but there is no evidence of his obtaining a PhD. He first travelled to Batavia (Jakarta), Java (1824) and gained a third class degree of apothecary (1825). The Dutch Government appointed him (1825) assistant to German physicist P. F. von Siebold. He and Siebold collected natural history specimens for the Leiden Museum (1920s), where Temminck and then Schlegel were Directors; they used much of the material Bürger collected for their *Fauna Japonica*. Bürger and Siebold collected together throughout the Dutch East Indies (Indonesia), then in Japan (1837–1838). Bürger was honourably discharged from government service (1843) and set up a rice and sugar business as well as mining and insurance.

Burger, M

Burger's Bright-eyed Frog *Boophis burgeri* **Glaw** & Vences, 1994

Marius Burger is a South African environmentalist and natural history tour guide. He completed (1990) a National Diploma in Nature Conservation. He worked (1987–1989) for Reserve Management and District Services, Cape Department of Nature Conservation, and (1989–1997) as a research technician in the same organisation. He then worked at the University of Cape Town as the National Coordinator of the South African Frog Atlassing Project and co-wrote *Progress and interim results of the Southern African Frog Atlas Project* (2000). He is currently working at the University of the Western Cape.

Burit

Treefrog sp. *Hypsiboas buriti* **Caramaschi** & **Cruz**, 1999

Buritis, Minas Gerais, Brazil was where the holotype was collected.

Burmeister

Burmeister's Treefrog *Hypsiboas prasinus* Burmeister, 1856

Burmeister's Leaf Frog *Phyllomedusa burmeisteri* **Boulenger**, 1882

Professor Karl Hermann Konrad Burmeister (1807–1892) was an ornithologist who was Director, Institute of Zoology, Martin Luther University, Halle Wittenberg, Germany (1837–1861). He made large collections during two expeditions to Brazil (1850–1852) and the La Plata region, Argentina (1857–1860). He lived in Argentina (1861–1892), being founding Director, Museo Nacional, Buenos Aires until retirement (1880). He was in the Prussian civil service but won his release by using the very inventive excuse that a persistent stomach complaint was caused by arsenic emissions in the museum, and by the drinking water in Halle that had high sulphate content! He wrote *Reise nach Brasilien* (1853). Three birds, a reptile and a mammal are named after him.

Burns

Unspotted Meadow Frog *Rana burnsi* Weed, 1922 NCR

[Now regarded as a morph of the Northern Leopard Frog, *Lithobates pipiens*]

F. J. & J. J. Burns were members of the Chicago produce and livestock dealers, F J Burns & Co. Among their product lines was a supply of frogs 'for educational purposes' (presumably for dissection rather than alimentation) and they were able to supply Weed with fresh frogs, caught locally.

Burrowes

Burrowes' Giant Glass Frog *Centrolene ballux* **Duellman** & Burrowes, 1989

Burrowes' Robber Frog *Pristimantis laticlavius* **Lynch** & Burrowes, 1990

Professor Dr Patricia A. Burrowes Gomez (b.1961) is an American biologist and herpetologist who is a professor in the Biology Department, Colegio Universitario de Cayey, Puerto Rico. Her master's degree (1987) and doctorate (1997) were both gained at the University of Kansas. She collected in Colombia with Duellman (1986) and they co-wrote *A new species of marsupial frog (Hylidae: Gastrotheca) from the Andes of Southern Colombia* (1986). She started working (2003) on a project on amphibians in Puerto Rico and the Dominican Republic, with a fellow professor from Puerto Rico, her husband Rafael L. Jogar.

Burrows

Burrows' Tree Frog *Litoria burrowsi* Scott, 1942
[Alt. Tasmanian Treefrog]

Miss M. Burrows collected the holotype in Cradle Valley, Tasmania. The binomial is sometimes 'corrected' to the feminine *burrowsae*. We have been unable to find further details of Miss Burrows.

Burton

Burton's Mayhay Frog *Choerophryne burtoni* **Richards**, **Dahl** & Hiaso, 2007

Dr Thomas 'Tom' C. Burton is a herpetologist at La Trobe University, Bendigo, Australia. A ligament found in frogs (Burton's ligament) is named after him.

Bustamente

Bustamente's Andes Frog *Bryophryne bustamantei* Chaparro et al., 2007

Javier Bustamante was honoured by the describers for '...his friendship and his support of taxonomic research and nature conservation in Peru.'

Butler

Butler's Narrow-mouthed Toad *Microhyla butleri* **Boulenger**, 1900
[Alt. Butler's Rice Frog, Butler's Pygmy Frog]

Arthur Lennox Butler (1873–1939) was born in Karachi to an Anglo-Irish family. He became Curator, Selangor Museum, Malaya, and then Superintendent of Game Preservation in the Sudan (1901–1915). He wrote *A list of the batrachians known to inhabit the Malay Peninsula, with some remarks on their habits, distribution, etc.* (1904). Five birds and four reptiles are named after him.

Bwana

Rio Chipillico Frog *Lithobates bwana* Hillis & de Sá, 1988
[Syn. *Rana bwana*]

'Bwana Jim' was a nickname given to James Ray **Dixon** (q.v.) by his students.

C

Cabrera

Slender-legged Treefrog sp. *Osteocephalus cabrerai* **Cochran** & **Goin**, 1970

Valle Robber Frog *Pristimantis cabrerai* Cochran & Goin, 1970

Isadore Cabrera was a Colombian naturalist and collector who, according to the describers, 'has contributed appreciably to the knowledge of his native country by both his botanical and his zoological collections.'

Cadena

Nutibara Robber Frog *Strabomantis cadenai* **Lynch**, 1986

Dr Augusto Alberto Cadena-García is a Colombian zoologist, Curator of Mammals, Institute of Natural Sciences, National University of Colombia. A prize named after him was established for the best zoology student in Colombia (2006). Two mammals are named after him.

Cainarachi

Cainarachi Poison Frog *Ameerega cainarachi* **Schulte**, 1989

Rio Cainarachi, Peru is the type locality.

Calabresi

Calabresi's African Bullfrog *Pyxicephalus obbianus* Calabresi, 1927

Professor Enrica Calabresi (1891–1944) was an Italian herpetologist, entomologist and teacher who graduated (1914) at the University of Florence, where she worked as Assistant, Department of Zoology and Comparative Anatomy. She qualified to teach (1924) and became Professor of Agricultural Entomology, University of Pisa (1936–1938). As she was Jewish, she lost her job (1938), then taught at the Jewish School, Florence (1939–1943). She was arrested (January 1944) and apparently committed suicide in custody.

Caldas

Caldas Snouted Treefrog *Scinax caldarum* **Lutz**, 1968

Poços de Caldas, Minas Gerais, Brazil is the type locality.

Caldwell

Caldwell's Frog *Babina caldwelli* **Schmidt**, 1925

Harry R. Caldwell (b.1876) was an American Methodist missionary at Yenping, Fukien, China (1900). He was a keen hunter and amateur naturalist who collected the frog holotype. He wrote *Blue Tiger – Strange Adventures of a Missionary in China* (1924), which includes his attempts to trap a blue-morph tiger, as well as *South China Birds* (1931).

Camerano

Transcaucasian Frog *Rana macrocnemis camerani* **Boulenger**, 1886

Professor Dr Lorenzo Camerano (1856–1917) was a herpetologist and entomologist. He started working life as a painter at the Museo Civico di Storia Naturale, Turin (1873), but became fascinated by zoology and signed on as a student, graduating from the Università degli Studi di Torino (1878) and becoming Professor (1880). He spent a short time at the Università degli Studi di Cagliari before returning to Turin as Professor of Comparative Anatomy until 1915. He was elected an Italian senator (1909). He wrote *Monografia degli Ofidi italiani* (1891). A bird is named after him.

Camp, CL

Inyo Mountains Salamander *Batrachoseps campi* Marlow, Brode & **Wake**, 1979

Professor Dr Charles Lewis Camp (1893–1975) was an American palaeontologist, historian and herpetologist. He was Assistant, Museum of Vertebrate Zoology, Columbia University (1908–1915 & 1916–1917) and gained a bachelor's degree at the University of California (1915). He served in the US Army (WW1) in Europe, China and South Africa (1918–1919). After this he returned to New York and worked as Assistant, AMNH (1919–1921), then returned to Columbia University, which awarded his master's degree (1921) and his doctorate (1923). He relocated (1922) to the University of California as Research Associate, staying there in various positions until retiring as Professor and Chairman, Department of Palaeontology (1940–1949) and Director, Museum of Palaeontology (1931–1941). He wrote *Classification of Lizards* (1923).

Camp, RD

Camp's Frog *Eleutherodactylus cystignathoides campi* **Stejneger**, 1915

[Alt. Rio Grande Chirping Frog]

Richard Denham Camp (1867–1929) was Secretary of the American Ornithologists Union (1888). The type specimen of the frog was uncovered in a collection presented by Camp to the Smithsonian.

Campbell

Campbell's Rainforest Toad *Incilius campbelli* **Mendelson**, 1994

Robber Frog sp. *Craugastor campbelli* Smith, 2005

Professor Dr Jonathan Atwood Campbell (b.1947) is Professor of Biology, University of Texas, Arlington. His main area of expertise is Guatemalan herpetofauna. He was awarded his master's degree at UTA (1977) and doctorate at the University of Kansas (1983). He returned to UTA as Curator of Herpetology and Assistant Professor (1983), progressing to Associate Professor (1988) and full Professor (1993). Five reptiles are named after him.

Camposseabra

Snouted Treefrog sp. *Scinax camposseabrai* **Bokermann**, 1968

(See **Seabra**, C)

Cannatella

Cannatella's Andes Frog *Hypodactylus lucida* Canatella, 1984

Cannatella's Plump Toad *Osornophryne talipes* Cannatella, 1986

Rocket Frog sp. *Aromobates cannatellai* Barrio-Amorós & Santos, 2012

Slender-legged Treefrog sp. *Osteocephalus cannatellai* Ron et al., 2012

Professor Dr David C. Cannatella is a zoologist, functional anatomist and herpetologist. He is Curator, Herpetology of Texas Natural History Collections. part of the Texas Natural Science Center, where he is Associate Professor, Section of Integrative Biology. The University of Southwestern Louisiana awarded his bachelor's degree in zoology (1976), and his master's (1979) and doctorate (1985) in systematics and ecology were both awarded by the University of Kansas. He wrote the article *An integrative phylogeny of Amphibia* (2007).

Cao Van Sung

Sung Toad *Leptolalax sungi* Lathrop et al., 1998 [Alt. Sung's Metacarpal-tubercled Toad]

Professor Dr Cao Van Sung (d.2002) was a Vietnamese zoologist and biologist who received his training in Russian universities. He worked at the University of Hanoi and was Emeritus Director of the Institute for Ecology and Biological Resources, Hanoi. A mammal and a reptile are named after him.

Capelle

Glass Frog sp. *Hyalinobatrachium cappellei* Lidth de Jeude, 1904

Dr Herman van Cappelle Jr (1857–1932) was a Dutch geologist and museum director. After completing his doctorate (Leiden, 1885) he became a secondary school teacher (1886) of geology, zoology and botany. He was part of an expedition to Surinam (1897) and another (1900) that he led, later publishing a book about it: *The interior of the Nickerie: Fate and general results of an expedition through the western part of the colony of Suriname in September and October of 1900* (1903). He also wrote numerous books and articles on geology, Suriname, myths and legends and his museum. He returned to teaching in the Hague and was involved in the founding of a museum there, 'The Museum for the Benefit of Education', later becoming its voluntary Director (1910) and eventually being paid (1916–1923).

Caramaschi

Treefrog sp. *Bokermannohyla caramaschii* **Napoli**, 2005

Caramaschi's Lime Treefrog *Sphaenorhynchus caramaschii* Toledo et al., 2007

Smooth Horned Frog sp. *Proceratophrys caramaschii* **Cruz**, Nunes & Junca, 2012

Professor Dr Ulisses Caramaschi is a Brazilian herpetologist, mainly interested in amphibians. He is Professor, Department of Vertebrates, National Museum, Federal University of Rio de Janeiro and Editor of Archives at the National Museum. His bachelor's degree in biology (1976) was from the University Estadual Paulista Júlio de Mesquita Filho, his master's in ecology (1981) from the University of Campinas, and doctorate in zoology (1989) from the University of São Paulo.

Carbone

Mahajanga Reed Frog *Heterixalus carbonei* Vences et al., 2000

Ermanno Carbone was the last Honorary Consul of Madagascar in Genoa.

Cardoso

Cardoso's Treefrog *Dendropsophus timbeba* **Martins** & Cardoso, 1987

Revolta Snouted Treefrog *Scinax cardosoi* **Carvalho** & **Peixoto**, 1991

Cycloramphid frog sp. *Rupirana cardosoi* **Heyer**, 1999

Rapids Frog sp. *Paratelmatobius cardosoi* **Pombal** & **Haddad**, 1999

Paradoxical Frog sp. *Pseudis cardosoi* Kwet, 2000
Tree Toad sp. *Hylodes cardosoi* Lingnau, Canedo & Pombal, 2008

Dr Adão José Cardoso (1951–1997) was a Brazilian herpetologist. His bachelor's degree (1975), his master's (1981) and his doctorate (1985) were all from the Universidade de Campinas (UNICAMP). The Zoological Museum at this university is now named after him.

Carles Vilà

Glass Frog sp. *Hyalinobatrachium carlesvilai* Castroviejo-Fisher et al., 2009

Professor Dr Carles Vilà is a Professor at the Estación Biológica de Doñana, Spain, and a lecturer in evolutionary biology at Uppsala Universitet, Sweden. The Universidad de Barcelona awarded his doctorate for work on wolf ecology and his current interests include the conservation biology and conservation genetics of vertebrates. He co-wrote *Detecting the vanishing populations of the highly endangered Darwin's fox,* Pseudalopex fulvipes (2004).

Carlos Sánchez

Robber Frog sp. *Pristimantis carlossanchezi* Arroyo, 2007

Carlos Alberto Sánchez is the author's grandfather.

Carmelita

Carmelita's Robber Frog *Pristimantis carmelitae* **Ruthven**, 1922

Myrtle Carmelita Carriker née Flye (b.1893) was the wife of Melbourne Armstrong Carriker Jr (q.v.). She was the daughter of an American engineer who became a coffee planter at Santa Marta, Colombia, which is where her husband met her whilst on a collecting expedition (1911). Ruthven visited the family and collected the frog holotype there, and also named an amphibian after Melbourne that the latter had collected. Myrtle returned to the USA (1927) and was later divorced. A bird is also named after her.

Carnevalli

Rio Doce Snouted Treefrog *Scinax carnevallii* **Caramaschi** & Kisteumacher, 1989

Professor Ney Eni Demas Carnevalli (1938–2002) was a zoologist and ornithologist, Federal University of Minas Gerais, Belo Horizonte, Brazil. He wrote (1978 unpublished) *Estudos da Ornitofauna do Parque Florestal do Rio Doce, Minas Gerais. Parte I.* He died of lung cancer.

Carnifex

Executioner Treefrog *Hyla carnifex* **Duellman**, 1969

Carnifex means public executioner or hangman. The original description says it is an allusion to John D. Lynch (q.v.) who collected many of the early specimens of this species and, presumably, is a private gallows humour joke between them in regard to hanging being a popular form of lynching.

Carqueja

Carqueja's Squeaker *Arthroleptis carquejai* Ferreira, 1906 [Alt. Cambondo Screeching Frog]

Professor Bento Carqueja (1860–1935) was a professor, journalist and philanthropist. He graduated (1882) in agriculture at the Porto Polytechnic Academy, Portugal, and then taught there (1882–1884). He transferred to Porto's Normal School, teaching agriculture and physical and natural sciences, and founding the school's botanical garden. He was Professor at the Porto Polytechnic Academy (1898–1915), later becoming full Professor at the Faculty of Sciences, Universidade do Porto, until his retirement (1928). He was also successful as a journalist (1880), eventually becoming co-owner of a newspaper. He was very interested in socio-economic matters, developed housing for the poor, and in later years donated his professorial salary to the Institute of Scientific Research in Socio-economic Sciences. The Portuguese botanist and explorer, Francisco Newton, was encouraged and helped by Carqueja, and Ferreira, the frog's describer, was a regular visitor to his home in Porto.

Carr

Slender-legged Treefrog sp. *Osteocephalus carri* **Cochran** & **Goin**, 1970
Cloud Forest Salamander *Bolitoglossa carri* **McCranie** & Wilson, 1993

Professor Dr Archibald (Archie) Fairly Carr Jr (1909–1987) was a conservationist who was Professor of Zoology, University of Florida, where he studied zoology before settling on a career as a herpetologist, becoming one of the world's leading experts on sea turtles. The Dr Archie Carr Wildlife Refuge, Costa Rica, was established and named in his honour. He wrote *Ulendo: Travels of a Naturalist in and out of Africa* (1954). Two reptiles are named after him.

Carrangueros

Robber Frog sp. *Pristimantis carranguerorum* **Lynch**, 1994

Los Carrangueros de Raquira is a popular Colombian folk-music band, founded (1980) by Jorge Velosa.

Carriker

Carriker's Harlequin Frog *Atelopus carrikeri* **Ruthven**, 1916
[Alt. Guajira Stubfoot Toad]

Melbourne Armstrong Carriker (1879–1965) was an American ornithologist and entomologist, specialising in Central and northern South America. He was sent on several expeditions by the Philadelphia Academy of Natural Sciences to Peru and Bolivia (1929–1938). He greatly enhanced the bird collections of the Carnegie Museum and the Smithsonian. He collected the frog holotype (1914). Nine birds and a mammal are named after him.

Carrillo

Ancash Water Frog *Telmatobius carrillae* Morales, 1988

Dr Nelly Carrillo de Espinoza (b. 1932) is a zoologist and herpetologist who has worked with the describer. She is an Honorary Researcher at the Natural History Museum, Universidad Nacional Mayor de San Marcos, Peru. She co-wrote *Lista taxonómica preliminar de los reptiles vivientes del Perú* (1995).

Carrizo

Carrizo's Treefrog *Hypsiboas varelae* Carrizo, 1992

Gustavo R. Carrizo is a zoologist and herpetologist at the Argentine Museum of Natural Sciences. He co-wrote *Catalogue of the voices of Argentine Amphibians Part 1* (1993).

Carvalho

Carvalho's Surinam Toad *Pipa carvalhoi* **Miranda-Ribiero**, 1937
Carvalho's Robber Frog *Pristimantis carvalhoi* **Lutz**, 1952
Carvalho's Silent Frog *Syncope carvalhoi* **Nelson,** 1975
Carvalho's Treefrog *Bokermannohyla carvalhoi* **Peixoto**, 1981
Carvalho's Button Frog *Cycloramphus carvalhoi* **Heyer**, 1983
Carvalho's Bug-eyed Frog *Zachaenus carvalhoi* **Izecksohn**, 1983
Carvalho's Tree Toad *Dendrophryniscus carvalhoi* Izecksohn, 1994
Central Humming Frog *Chiasmocleis carvalhoi* **Cruz, Caramaschi** & Izecksohn, 1997
Oval Frog sp. *Elachistocleis carvalhoi* Caramaschi, 2010

Dr Antenor Leitão de Carvalho (1910–1985) was a Brazilian ichthyologist and herpetologist specializing in frogs. He became a pilot in the merchant marine (1927–1932) and, whenever in port (Rio de Janeiro), he volunteered to help out at the Museum. He became Field Collector for the Museum of Rio de Janeiro (1933), undertaking a number of expeditions in Brazil. He went on to become Curator of Herpetology (1941) and, eventually, the Museum's Vice-Director. He collected specimens of many different taxa all over Brazil (1930s–1940s). He wrote *A Preliminary Synopsis of the Genera of American Microhylid Frogs* (1954). Five reptiles are named after him. (See **Antenor**)

Caspari

Robber Frog sp. *Eleutherodactylus casparii* **Dunn**, 1926

Mr Caspari of Mina Carlota, Cuba, was thanked by Dunn for his help. However, Dunn neglected to provide forename(s) or other details.

Castroviejo

Temperate Snouted Treefrog *Scinax castroviejoi* **De La Riva**, 1993
Glass Frog sp. *Vitreorana castroviejoi* **Ayarzagüena** & **Señaris**, 1997

Dr Javier Castroviejo Bolibar is a Spanish zoologist and ecologist who created a number of biological stations, including one in Doñana, Spain. He is Presidente del Comité Español del Programa Hombre y Biosfera Red IberoMaB de la UNESCO. He wrote *Premières données sur l'ecologie hivernale des vertébrés de la Cordillière Cantabrique* (1970). A mammal and a reptile are named after him.

Catalina

Robber Frog sp. *Craugastor catalinae* **Campbell** & **Savage**, 2000

Dr Karen R. Lips is an American biologist and zoologist who is an Associate Professor of Biology at the University of Maryland. The etymology tells us that the binomial is from the Spanish personal name, Catalina: the equivalent of Karen. The University of South Florida, Tampa, awarded her bachelor's degree in zoology (1988) and the University of Miami, Coral Gables, her doctorate (1995). She co-wrote *Emerging infectious disease and the loss of biodiversity in a Neotropical amphibian community* (2006). She collected some of the type-material.

Catesby

American Bullfrog *Rana catesbeiana* Shaw, 1802
[Syn. *Lithobates catesbeianus*]

Mark Catesby (1683–1749) was an English naturalist, artist and traveller. He made two journeys to the Americas (1712–1719 & 1722–1726). During his travels Catesby observed that birds migrate and published his observations in *On the passage of birds* (1747). This discovery was entirely contrary to the prevailing view that birds hibernated in caves or at the bottom of ponds. He used to ship his snake specimens back to

England in jars of rum, which the sailors sometimes drank – thus ruining his specimens! He observed the similarity in the features of the Native Americans and peoples of Asiatic origin and was the first person to hypothesize the existence in the distant past of a land bridge between Asia and the Americas. He wrote *The Natural History of Carolina, Florida and the Bahama Islands: Containing the figures of birds, beasts, fishes, serpents, insects and plants* (1731–1743). Five birds and two reptiles are named after him.

Catharina

Catharina Snouted Treefrog *Scinax catharinae* **Boulenger**, 1888

Serra do Catharina, Santa Catarina, Brazil is the type locality.

Cei

Cei's White-lipped Frog *Leptodactylus chaquensis* Cei, 1950

Catamarca Water Frog *Telmatobius ceiorum** **Laurent**, 1970

Patagonia Frog sp. *Atelognathus ceii* Basso, 1998

Professor Dr José Miguel Cei (1918–2007) was a biologist, ecologist and herpetologist at the National University, Córboba, Argentina. He wrote *Batracios de Chile* (1962). Five reptiles are named after him.

* The plural form *ceiorum* honours both the Professor and his son, Roberto, the latter having collected the holotype.

Celsa

Glass Frog genus *Celsiella* Guayasamin et al., 2009

Treefrog sp. *Tepuihyla celsae* Mijares-Urrutia, Manzanilla & **La Marca**, 1999

Dr Josefa Celsa Señaris (see **Señaris**).

Cepeda

Rocket Frog sp. *Allobates cepedai* Morales, 2002

Professor Dr Jorge Cepeda-Pizarro is a Chilean biologist and entomologist who is Professor of Biological Sciences at the Faculty of Sciences, Universidad de la Serena, part of the Universidad de Chile. New Mexico State University awarded both his master's degree and his doctorate. His main area of study is arthropods. He and Morales are colleagues at the Universidad de la Serena and co-wrote *Limnología del humedal Tambo-Puquíos* (2006). A scorpion is named after him.

Cesari

Oval Frog sp. *Elachistocleis cesarii* **Miranda-Ribeiro**, 1920

The original description does not explain the etymology and we are unable to identify who Cesari was.

Cevallos

Palanda Rocket Frog *Hyloxalus cevallosi* **Rivero**, 1991

Dr Gabriel Cevallos García (1913–2004) was honoured by Rivero who wrote that he was a '...distinguished Ecuadorean writer and thinker, and great student of the history of his country...' Cevallos had been a professor, Faculty of Arts, University of Cuenca, Ecuador, and had served as that university's Chancellor (1964–1968).

Chabanaud

Chabanaud's Reed Frog *Hyperolius chabanaudi* **Ahl**, 1931 NCR

[JrSYN *Hyperolius phantasticus*]

Dr Paul Chabanaud (1876–1959) was a French ichthyologist and herpetologist. He took his first degree at Poitiers (1897). He volunteered (1915) his services to the Muséum National d'Histoire Naturelle, Paris under Louis Roule, who asked him to identify herpetological specimens and sent him on a scientific expedition to French West Africa (1919). He travelled to Senegal and Guinea before walking 1,200 km through southern Guinea and Liberia, returning to France (1920), when he became a Preparator of Fishes at the Museum with a special interest in flatfish. He took his doctorate at the Sorbonne (1936). He wrote 40 papers on herpetology (1915–1954). Three reptiles are named after him.

Chac

Chac's Rainfrog *Craugastor chac* **Savage**, 1987

Savage wrote that 'The name 'chac' is derived from the Mayan name for the rain god, whose beneficent downpours are essential to the lives of both rainfrogs and man in Central America.'

Chakrapani

Chakrapani's Narrow-mouthed Frog *Microhyla chakrapanii* **Pillai**, 1997

[Alt. Mayabunder Rice Frog]

S. Chakrapani led a surveying expedition to the Andaman Islands (1972) and collected this frog at Mayabunder, Middle Andaman Island.

Chamberlain

Chamberlain's Dwarf Salamander *Eurycea chamberlaini* Harrison & Guttman, 2003

Edward Burnham Chamberlain (1895–1986), former Curator of Vertebrates at the Charleston Museum. After military service and a year at the Smithsonian, he started at the Charleston Museum as Head of Preparation (1924), eventually rising to Curator and retiring (1952).

Chami

Glass Frog sp. *Nymphargus chami* **Ruiz-Carranza** & **Lynch**, 1995

The Chami are a Colombian indigenous group within whose territory this frog is found.

Chamula

Chamula Mountain Brook Frog *Duellmanohyla chamulae* **Duellman**, 1961

This frog is named after the Chamula Indians of the highlands of Chiapas, southern Mexico, where the species is endemic.

Chancas

Peru Cochran Frog *Nymphargus chancas* **Duellman** & **Schulte**, 1993

The Chancas are a Peruvian indigenous people.

Chang

Chinese Newt sp. *Pachytriton changi* Nishikawa, **Matsui**, & Jiang, 2012

Dr Mangven L. Y. Chang (b.1902) was a Professor herpetologist at East China Normal University, Shanghai. The etymology reads: 'The specific epithet is dedicated to the late Dr Mangven L. Y. Chang, who pioneered the modern scientific study of Chinese urodeles in the 1930s.' He wrote several monographs and longer works including (co-written) *On the Salamanders of Chekiang* (1933) and *Contribution à l'étude Morphologique, Biologique et Systématique des Amphibiens urodèles de la Chine* (1936).

Channing

Toad sp. *Amietophrynus channingi* Barej et al., 2011 [Syn. *Bufo channingi*]

Professor Dr Alan Channing is a herpetologist at the Department of Biodiversity & Conservation Biology, University of the Western Cape, South Africa, where he is a professor. He wrote *Amphibians of Central and Southern Africa* (2001).

Chapin

Chapin's Frog *Amietia angolensis chapini* **Noble**, 1924 [Taxonomy unclear: sometimes viewed as a synonym of *A. angolensis*]

Dr James Paul Chapin (1889–1964) was an American ornithologist and co-leader of the Lang-Chapin Expedition, which made the first comprehensive biological survey of the then Belgian Congo (1909–1915). He worked at the AMNH, New York, and was President of the Explorers' Club (1949–1950). He wrote *Birds of the Belgian Congo* (1932). Fourteen birds, four reptiles and two mammals are named after him.

Chappuis

Turkana Toad *Bufo chappuisi* **Roux**, 1936 NCR [JrSYN *Poyntonophrynus lughensis*]

Dr Pierre-Alfred Chappuis, (1891–1960) was a French-born Swiss zoologist and biospeleologist who specialized in isopods. He led the 1933 Mission Scientifique de l'Omo, Ethiopia, during which the toad holotype was collected.

Charazani

Charazani Treefrog *Hyloscirtus charazani* **Vellard**, 1970

Charazani, Departamento La Paz, Bolivia is the type locality.

Chari

Chari's Bush Frog *Raorchestes charius* **Rao**, 1937 [Alt. Seshachar's Bush Frog; Syn. *Philautus charius*]

Chari is a diminutive for **Seshachar** (q.v.).

Charlotte

Charlotte's Madagascar Frog *Mantidactylus charlotteae* Vences & **Glaw**, 2004

Charlotte Richter-Pfeil gives financial support to BIOPAT.

Cheesman

Kokoda Rainforest Frog *Cophixalus cheesmanae* **Parker**, 1934

Cheesman's Papua Frog *Barygenys cheesmanae* Parker, 1936

Cheesman's Wrinkled Ground Frog *Platymantis cheesmanae* Parker, 1940

Cheesman's Treefrog *Litoria cheesmani** **Tyler**, 1964

Miss Lucy Evelyn Cheesman (1881–1969) wanted to train as a veterinary surgeon, but in her time the restrictions on the education of women precluded it. Instead she became an entomologist, explorer and traveller, making a number of expeditions to the Galapagos Islands, the Marquesas, New Guinea, the New Hebrides and the Solomon Islands (1924–1936). She worked for many years as a volunteer at the British Museum (Natural History), London, and was the first female curator at London Zoo. She wrote *The Cyclops Mountains of Dutch New Guinea* (1938). A reptile is also named after her.

*The binomial of the treefrog, *cheesmani*, is sometimes 'corrected' to the feminine *cheesmanae*.

Chen-fu

Chenfu's Treefrog *Rhacophorus chenfui* **Liu**, 1945

Dr Chen-fu Wu was an entomologist of Yen Ching University (University of Beijing) in the 1940s.

Chermock

Salamander sp. *Desmognathus chermocki* **Bishop** & Valentine, 1950

[JrSYN *Desmognathus aeneus*]

Dr Ralph L. Chermock (1918–1977) was a lepidopterist. He was Professor of Biology at Alabama University (1947–1966) and Director, Alabama Natural History Museum (1960–1966). He was one of the collectors of the type series.

Chirio

Wolterstorff Toad sp. *Wolterstorffina chirioi* Boistel & **Amiet**, 2001

Laurent Chirio is a herpetologist at the Muséum National d'Histoire Naturelle, Paris. He co-wrote *Biogeography of the reptiles of the Central African Republic* (2006).

Chirripo

Suretka Glass Frog *Hyalinobatrachium chirripoi* **Taylor**, 1958

The species is named for the local inhabitants of the area where the frog occurs: the Chirripó of Costa Rica.

Chisholm

Salado Salamander *Eurycea chisholmensis* Chippindale et al., 2000

[Alt. Chisholm Trail Salamander]

The Chisholm Trail, which was a route used for cattle drives from Texas to Kansas.

Christian

Christian's Marsupial Frog *Gastrotheca christiani* **Laurent**, 1967

Dr Christian P. J. Halloy is Adjunct Professor of Physics, University of Tennessee. He is a Belgian who was brought up in Africa, USA and Argentina. His master's degree in physics was earned (1977) at the National University of Tucumán, Argentina, and his doctorate in physical chemistry was awarded by the Ecole Nationale Supérieure des Industries Chimiques, France. He and his brother Stéphan (q.v.), when they were teenagers in Argentina, used to help their stepfather Raimond Laurent to collect specimens, and they discovered this species (1967).

Christy

Christy's Tree Frog *Leptopelis christyi* **Boulenger**, 1912

Christy's Ridged Frog *Ptychadena christyi* Boulenger, 1919

Dr Cuthbert Christy (1863–1932) qualified as a physician at Edinburgh. He travelled in the West Indies and South America (early 1980s), subsequently joining the army as a doctor, and was in northern Nigeria (1898–1900) and afterwards for a short time in India. He was then in Uganda and the Congo (1902–1903), and visited Ceylon (Sri Lanka) (1904–1914), Kenya, Uganda, the Gold Coast (Ghana), Cameroon and Sudan. He served in Africa and Mesopotamia (Iraq) (WW1), then explored in the Sudan, Nyasaland (Malawi) and Tanganyika (Tanzania), and was a member of a League of Nations Commission enquiring into slavery and forced labour in Liberia. He was Director, Congo Museum, Tervuren, Belgium. While on a zoological expedition to the Congo (1932) he was killed when gored by a buffalo. A mammal, a bird and three reptiles are named after him.

Chryses

Golden Treefrog *Plectrohyla chryses* **Adler**, 1965

In Greek mythology, Chryses was a priest of Apollo at Chryse, near the city of Troy. The Greek word *chrȳsós* means 'gold'. The treefrog is described as having 'dorsum of body, head, and limbs, and sides of body metallic golden yellow', so the name also reflects the frog's appearance, but Adler specifically says that the name 'refers to Chryses, one of the priests of Apollo.'

Chudeau

Bata Marsh Toad *Amietophrynus chudeaui* **Chabanaud**, 1919

René Chudeau (1864–1921) was a French explorer and geologist who worked in French West Africa, especially in the Sahara and Sahel (1904–1914), discovering the first dinosaur bones in Niger. He was summarily dismissed (1890) from his post as lecturer at the University of Besançon for living with an alleged prostitute. After returning to France, he worked as a consulting engineer for the Industrial Bank of China, which went bankrupt; he lost his job and died miserable and alone. The day after his death he was elected to a position at the Muséum National d'Histoire Naturelle, Paris. He wrote *Le Sahara Soudanais* (1909). A mammal is named after him.

Cisternas

Brown Midwife Toad *Alytes cisternasii* **Bosca**, 1879

[Alt. Iberian Midwife Toad]

Professor Dr Rafael Cisternas y Fonseré (1818–1876) was a Spanish physician, naturalist and ichthyologist. He received a doctorate in natural sciences (1846), having already qualified as both a physician and a surgeon. He taught mineralogy and zoology at the University of Barcelona (1845–1848). He became Professor of Natural History, Oviedo and Valladolid (1848), and at the University of Salamanca (1851). He was Professor of Natural Sciences, University of Valencia, responsible for the teaching of mineralogy, botany and zoology (1861–1876). He was also in charge of the University Natural History Museum and Botanical Gardens. He took over the Colleges of Agriculture and Veterinary Medicine and the teaching of mineralogy and applied chemistry at the School of Architecture (1869). He was the first teacher in Valencia to teach Darwinism. He wrote *Ensayo de un catálogo descriptivo de los peces de agua dulce que habitan en la provincia de Valencia* (1877).

Clapp

Rainforest Frog sp. *Cophixalus clapporum* Kraus, 2012

George Edward Clapp (b.1941) in England and Eleanor Elizabeth Christina Keisa Clapp (b.1984) in Papua New Guinea are father and daughter. George has a bachelor's degree from the University of Queensland and worked in Papua New Guinea for over 30 years as a patrol and district officer and for oil and resource companies. He became Woodlark Mining's Community Relations and Environment Manager (2007) and is now retired. Eleanor is an environmental scientist who has worked for Woodlark Mining's parent company, Kula Gold, as Environmental Database Manager since 2009. She has a bachelor's degree awarded by the University of the Sunshine Coast, Queensland and a master's from James Cook University, Queensland.

Clark, HC

Mexican Caecilian *Gymnophis clarkii* **Barbour**, 1926 NCR

[JrSYN *Dermophis mexicanus*]

Dr Herbert Charles Clark (1877–1960) was the first Director of Gorgas Memorial Laboratory, Panama (1929–1954). The United Fruit Company previously employed him as Director of Laboratories and Preventive Medicine. He also organized an annual census of the snake population of Panama (1929–1953). Five reptiles are named after him.

Clark, JH

Clark's (Striped) Tree Frog *Pseudacris clarkii* **Baird**, 1854

[Alt. Spotted Chorus Frog]

Lieutenant John Henry Clark (1830–1885) was an American surveyor, naturalist and collector. He was a student at Dickinson College (c.1844). He was a zoologist on the United States and Mexican Boundary Survey (1850–1855). Under the auspices of the Smithsonian, he conducted the Texas Boundary Survey (1860). A bird and two reptiles are named after him.

Clarke

Clarke's Banana Frog *Afrixalus clarkei* **Largen**, 1974

[Binomial sometimes amended to the plural, *clarkeorum*]

Mr and Mrs R. O. S. Clarke, who lived in Ethiopia, were Largen's generous hosts. Mr Clarke helped Largen with his collecting.

Claudia

Bocas Poison Frog *Ranitomeya claudiae* Jungfer, Lötters & Jörgens, 2000

Claudia Vlasimsky's father, Stan M. Vlasimsky, has been a big financial supporter of BIOPAT.

Clemson

Clemson Salamander *Plethodon clemsonae* **Brimley**, 1927 NCR

[JrSYN *Plethodon metcalfi*]

This is named after Clemson College, South Carolina, where the holotype was collected.

Clivio

Eritrean Clawed Frog *Xenopus clivii* **Peracca**, 1898

[Alt. Peracca's Clawed Frog]

Lieutenant P. Clivio was an Italian army veterinary surgeon who collected the holotype in Eritrea and presented it to the Turin Museum.

Cochran

Glass Frog genus *Cochranella* **Taylor**, 1951

Cochran's Robber Frog *Eleutherodactylus cochranae* **Grant**, 1932

Cochran's Running Frog *Kassina cochranae* **Loveridge**, 1941

Cochran's False Brook Salamander *Pseudoeurycea cochranae* Taylor, 1943

Cochran's Stream Salamander *Batrachuperus cochranae* **Liu**, 1950 NCR

[JrSYN *Batrachuperus pinchonii*]

Cochran's Snouted Treefrog *Scinax similis* Cochran, 1952

Canebrake Treefrog *Aplastodiscus cochranae* **Mertens**, 1952

Cochran's Lime Treefrog *Sphaenorhynchus surdus* Cochhran, 1953

Cochran Frog *Nymphargus cochranae* **Goin**, 1961

Chocolatefoot Leaf Frog *Phasmahyla cochranae* **Bokermann**, 1966
Cochran's Guanabara Frog *Euparkerella cochranae* **Izecksohn**, 1988

Dr Doris Mable Cochran (1898–1968) was an American herpetologist. She was educated in Washington DC and after graduating from high school worked for the War Department. She took evening classes at George Washington University, where she obtained her master's degree in science (1921). She received a master's degree in zoology (1928) from Johns Hopkins University and a doctorate from the University of Maryland (1933). She also trained as, and became, a highly skilled scientific illustrator. She worked at the Smithsonian (1919–1968), first as an assistant to Leonhard Stejneger (q.v.) in the Division of Reptiles and Amphibians, becoming Assistant Curator, Reptiles and Amphibians (1927), Associate Curator (1942) and Curator (1956). After Stejneger's death (1943) she became acting Head of the Division. Her main interest was the herpetofauna of the West Indies and Tropical America, making field trips (1935 & 1962–1963) to work with the Brazilian herpetologists Adolpho and Bertha Lutz. She wrote *The Frogs of Southeastern Brazil* (1955). Nine reptiles are named after her. (See **Doris** and **Mable**).

Cofan

Andean Toad sp. *Osornophryne cofanorum* Mueses-Cisneros, Yanez-Munoz & Guayasamin, 2010

The Cofan are an indigenous people, Amazonian Indians of Ecuador.

Cogger

Treefrog genus *Coggerdonia* Wellington & Wells, 1985 NCR
[Not widely accepted. In *Litoria*]
Kaironk Land Frog *Oxydactyla coggeri* **Zweifel**, 2000
Cogger's (Barred) Frog *Mixophyes coggeri* **Mahony** et al., 2006
[Alt. Pilgrim Barred Frog]

Dr Harold 'Hal' G. Cogger (b.1935) is an Australian herpetologist who spent his entire career at the Australian Museum, Sydney. Both his bachelor's (1959) and master's (1961) degrees are from the University of Sydney and his doctorate (1970) from Macquarie University. He was Curator of Reptiles and Amphibians (1960–1975) and Deputy Director of the Museum (1976–1995). He was Conjoint Professor, University of Newcastle, NSW (1997–2001). He has travelled widely in the cause of research, visiting most parts of Australia and New Guinea, Japan, Indonesia and many islands in the western Pacific. He wrote *Reptiles & Amphibians of Australia* (1975). Seven reptiles are named after him.

Collett

Collett's Tree Frog *Polypedates colletti* **Boulenger,** 1890
[Alt. Collett's Whipping Frog]

Professor Dr Robert Collett (1842–1913) was a Norwegian zoologist and ichthyologist. He worked at the Christiania Museum (1871–1913), first as an Assistant Curator, then Curator (1874) and Director (1892), and became Professor of Zoology at the University (1884). Christiania is now called Oslo – Norway became independent from Sweden in 1905. A mammal, two reptiles and five birds are named after him.

Collins

Collins' Poison Frog *Ranitomeya abdita* **Myers** & **Daly**, 1976

Joseph T. Collins (1939–2012) was Herpetologist Emeritus at the Museum of Natural History, University of Kansas. He wrote *Amphibians and Reptiles in Kansas* (1974). Although his primary interest was herpetology, he also published on mammalogy, ichthyology, systematics and taxonomy. Along with William Duellman (q.v.) he collected the paratypes of this frog in Ecuador (1971).

Coloma

Carchi Andes Toad *Andinophryne colomai* **Hoogmoed,** 1985
Robber Frog sp. *Pristimantis colomai* **Lynch** & **Duellman**, 1997
Strabomantid frog sp. *Noblella coloma* Guayasamin & Terán-Valdez, 2009

Professor Dr Luis Aurelio Coloma Román (b.1962) is a zoologist and herpetologist. He is Curator, Herpetology Section, Museum of Zoology, Pontificia Universidad Católica del Ecuador. He took his initial degrees at universities in Ecuador and followed up with a doctorate from University of Kansas. A reptile is also named after him.

Colonnello

Ayarzagüena's Frog *Dischidodactylus colonnelloi* **Ayarzagüena**, 1985

Dr Giuseppe Colonnello Bertoli is a Venezuelan botanist and hydrobiologist who is Curator of the Herbarium at the Museo de Historia Natural La Salle. He wrote *Aquatic vegetation of the Orinoco River Delta (Venezuela). An overview* (1996).

Conant

Spotted Dusky Salamander *Desmognathus conanti* Rossman, 1958

Conant's False Brook Salamander *Pseudoeurycea conanti* **Bogert**, 1967

Conant's Mushroomtongue Salamander *Bolitoglossa conanti* **McCranie** & Wilson, 1993

Roger Conant (1909–2003) was a herpetologist, author and conservationist. While still a teenager he took a job at Philadelphia Zoo to earn money to help his widowed mother. The result was a life-long passion for reptiles. He was Curator of Reptiles at Toledo Zoo, Ohio (1929–1935), leaving to return to Philadelphia Zoo, becoming its Director (1967–1973). In addition to fieldwork in USA and Mexico, he visited Asia and Africa. He wrote and presented a 15-minute weekly educational programme for a Philadelphia radio station called 'Let's Visit the Zoo' (1936–1969). His wife was the animal artist and photographer Isabelle de Peyster Hunt (1901–1976) and they created *A Field Guide to Reptiles & Amphibians of the United States & Canada East of the 100th Meridian* (1958). In retirement at Albuquerque, New Mexico, he continued to research and write. He was Director Emeritus, Philadelphia Zoo, and an Adjunct Professor, University of New Mexico. He died of cancer. Three reptiles are named after him.

Condorir

Andes Frog sp. *Psychrophrynella condoriri* **De la Riva**, Aguayo & Padial, 2007

Condorir is a mountain in Bolivia – it has the appearance of a giant condor with outstretched wings.

Contreras

Contreras' Water Frog *Telmatobius contrerasi* **Cei,** 1977

Professor Dr Julio Rafael Contreras Roque (1933-fl.2012) is an Argentine biologist who was at Museo Argentino de Ciencias Naturales 'Bernardino Rivadavia', Buenos Aires and President, Buenos Aires Foundation of Natural History Felix de Azara. He was Director of the Biological Station, Isla Victoria, Bariloche (1975). Since 2003 he has lived in Paraguay, where he teaches and researches at National University of Pilar. A mammal and a bird are named after him.

Cook

Cook's Robber Frog *Eleutherodactylus cooki* **Grant**, 1932

Dr Melville 'Mel' Thurston Cook (1869–1952) was a botanist and entomologist who became Director, Insular Experiment Station, Rio Piedras, part of the Universidad de Puerto Rico. He joined the faculty of Rutgers University (1910) to teach botany. He was a Professor and seconded to help teach at New Jersey College for Women (1918). He spent some months in Puerto Rico with a party of his students (1926). He wrote *Snake Killing Mongooses a Plague to Puerto Rico* (1929). A reptile is named after him.

Cooper

Cooper's Black Caecilian *Praslinia cooperi* **Boulenger**, 1909

Sir Clive Forster-Cooper (1880–1947) was educated at Trinity College, Cambridge. He joined J. Stanley Gardiner's expedition to the Maldives and Laccadives islands (1900), then joined the staff of the North Sea Fisheries Commission for Scientific Investigation (1902–1903). He returned to Cambridge and went on an expedition to the Seychelles with J. Stanley Gardiner on 'HMS Sealark' (1906). He met Dr C. W. Andrews at the BMNH, who introduced him to palaeontology. He went with Andrews to Egypt (1907). He spent a year (1909–1910) at the AMNH, New York, before returning to Cambridge. He was appointed Director, Cambridge University Museum of Zoology (1914). He worked (1914–1918) on malaria at the School of Tropical Medicine, Liverpool. He became Cambridge University Reader in Vertebrates, Fellow and Bursar of Trinity Hall, and also became a Fellow of the Royal Society (1936). He was Director of the BMNH (1938–1947), actually living in the Museum, ensuring the survival of the collections despite severe bomb damage (1940 & 1945). He was knighted (1946), but died two months before he was due to retire (1947). He was described as modest and shy, appreciated paintings, and was a watercolorist and draughtsman of some skill.

Cooper (Omer-)

Cooper's Grassland Frog *Ptychadena cooperi* **Parker**, 1930

Joseph Omer-Cooper (1893–1972) was a British-born zoologist and entomologist whose bachelor's degree (1925) and master's (1927) were awarded by Cambridge, where he was on the staff of the Zoology Department (1924–1927), followed by a Senior Lectureship at Durham University (1927–1935). He was at Rhodes University, South Africa (1936–1954), becoming Head of Department (1937) and Professor of Zoology (1940). He collected in Egypt and Ethiopia (1926–1935). He made one final expedition after retirement to Chad and Nigeria (1963). He wife, Joyce, was also a distinguished zoologist and lecturer at Rhodes University. He collected the frog holotype.

Cope

Cope's Caribbean Toad *Peltophryne empusa* Cope, 1862
[Alt. Cuban Small-eared Toad]
Cope's Toad *Rhinella diptycha* Cope, 1862
Cope's Spea *Spea bombifrons* Cope, 1863
[Alt. Plains Spadefoot Toad]
Cope's Hyla *Hyla arenicolor* Cope, 1866
[Alt. Canyon Treefrog]
Cope's Toad-frog *Phrynobatrachus acridoides* Cope, 1867
[Alt. Eastern Puddle Frog, Zanzibar Puddle Frog]
Cope's (Assam) Frog *Hylarana leptoglossa* Cope, 1868
Cope's Mushroomtongue Salamander *Bolitoglossa morio* Cope, 1869
Cope's Eastern Paraguay Treefrog *Hypsiboas polytaenius* Cope, 1870
Cope's White-lipped Frog *Leptodactylus labialis* Cope, 1877 NCR
[JrSYN *Leptodactylus mystacinus*]
Cope's (Grey) Treefrog *Hyla chrysoscelis* Cope, 1880
Cope's Salamander *Ambystoma copeianum* **Hay**, 1885 NCR
[JrSYN *Ambystoma tigrinum*]
Cope's Brown Treefrog *Ecnomiohyla miliaria* Cope, 1886
Cope's Swamp Frog *Pseudopaludicola mystacalis* Cope, 1887
Cope's Frog *Lithobates virgatipes* Cope, 1891
[Alt. Carpenter Frog; Syn. *Rana virgatipes*]
Cope's Giant Salamander *Dicamptodon copei* **Nussbaum**, 1970
El Copé Giant Salamander *Bolitoglossa copia* **Wake**, **Hanken** and Ibanez, 2005
Cope's Tree Frog *Chorophilus triseriatus corporalis* Cope, 1875 NCR
[No designated type specimen: taxon now known as *Pseudacris kalmi*]
Cope's Toad *Anaxyrus americanus copei* Yarrow & Henshaw, 1878
[Alt. Hudson Bay Toad; Syn. *Bufo americanus copei*]
Cope's Cave Salamander *Eurycea longicauda melanopleura* Cope, 1894
[Alt. Dark-sided Salamander]

Edward Drinker Cope (1840–1897) was an American palaeontologist, anatomist, herpetologist and ichthyologist. He studied under Baird at the Smithsonian (1859), at the British Museum, London, and the Jardin des Plantes, Paris (1863–1867). He was Professor of Comparative Zoology and Botany, Haverford College, Pennsylvania (1864–1867), and was appointed Curator, Philadelphia Academy of Natural Sciences (1865). He was Palaeontologist on the Wheeler Survey (1874–1877) west of the 100th meridian in New Mexico, Oregon, Texas and Montana. He was Professor at the University of Pennsylvania: of Geology and Mineralogy (1889–1895) and Zoology and Comparative Anatomy (1895–1897). He was senior naturalist (1878) on the periodical *American Naturalist*, which he co-owned. He wrote *Systematic Arrangement of the Extinct Batrachia, Reptiles and Aves of North America* (1869–1870). In his will he donated his body to science. His cause of death was listed as uremic poisoning, but it was rumoured that he died of syphilis. However, in 1995 permission was granted to have Cope's skeleton medically examined; there was no evidence of bony syphilis. The etymology for the El Copé Giant Salamander makes it clear that it is named in honour of both the town of El Copé and Edward Drinker Cope. 59 reptiles are named after him!

Copland

Copland's Rock Frog *Litoria coplandi* **Tyler**, 1968
[Alt. Sandstone Frog, Saxicoline Treefrog]

Stephen John Copland (fl.1962) was a herpetologist at Department of Zoology, University of Sydney, where he catalogued the herpetological collection at Macleay Museum (1945–1947). He was awarded his master's by the University of Sydney (1949). He worked largely in isolation and wrote very detailed descriptions. He travelled within New South Wales amassing a large collection. He became President of the Linnean Society of New South Wales (1957). His extensive collection, including many holotypes, is in the Australian Museum. He wrote *Reptiles Occurring Above the Winter Snowline at Mt. Kosciusko* (1947). Four reptiles are named after him.

Coppinger

Spiny-chest Frog sp. *Alsodes coppingeri* **Günther**, 1881

Dr Richard William Coppinger (1847–1910) was a naval surgeon who served on HMS Alert (1881–1882) and was Inspector-General, Hospitals and Fleets (1901–1904). Two birds are named after him

Cordeiro

Cordeiro's Humming Frog *Chiasmocleis cordeiroi* **Caramaschi** & Pimenta, 2003

Paulo Henrique Chaves Cordeiro is a Brazilian biologist and field collector who has worked with the junior author.

Cordova

Robber Frog sp. *Pristimantis cordovae* **Lehr** & **Duellman**, 2007

Dr Jesús 'Jess' H. Córdova Santa Gadea is a biologist and herpetologist at the Natural History Museum in Lima, Peru. He worked at the Laboratory of Human Genetics, Universidad Nacional Mayor de San Marcos, Lima (1992). Both Lehr and Duellman have worked with him. He and Lehr wrote a paper (2002) on Peruvian amphibians and he co-wrote *Conservation priorities for harlequin frogs (*Atelopus *ssp.) of Peru* (2005).

Cortés/Cortez

Cortez' Hidden Salamander *Cryptotriton nasalis* **Dunn**, 1924

[Alt. Cortes Salamander]

Cortés Department, Honduras. The spelling 'Cortez' and use of the apostrophe are errors.

Cosenza

Snouted Treefrog sp. *Scinax cosenzai* Lacerda, **Peixoto** & **Feio**, 2012

Professor Braz Antônio Pereira Cosenza is a Brazilian biologist, botanist and conservationist who is courses co-ordinator at Faculdades Vale do Carangola, Minas Gerais, Brazil. His bachelor's degree was awarded by the Universidade Federal de Juiz de Fora (1989), and his master's (2003) and doctorate (2012) were awarded by the Universidade Federal de Viçosa. He is particularly noted and celebrated for his contributions to the studies of flora and fauna of Serra do Brigadeiro.

Couch

Couch's Spadefoot Toad *Scaphiopus couchii* **Baird**, 1854

General Darius Nash Couch (1822–1897) was a US Army officer, a general during the Civil War. He was also an explorer, who took leave of absence to lead a zoological expedition in Mexico. He was a lieutenant (1846) in Mexico and fought at the Battle of Buena Vista (1847). He returned to Washington (1854), resigned his commission (1855) and became (1855–1861) a merchant and manufacturer in New York and Massachusetts. On the outbreak of the American Civil War he rejoined the army as a colonel. He offered to resign on the grounds of ill health (1863), but was persuaded to stay on by being promoted to major-general. He was in charge of all the ceremonies associated with the consecration of the National Cemetery at Gettysburg (1865), the occasion of Abraham Lincoln's Gettysburg Address. After the Civil War he again resigned from the army and was Collector, Port of Boston (1866–1867), President of a Virginia mining and manufacturing concern (1867–1877), and an administrator in Connecticut (1877–1884). Two birds and two reptiles are named after him.

Coudreau

Coudreau's Frog *Rhombophryne coudreaui* **Angel**, 1938

[Alt. Betampona Digging Frog]

Jean Coudreau (fl.1961) was a colonial forestry administrator in Madagascar (1935–1961) who also collected botanical specimens.

Courtois

Spiny Frog sp. *Quasipaa courtoisi* **Angel**, 1922

The Reverend Fréderic Courtois (1860–1928) was a French missionary to China (1901–1928). He was an amateur naturalist and was appointed Director of the Natural History Museum in Sikawei, near Shanghai (1903). He wrote *Les Oiseaux du Musée de Zi-Ka-Wei* (1912). Two birds are named after him.

Covacevich

Magnificent Brood Frog *Pseudophryne covacevichae* Ingram & **Corben**, 1994

Jeanette Adelaide Covacevich (b.1945) was, until retirement (2002), Senior Curator, Reptiles and Amphibians, Queensland Museum. She remains a prominent figure in conservation in Queensland and is recognized for her efforts to preserve the unique character of North Stradbroke Island. She wrote *The Snakes of Brisbane* (1980). A reptile is named after her.

Cowan

Betsileo Giant Treefrog *Platypelis cowanii* **Boulenger**, 1882

Cowan's Mantella *Mantella cowani* Boulenger, 1882

The Reverend William Deans Cowan (1844–1923) was a missionary in Madagascar for ten years (late 1800s). He made a geological expedition in south-central Madagascar and was the author of *The Bara Land: A Description of the Country and People*. A mammal and a bird are named after him.

Cox

Cox's Stefania Treefrog *Stefania coxi* MacCulloch & Lathrop, 2002

Carter J. Cox was a field assistant on an expedition with the describers to Mount Ayanganna, Guyana, and collected the holotype.

Coyne

Coyne's Stubfoot Toad *Atelopus coynei* **Miyata**, 1980
[Alt. Rio Faisanes Stubfoot Toad]

Professor Dr Jerry Coyne (b.1949) is an American biologist who is Professor at the Department of Ecology and Evolution, University of Chicago. He graduated (1971) from the College of William & Mary, Williamsburg, Virginia and later did his doctorate at Harvard. He has been Associate Editor, *Evolution* (1985–1988 and 1994–2000) and *The American Naturalist* (1990–1993). He is a critic of Intelligent Design and has noted it as '...the latest pseudoscientific incarnation of religious creationism, cleverly crafted by a new group of enthusiasts to circumvent recent legal restriction.' He is also opposed to scientific racism and stated in his review of *Darwin's Black Box: The Biochemical Challenge to Evolution* (1996) that the Bell Curve was pseudoscientific. He is clearly a generous man as Miyata wrote: 'The specific epithet is a patronym for Dr Jerry Coyne, whose timely financial assistance helped stave off the wolves on several occasions and allowed me to complete the description.'

Cranwell

Cranwell's Horned Frog *Ceratophrys cranwelli* **Barrio**, 1980

Dr Jorge A. Cranwell was in charge of the herpetology section, Museo Argentino de Ciencias Naturales 'Bernardino Rivadavia', Buenos Aires, Argentina. He wrote *Para la herpetología de Misiones* (1943). A reptile is named after him.

Cristina (Ardila-Robayo)

Cristina's Robber Frog *Pristimantis cristinae* **Lynch** & **Ruiz-Carranza**, 1985
Glass Frog sp. *Nymphargus cristinae* Ruiz-Carranza & Lynch, 1995
Beaked Toad sp. *Rhinella cristinae* Vélez-Rodríguez & Ruiz-Carranza, 2002

Maria Cristina Ardila-Robayo works and writes with both Lynch and Ruiz Carranza. She and Ruiz-Carranza are both at the Museo de Historia Natural, Bogotá, Colombia where she is an associate professor. The etymology for the toad calls her '...a tireless investigator, and one of the best collectors of Colombian amphibians.' (See **Ardila**)

Cristina (Braun)

Cristina's Smooth Horned Frog *Proceratophrys cristinae* **Braun**, 1973 NCR
[JrSYN *Proceratophrys bigibbosa*]

Cristina Assunção Sirangelo Braun was the wife of the describer (see **Braun, PC & CAS**).

Crombie

Crombie's Dwarf Frog *Physalaemus crombiei* **Heyer** & **Wolf**, 1989
Rocket Frog sp. *Allobates crombiei* Morales, 2002

Ronald Ian Crombie (b.1949) is a herpetologist at the Division of Amphibians and Reptiles, National Museum of Natural History, Washington DC. He collected on the tiny island of Catalinita, off the coast of the Dominican Republic (1975). He wrote *Herpetological publications of the National Museum of Natural History* (1994). A reptile is named after him.

Crump

Crump's Treefrog *Dendropsophus brevifrons* **Duellman** & Crump, 1974

Dr Martha L. Crump is a field biologist and herpetologist who is an Adjunct Professor of Biological Sciences at Northern Arizona University. The University of Kansas, Lawrence awarded her degrees – bachelor's (1968), master's (1971) and doctorate (1974). She wrote *In search of the Golden Frog* (2000) and *Headless Males Make Great Lovers* (2007).

Cruz

Leaf Frog genus *Cruziohyla* **Faivovich** et al., 2005
Cruz Robber Frog *Craugastor cruzi* **McCranie, Savage** & Wilson, 1989
Treefrog sp. *Dendropsophus cruzi* **Pombal** & Bastos, 1998
Mangaratiba Tree Frog *Phasmahyla cruzi* Carvalho-e-Silva, Silva & Carvalho-e-Silva, 2009

Dr Carlos Alberto Gonçalves da Cruz (b.1944) is a Brazilian herpetologist who was at the Universidade Federal Rural do Rio de Janeiro (1992), and is now a Research Associate at the Museu Nacional, Rio de Janeiro. He co-wrote *Phyllomedusa: posição taxonômica, hábitos e biologia* (Amphibia, Anura, Hylidae) (2002).

Cuc

Nam Tha Frog *Amolops cucae* Bain, **Stuart** & **Orlov,** 2006

Dr Ho Thu Cuc is Senior Herpetologist at the Institute of Ecology and Biological Resources, Hanoi. She co-wrote *Herpetofauna of Vietnam* (2009). A reptile is named after her.

Cundall

Cundall's Robber Frog *Eleutherodactylus cundalli* **Dunn**, 1926

Frank Cundall (1858–1937) was an Englishman who went to Jamaica (1891) and was Secretary and Librarian of the Institute of Jamaica (1891–1937). He was the doyen of Jamaica's scholars and historians and published on virtually every aspect of the island's life and history. He had an interest in herpetology and kept a Jamaican Boa, *Epicrates subflavus*, as a pet. Dunn spent ten days in Jamaica (1925) and in that time discovered seven new species of the genus *Eleutherodactylus*, all endemic to Jamaica, naming most of them after Jamaican naturalists or worthies.

Cuentas

Robber Frog sp. *Pristimantis cuentasi* **Lynch**, 2003

Daniel Cuentas Montalvo collected the frog holotype (2001) in Colombia. He co-wrote *Anuros del departamento del Atlántico y Norte de Bolívar* (2002).

Cunha

Cunha's Caecilian *Typhlonectes cunhai* Cascon, Lima-Verde, & Benevides Marques 1991 NCR

[JrSYN *Typhlonectes compressicauda*]

Osvaldo Rodrigues da Cunha (1916–2007) was a zoologist and herpetologist at the Museu Paraense Emilio Goeldi, Belém, Brazil. He wrote *O Naturalista Alexandre Rodrigues Ferreira: Uma Analise Comparativa de sua Viagem Filosofica (1783–1793) Pela Amazonia e Mato Grosso com a de Outros Naturalistas Posteriores* (1989). Three reptiles are named after him.

Cuvier

Cuvier's Foam Froglet *Physalaemus cuvieri* **Fitzinger**, 1826

Georges Léopold Chrétien Frédéric Dagobert Baron Cuvier (1769–1832), better known by his pen name Georges Cuvier, was a French naturalist and one of the scientific giants of his age. He supported the geological school of thought called 'catastrophism', according to which. paleontological discontinuities were evidence of sudden and widespread catastrophes, and extinctions took place suddenly. The harshness of his criticism towards scientific opponents, and the strength of his reputation, discouraged other naturalists from speculating about the gradual transmutation of species, right up until Darwin's time. Cuvier is also famed for having stayed in a top government post, as Permanent Secretary, Academy of Sciences, through three regimes, including that of Napoleon Bonaparte. Among his writings are *Tableau élémentaire de l'histoire naturelle des animaux* (1798) and *Le Règne animal distribué d'après son organisation* (4 volumes, 1817). Seven birds, six reptiles and three mammals are named after him.

D

Dagmar

Andes Frog sp. *Phrynopus dagmarae* **Lehr**, **Aguilar** & **Köhler**, 2002

Dagmar Schramm lives in Eberbach, Germany. The describers write: 'We are indebted to... ...H. Wölfel-Schramm (Eberbach, Germany)....for supporting taxonomic research and nature conservation in Peru.' Dagmar Schramm is presumably related to this person, but the relationship is not mentioned in the original description.

Dahl, C

Treefrog sp. *Litoria chrisdahli* Richards, 2007

Chris Dahl is a biologist and herpetologist who was a student at the Department of Biology, University of Papua New Guinea, Madang when he collected the type series (2007). He is now a researcher and Deputy Director at the New Guinea Binatang Research Centre. He co-wrote *Beta diversity of frogs in the forests of New Guinea, Amazonia and Europe: contrasting tropical and temperate communities* (2008).

Dahl, K

Dahl's (Olive) Treefrog *Litoria dahlii* **Boulenger**, 1896
[Alt. Dahl's Aquatic Frog]

Professor Dr Knut Dahl (1871–1953) was a Norwegian naturalist, explorer and collector. He was in Australia (1894–1896), collecting specimens for the Zoological Museum, Norwegian University. He wrote *In Savage Australia: An account of a hunting and collecting expedition to Arnhem Land and Dampier Land* (English publication, 1926). Two mammals are named after him.

Dalcq

Kivu River Frog *Phrynobatrachus dalcqi* **Laurent**, 1952

Dr Albert M. Dalcq (1893–1973) was an embryologist who used amphibians for his experiments. He qualified as a physician and became Assistant at the Department of Anatomy and Embryology, Faculty of Medicine, Brussels (1919), becoming its Head and Professor (1930–1963). The Second World War interrupted his studies as the University of Brussels refused to bow to the demands of the occupying Germans, and instead closed down (1941–1945).

Dämel

Australian Wood Frog *Hylarana daemeli* **Steindachner**, 1868

Edward Dämel (1821–1900) was a collector, mainly of arachnids and Pacific entomological specimens (1860–1874). He collected in Queensland and New South Wales (1867–1874) for the Godeffroy Museum, Hamburg. A reptile is named after him.

Dammerman

Dammerman's Wart Frog *Limnonectes dammermani* **Mertens**, 1929

Dr Karel Willem Dammerman (1888–1951) was a Dutch field zoologist, botanist and collector who worked in the East Indies. He wrote *Preservation of Wildlife and Nature Reserves in the Netherlands Indies* (1929). He also wrote an article (1929), *The Orang Pendek or Ape-man of Sumatra*. The existence of this 'mini-Yeti' is still questioned by zoologists, whilst cryptozoologists seek evidence for it. Five birds and a mammal are named after Dammerman.

Dana

Cuzco Robber Frog *Pristimantis danae* **Duellman**, 1978
Bolivar Snouted Treefrog *Scinax danae* Duellman, 1986

Dana Trueb Duellman is the daughter of the describer and his wife Linda Trueb. He wrote of the treefrog: 'This species is named for Dana Trueb Duellman, whose youthful enthusiasm in the search for frogs greatly enriched our collections'.

Daniel, G (Patiño)

Santa Rita Rocket Frog *Colostethus fraterdanieli* **Silverstone**, 1971
Antioquia Leaf Frog *Agalychnis danieli* **Ruiz-Carranza** et al., 1988

Brother Daniel González Patiño (1909–1988) was a Colombian monk who joined the Museo de Historia Natural, Instituto de La Salle, Bogotá (1937). He was Director of the Museum (1980s). His major interests were botany, herpetology and mineralogy. Four reptiles are named after him.

Daniel, JC

Daniel's Frog *Rana danieli* **Pillai** & Chanda, 1977 NCR
[JrSYN *Hylarana garoensis*]
Daniel's Caecilian *Gegeneophis danieli* Giri, Wilkinson & Gower, 2003
[Alt. Amboli Caecilian]
Daniel's Night Frog *Nyctibatrachus danieli* **Biju** et al., 2011

Jivanayakam Cyril aka 'JC' Daniel (1927–2011) was an Indian naturalist particularly interested in herpetology. He joined the Bombay Natural History Society as an assistant to Dr Salim Ali (1950), eventually becoming its Director until retirement (1991), when he was elected an Honorary Member. He initiated the 'Hornbill' magazine. He wrote *The Book of Indian Reptiles* (1983). A reptile is named after him.

Daniela

Ivory Coast Toad *Amietophrynus danielae* **Perret**, 1977
[Syn. *Bufo danielae*]

Dr Danièle Murith was a parasitologist who worked at the Swiss Centre of Research, Ivory Coast, in the 1970s, In the 1980s she was a post-doctoral Fellow at the Department of Zoology, University of Guelph, Ontario, Canada. She wrote *Examples of the close relationship between the mode of transmission of some parasites and the niche of the host amphibians* (1981).

Daniels

Blue Ridge Spring Salamander *Gyrinophilus porphyriticus danielsi* Blatchley, 1901

Lorenzo E. Daniels (1851–1918) of Laporte, Indiana was an amateur naturalist. Blatchley relates in his description of the salamander that Daniels was '...collecting mollusca in southeastern Tennessee' (1900) and 'picked up six specimens of salamanders and placed them in a flask of alcohol which by chance he had with him.' Daniels also had a collection of fossils, which he had accumulated by 1900. He is recorded collecting around Lily Lake, Indiana (1902). He collected 'land shells and ferns' in Arizona (1907) with a small party including ornithologist Harry Swarth, who collected birds there. Four reptiles are named after him.

Danko

Water Frog sp. *Telmatobius dankoi* Formas et al., 1999

Dr Danko Brncic Juricic (1925–1998) was a Chilean geneticist and naturalist of Croatian descent. He was working at the Institute of Biosciences, Universidade de São Paulo (1952), and at Austin, Texas as a Guggenheim Fellow (1969). He was a member of the Chilean Academy of Sciences and a professor at Universidad de Chile.

Dantas

Feijó White-lipped Frog *Hydrolaetare dantasi* **Bokermann**, 1959
[Syn. *Leptodactylus dantasi*]

R. P. Edison de Souza Dantas of Tarauacá, State of Acre, Brazil, gave the describer a lot of help during his stay in the area.

Dante

Murici Spinythumb Frog *Crossodactylus dantei* Carcerelli & **Caramaschi**, 1993

Dante Buzzetti is a biologist and ornithologist who is an expert on the birds of Murici, Brazil.

Dao

Dao Frog *Amolops daorum* Bain et al., 2003

The Dao are an ethnic minority group in Vietnam.

Darling

Darling's Frog *Hylarana darlingi* **Boulenger**, 1902
[Alt. Golden-backed Frog]

James ffolliott Darling was an Irish zoological collector in Rhodesia (Zimbabwe) at the end of the nineteenth century and beginning of the twentieth. Boulenger wrote *A List of the Fishes, Batrachians and Reptiles Collected by Mr. J. ffolliott Darling in Mashonaland, with Descriptions of New Species* (1902). Darling gave collections of insects and arachnids to the South Africa Museum (1896–1897). Two mammals are named after him.

Darlington

Australian Marsupial Frog *Assa darlingtoni* **Loveridge**, 1933
[Alt. Hip-pocket Frog, Pouched Frog]
Darlington's Robber Frog *Eleutherodactylus darlingtoni* **Cochran**, 1935
Darlington's Madang Treefrog *Litoria darlingtoni* Loveridge, 1945
Darlington's Rainforest Frog *Albericus darlingtoni* Loveridge, 1948

Dr Philip Jackson Darlington Jr (1904–1983) was an evolutionary biologist, zoogeographer and beetle taxonomist. He collected in Colombia, Puerto Rico, Haiti, Cuba, New Guinea, Australia and Tierra del Fuego. He was Assistant Curator of Insects, Museum of Comparative Zoology, Harvard (1940), where he later became Curator, and Professor of Zoology. He took a year off (1956–1957) to live out of the back of a truck with his family in Australia. He wrote *Zoogeography: The Geographical Distribution of Animals* (1957). Four reptiles and a mammal are named after him.

Darwin

Darwin's Frog *Rhinoderma darwinii* **Duméril** & **Bibron**, 1841
Chile Darwin's Frog *Rhinoderma rufum* Philippi, 1902
Charles Darwin's Frog *Ingerana charlesdarwini* **Das**, 1998

Charles Robert Darwin (1809–1882) was the prime advocate, together with Wallace, of natural selection as the way in which speciation occurs. To quote from his most famous work, *On the Origin of Species by Means of Natural Selection* (1859), 'I have called this principle, by which each slight variation, if useful, is preserved, by the term Natural Selection.' Darwin was the naturalist on 'HMS Beagle' on her scientific expedition round the world (1831–1836). In South America he found fossils of extinct animals that were similar to extant species. On the Galápagos Islands he noticed many variations among plants and animals of the same general type as those in South America. On his return to London he conducted thorough research of his notes and specimens. Out of this study grew several related theories: evolution did occur; evolutionary change was gradual, taking thousands or even millions of years; the primary mechanism for evolution was a process called natural selection; and the millions of species alive today arose from a single original life-form through a branching process called 'speciation'. Nine reptiles, four mammals and twenty-three birds are named after him.

Darwin & Wallace

Darwin/Wallace Poison Frog *Epipedobates darwinwallacei* Cisneros-Heredia & Yánez-Muñoz, 2010

(See entries for **Darwin** and **Wallace, AR**)

Dary

Ford's Robber Frog *Craugastor daryi* **Ford** & **Savage**, 1984

Mushroomtongue Salamander sp. *Bolitoglossa daryorum* **Campbell** et al., 2010

Mario Dary Rivera (1928–1981) was a Guatemalan conservationist and biologist. He was President of the University of San Carlos, where he had founded the School of Biology (1973). He was instrumental in establishing the *biotopo del quetzal* (essential to the survival of the Quetzal) in the Guatemalan cloud forest (1976). His actions were not universally popular: some considered him to be a subversive. A death squad assassinated him (1981). His son, Juan Mario Dary Fuentes, followed strongly in his father's path, rising to become the Guatemalan Minister of Environment and Natural Resources. The frog is named only after Mario Dary; the plural *daryorum* ('of the Darys') for the salamander also honours Juan Mario Dary and other members of this exceptional family. A reptile is named after him.

Das

Das' Dwarf Toad *Adenomus dasi* Manamendra-Arachchi & Pethiyagoda, 1998

Dr Indraneil Das (b.1964) is an Indian herpetologist, now based at University Malaysia Sarawak, where he is an Associate Professor (1998). After early education in India, he received his doctorate in animal ecology from Oxford. Among his many publications is *Biogeography of the Reptiles of South Asia* (1996). Two reptiles are named after him. (See also **Indraneil**)

David

Chinese Giant Salamander *Andrias davidianus* **Blanchard**, 1871

Fr Jean Pierre Armand David (Père David) (1826–1900) was a French Lazarist priest and zoologist who went as a missionary to China (1862) and was the first Westerner to observe many animals, including the eponymous Père David's Deer. He co-authored *Les Oiseaux de Chine* (1877). He collected thousands of natural history specimens and many taxa are named after him, including a mammal, fifteen birds, and four reptiles.

David (Gower)

Chorla Giant Striped Caecilian *Ichthyophis davidi* Bhatta et al., 2011

Dr David J. Gower is a zoologist and herpetologist. University of Reading awarded his bachelor's degree in zoology and his doctorate was from the University of Bristol (1994). After postdoctoral work at Universität Tübingen, Germany and at Bristol, he joined the Department of Zoology, Natural History Museum, London (1999) and was appointed as a Herpetology Researcher (2004). He co-wrote *The conservation biology of caecilians* (2005).

Davies

Davies' Tree Frog *Litoria daviesae* **Mahony** et al., 2001

Davies' Toadlet *Uperoleia daviesae* **Young, Tyler** and Kent, 2005

[Alt. Howard River Toadlet]

Dr Margaret M. Davies (b.1944) is a zoologist and herpetologist at the Department of Zoology, University of Adelaide, which awarded her doctorate (1987). The etymology for the toadlet reads: 'The specific epithet *daviesae* honors Margaret Davies, whose published contributions have substantially expanded knowledge of the genus *Uperoleia*.'

Day

Day's Big-eyed Treefrog *Litoria dayi* **Günther**, 1897

[Alt. Australian Lace-lid; Syn. *Nyctimystes dayi*]

William S. Day was a naturalist and traveller who collected in Australia and New Guinea for Lord Rothschild's Tring Museum at the end of the nineteenth century.

Dayman

Dayman Rainforest Frog *Cophixalus daymani* **Zweifel**, 1956

Dayman Big-eyed Treefrog *Litoria daymani* Zweifel, 1958

These amphibians are named after their type locality, Mount Dayman, Papua New Guinea.

Debussy

Battak Frog *Hylarana debussyi* **Van Kampen**, 1910

Dr L. P. Debussy was a biologist working for Tobacco Growers Association, Deli, Sumatra. He travelled in the USA (1910), returned to Sumatra and was then appointed (1917) Director, Zoological Section, Colonial Museum, Amsterdam. He travelled back to Holland via San Francisco and New York, after having trouble finding a vessel sailing to Holland (neutral in the First World War). He collected the frog holotype.

Decary

Decary's Madagascar Frog *Gephyromantis decaryi* **Angel**, 1930

Raymond Decary (1891–1973) was a colonial administrator in Madagascar (1916–1944). He was a zoologist, botanist, geologist, ethnographer and was interested in everything to do with Madagascar, contributing over 40,000 specimens of Malagasy flora to the Paris Herbarium. He qualified in law (1912). He was seriously wounded in the Battle of the Marne (1914) and was unable to return to active service. He went to Madagascar (1916) as an officer in the Reserve, which released fully fit officers for active service in France. He trained as a colonial administrator (1921), returning to Madagascar (1922), and undertook seven scientific expeditions in the island (1923–1930). He became Director of Scientific Research, Madagascar (1937). He was again in the French Army in Madagascar (1939–1944), returning to France after the Liberation. Demobilized (1945), he retired to private life to continue his research. He wrote *Malagasy fauna* (1950). Three reptiles and a bird are named after him.

Defler

Poison Frog sp. *Ranitomeya defleri* Twomey & Brown, 2009

Dr Thomas Defler is an American primatologist who has worked in Colombia (1977) after being awarded his doctorate (1976) by the University of Colorado, Denver. He founded (1983) Estación Biológica Caparú, but was forced to abandon it (1998) by FARC guerillas and has developed another Amazonian research station, Estación Ecológica Omé, under the aegis of Universidad Nacional de Colombia, Bogotá, at which he also teaches. He wrote *Primates de Colombia* (2003).

Degen

Degen's Toad *Amietophrynus vittatus* **Boulenger**, 1906

[Alt. Lake Victoria Toad]

Edward J. E. Degen (1852–1922) was born in Basel and died in London. He collected reptiles, mammals and fishes in East Africa (ca. 1895–1905). He was in Abyssinia (Ethiopia) (1902). After leaving Africa he worked as an articulator / taxidermist at the British Museum (Natural History), London. A reptile and a bird are named after him.

De Granville

Degranville's Rocket Frog *Anomaloglossus degranvillei* **Lescure**, 1975

Dr Jean-Jacques de Granville (b.1943) is a botanist who works for Centre ORSTOM de Cayenne and is Curator, Herbarium, Cayenne, French Guiana. He is an expert on the flora of French Guiana. He was one of the first scientists to explore Pic Matécho and has led subsequent expeditions to study the region and its various habitats. He wrote *Phytogeographical characteristics of the Guianan forests* (1988).

De Jong

Madagascar Frog sp. *Blommersia dejongi* Vences et al., 2010

Wilfried W. de Jong is Professor Emeritus, Department of Biochemistry, Radboud University, Nijmegen, Netherlands. He has written many papers, both alone and with others, including *Molecules remodel the mammalian tree* (1998). The etymology for the frog says: 'The species name is dedicated to Wilfried W. de Jong, Nijmegen, in recognition of his outstanding contributions to understanding the molecular phylogenetics among the main mammal lineages.'

Delacour

Tonkin Asian Frog *Nanorana delacouri* **Angel**, 1928

[Syn. *Annandia delacouri, Quasipaa delacouri*]

Dr Jean Theodore Delacour (1890–1985) was a French-American ornithologist renowned for discovering, keeping and breeding some of the rarest birds in the world. He was born in Paris and died in Los Angeles. He created (1919–1920) the zoological gardens at Clères and donated them to the Muséum National d'Histoire Naturelle, Paris (1967). He undertook several expeditions to Indo-china, particularly Vietnam, and collected specimens, especially pheasants. He co-wrote *Birds of the Philippines* (1946). Among the taxa named after him are three birds, three mammals and a reptile.

Delalande

Delalande's Sand Frog *Tomopterna delalandii* **Tschudi**, 1838

[Alt. Cape Sand Frog, Delalande's Dwarf Bullfrog]

Pierre Antoine Delalande (1787–1823) was a French naturalist and explorer. He worked at the Muséum National d'Histoire Naturelle in Paris. He collected in the region around Rio de Janeiro, Brazil (1816) with Auguste Saint-Hilaire, and in the African Cape with his nephew Jules Verreaux and Andrew Smith (1818). He returned to France with a huge natural history collection, but with his health badly damaged by tropical diseases. Five birds and four reptiles are named after him.

De la Riva

Treefrog sp. *Dendropsophus delarivai* **Köhler** & **Lötters**, 2001

Ignacio de la Riva is a zoologist, evolutionary biologist and herpetologist at the Museo Nacional de Ciencias Naturales, Madrid. The Universidad Complutense, Madrid awarded his bachelor's degree (1986) and doctorate (1993). He was a post-doctoral fellow (1994) at the Department of Herpetology, Natural History Museum, University of Kansas. He wrote *Taxonomic status of* Bufo simus *O. Schmidt, 1857* (Anura: Bufornidae) (2004).

De la Torre

Rocket Frog sp. *Hyloxalus delatorreae* **Coloma**, 1995

Dr Stella de la Torre is an Ecuadorian biologist and ecologist at the College of Biological and Environmental Sciences, Department of Ecology, Universidad San Francisco de Quito, where she is a professor. The University of Wisconsin, Madison awarded her doctorate. She collected paratypes of the frog.

Deloitte

Forest Toad sp. *Nectophrynoides deloittei* ?Awaiting formal description

Deloitte is a well-known firm of accountants. The binomial was chosen by ARC (Africa Rainforest Conservancy), a US-based organisation devoted to the preservation of Tanzanian rain forests, which Deloitte supports.

Deloustal

Tam Dao Salamander *Paramesotriton deloustali* **Bourret**, 1934

[Alt. Tam Dao Newt, Vietnam Warty Newt]

Eugene Deloustal (1881–1942) was a member of a French family who lived at Tam-Dao, Tonkin (Vietnam) (1886). He studied engineering in Tonkin where Bourret was in the same school. They became life-long friends and Deloustal pointed out the salamander to Bourret, who was staying with the Deloustal family at Tam-Dao at the time. Deloustal left Tonkin for France (1939) and commanded the sixth batallion of Indochinese labourers (1942). He died at La Ciotat of malaria.

Demarchi

Eritrea Pond Frog *Pelophylax demarchii* **Scortecci**, 1929

Dr Marco de Marchi (1872–1936) was an Italian naturalist, hydrobiologist, industrialist and patron of the sciences. On the death of his father (1893) he left the Università degli Studi di Pisa and went to Argentina to run the family chemical business there. He studied Argentine hummingbirds and wrote a thesis on them that gained him a degree (1901) from the Università degli Studi di Pavia. He funded many scientific expeditions, libraries, museums, institutions, scholarships etc. and on his death he left to the state the Institute of Hydrology at Pallanza. Like many naturalists, he was an ardent philatelist.

Dendy

Dendy's Toadlet *Pseudophryne dendyi* **Lucas**, 1892

[Alt. Southern Toadlet]

Dr Arthur Dendy (1865–1925) was an English zoologist and a world expert on sponges. His bachelor's degree in zoology (1885), master's (1887), and doctorate (1891) were awarded by Victoria University, Manchester. He worked on the team evaluating the reports of the Challenger Expedition (1885) and after that joined the staff of the British Museum. He was appointed to be Demonstrator and Assistant Lecturer in biology at the University of Melbourne, Australia (1888). He was Professor of Biology, Canterbury College, University of New Zealand (1893–1902) and (1903–1905) he held the identical post at the University of Cape Town, South Africa. He returned to England (1905) as Professor of Biology, King's College, London. He died following an operation for chronic appendicitis. Dendy collected the toadlet holotype (1891) during an expedition with Lucas. Three birds are named after him.

Denhardt

Denhardt's African Caecilian *Boulengerula denhardti* **Nieden**, 1912

Clemens Andreas Denhardt (1852–1928) and Gustav Denhardt (1856–1917) were brothers who explored in Africa. They were friends of Dr Gustav Adolf Fischer (1848–1886), with whom (1878) they explored the Tana River region, East Africa. The brothers explored from Lamu to Vita, where the Sultan of Witu asked for a treaty with Germany. Clemens acquired territory for Germany, which he transferred to a German colonial society, Deutsche Witugesellschaft, for which he remained as agent as well as being adviser to the Sultan – in whose name he set up a customs post on the Tana River and succeeded in severely annoying the British. All the land he had acquired was ceded by Germany to Great Britain (1890) in exchange for Heligoland (an island in the North Sea). He wrote *Anleitung zu geographischen Arbeiten bei Forschungsreisen* (1883). He appears in museum records as a collector of specimens, so we suspect that he rather than Gustav is honoured in the caecilian's binomial, but we cannot be sure as the original description has no first names or initials to clarify which brother was intended.

Dennis

Dennis' Chirping Frog *Eleutherodactylus dennisi* **Lynch**, 1970

David M. Dennis was an illustrator and photographer. He was Head, Biological Illustration Unit, College of Biological Sciences, Ohio State University for 30 years. Lynch wrote of Dennis that his '... drawings greatly enhance the worth of this paper.'

Dennys

Dennys' Whipping Frog *Rhacophorus dennysi* **Blanford**, 1881

[Alt. Blanford's Whipping Frog; Syn. *Polypedates dennysi*]

Dr Nicholas Belfield Dennys (d.1900) joined the Admiralty (1855) and was appointed (1863) as Student Interpreter in Peking (Beijing). He became an author and journalist and was (1866–1876) editor of the *China Mail*, Hong Kong and Curator of the Hong Kong Museum. He became (1877) Assistant Protector of the Chinese in Singapore, and Librarian and Curator of the Museum in Singapore. He was Honorary Secretary and first editor of the journal of the Malayan branch of the Royal Asiatic Society and contributed articles (1878–1886) to it. He was a magistrate in Singapore and Gopeng (1879–1888). Invalided home (1889), he resigned (1890). He became (1894) Protector of the Chinese and Magistrate in British North Borneo (Sabah, Malaysia), editor of the *British North Borneo Herald* and (1899) Acting Judge and Member of Council. He wrote *Treaty ports of China and Japan* (1867). He collected botanical and zoological specimens, which he sent to a number of museums. He was in Hong Kong when he died.

Dente

Amapa Treefrog *Hypsiboas dentei* **Bokermann**, 1967

Emilio Dente was a highly skilled taxidermist. He was at the Zoology Department, Universidade de São Paulo, Brazil (1947) when the 17-year old Bokermann joined the staff as a supernumerary servant labourer. He took Bokermann, who was indigent, under his wing and fed him, as well as teaching him laboratory techniques. They collected together on a number of expeditions, such as in Pará (1955). Dente was at the Belém Virus Laboratory (1963) and worked at the Adolfo Lutz Institute (1974).

Derby

Derby Toadlet *Uperoleia aspera* **Tyler**, **Davies** & **Martin**, 1981

This amphibian is named after a town in Western Australia, near which the type specimen was collected.

De Roo

Togo Slippery Frog *Conraua derooi* Hulselmans, 1972

Antoon Emeric Marcel de Roo (1936–1971) was a Belgian naturalist. He co-wrote *Contribution à l'étude des Chiroptères de la République du Togo* (1969). A mammal and a bird are named after him.

De Saeger

Byangolo Frog *Amietia desaegeri* **Laurent**, 1972

Henri Jules de Saeger (1901–1994) conducted a survey (1949–1952) of the Garamba National Park, Belgian Congo (DRC). Verschuren, in his obituary of de Saeger, described him as 'Un pionnier de la conservation de la nature'.

Despax

Despax's Robber Frog *Pristimantis riveti* Despax, 1911

Raymond Justin Marie Despax (1886–1950) was a French entomologist and herpetologist. He wrote *Sur trois collections de reptiles et de batraciens provenant de l'Archipel Malais* (1912). Three reptiles are named after him.

De Vilhena

De Vilhena's Reed Frog *Hyperolius vilhenai* **Laurent**, 1964
[Alt. Luita Reed Frog]

E. de Vilhena was the deputy administrator of the Diamond Company of Angola.

DeVille

DeVille's Robber Frog *Pristimantis devillei* **Boulenger**, 1880

Émile de Ville (1824–1853) was a French naturalist who collected in Latin America (1843–1847). King Louis-Philippe of France ordered Castelnau, accompanied by Deville (among others), to explore Brazil and Peru during this period. Nine birds are named after him.

De Villiers

De Villiers' Moss Frog *Arthroleptella villiersi* **Hewitt**, 1935

Dr Cornelius 'Con' Gerhaardus Stephanus de Villiers (1894–1978) was a South African zoologist. His bachelor's (1914) and master's (1916) degrees were awarded by Stellenbosch University, where he became Professor of Zoology (1923–1959). His doctorate (1922) was from Zurich; Stellenbosch awarded him an honorary doctorate (1967). He was Visiting Professor in Rome (1935–1936) and at American and German universities. He spoke six languages, wrote short stories, and was an authority on music and the genealogy of old South African families. He wrote *Studies on the cranial anatomy of* Ascaphus truei *Stejneger, the American 'leiopelmid'* (1934).

Devincenzi

Rivera Redbelly Toad *Melanophryniscus devincenzii* **Klappenbach**, 1968

Dr Garibaldi José Devincenzi (1882–1943) was a Uruguayan naturalist and ichthyologist, who was qualified as a physician (1909). He was Director, Uruguay Museum, Montevideo (1912–1942). He produced the first systematic catalogue of Uruguayan fish, reptiles, mammals and birds, and wrote *Peces del Úruguay* (1924).

DeVries

Humming Frog sp. *Chiasmocleis devriesi* Funk & **Cannatella**, 2009

Dr Philip J. DeVries (b.1952) is an entomologist and biologist who researches into the tropical ecology and biology of the Amazon basin. After a bachelor's degree at the University of Michigan, he was a Peace Corps volunteer in Costa Rica, and a curator of Lepidoptera at the Museo Nacional de Costa Rica (1975–1980). The University of Texas, Austin (1987) awarded his doctorate and he was post-doctoral fellow (1987–1988) at the Smithsonian Tropical Research Institute, Panama. He was an assistant professor at the University of Oregon (1994–2000), Director of the Center for Biodiversity Studies and Curator of Lepidoptera, Milwaukee Public Museum, Wisconsin (2000–2004), becoming (2005) professor at the University of New Orleans, Louisiana. He wrote the 2-volume *The Butterflies of Costa Rica and their Natural History* (1987 & 1997).

De Witte

De Witte's River Frog *Afrana wittei* **Angel**, 1924
[Alt. Molo Frog; Syn. *Amietia wittei*]
De Witte's Clicking Frog *Kassinula wittei* **Laurent**, 1940
De Witte's Leaping Frog *Afrixalus wittei* Laurent, 1941
De Witte's Snout-burrower *Hemisus wittei* Laurent, 1963
De Witte's Clawed Frog *Xenopus wittei* Tinsley, Kobel & Fischberg, 1979

Dr Gaston-François de Witte (1897–1980) was in the Belgian Congo (1933–1935 and 1946–1949). He was originally a colonial administrator, but worked as a naturalist and collector for the Institut des Parcs Nationales Congo-Belge, Tervuren from 1938. Two reptiles are named after him. (See also **Witte** & **Gaston**)

Díaz

Harlequin Frog sp. *Atelopus eusebiodiazi* Venegas et al., 2008

Eusebio Díaz is a taxidermist at the Natural History Museum, University of San Marcos, Peru. He collected the holotype (1987).

Di-Bernardo

Casque-headed Treefrog sp. *Trachycephalus diberbardoi* Kwet & Solé, 2008

Dr Marcos Di-Bernardo (1963–2006) was Professor of Zoology at Pontifícia Universidade Católica do Rio Grande do Sul, Porto Allegre, Brazil, which had awarded his bachelor's (1985) and master's

(1991) degrees. Universidade Estadual Paulista, São Paulo, awarded his doctorate (1998). He died of skin cancer aged 42. He co-wrote, with Axel Kwet (who wrote Di-Bernado's obituary), *Pró-Mata-Anfíbios. Amphibien. Amphibians* (1999).

Diesmos

Wrinkled Ground Frog sp. *Platymantis diesmosi* Brown & Gonzalez, 2007

Arvin Cantor Diesmos is a Filipino biologist, herpetologist and ecologist who is a researcher at the National Museum of the Philippines, and a Board Member of the Wildlife Conservation Society of the Philippines. His bachelor's degree in biology and master's (wildlife ecology) were awarded by the University of the Philippines. He is doing postgraduate studies at the National University of Singapore.

Dintelmann

Reed Frog sp. *Hyperolius dintelmanni* **Lötters** & Schmitz, 2004

Horst Dintelmann lives in Bonn, Germany, and has given the describers support in their work in Cameroon. Other authors, describing other taxa, have also thanked him for support so we assume he is a generous sponsor of natural history research.

Diogenes

Robber Frog sp. *Pristimantis diogenes* **Lynch** & **Ruiz-Carranza,** 1996

Diogenes (412–323 BC) was a philosopher of the Cynic school and most famous for telling Alexander the Great to get out of his light! He slept in a large jar in the Athens marketplace. Why the describers think this frog might be like him, we do not know.

Diringshofen

Sao Bento Button Frog *Cycloramphus diringshofeni* **Bokermann**, 1957

Ricardo von Diringshofen (1900–1986) was a Brazilian entomologist and an expert on Coleoptera, whose massive collection was acquired by the museum in São Paulo (1960s).

Dixon

White-lipped Peeping Frog *Eleutherodactylus dixoni* **Lynch**, 1991 NCR

[JrSYN *Eleutherodactylus albolabris*]

Dr James Ray Dixon (b.1928) took his bachelor's (1950), his master's (1957) and his doctorate (1961) in zoology, all at Texas A&M University. He was Curator of Reptiles at Ross Allen Reptile Institute (1954–1955) and Associate Professor, Veterinary Medicine, Texas A&M University (1951–1961). He was Assistant Professor, Wildlife Management, New Mexico State University (1961–1965), then became Curator of Herpetology, Los Angeles County Museum. He was Professor of Wildlife and Fishery Sciences, Texas A&M University (1971) and he is Professor Emeritus and Curator Emeritus, Texas Cooperative Wildlife Collection. He wrote *Amphibians and Reptiles of Texas* (1987). Eight reptiles are named after him. (Also see **Bwana**)

Dobson

Dobson's Burrowing Frog *Sphaerotheca dobsonii* **Boulenger**, 1882

George Edward Dobson (1848–1895) was an Irish zoologist and an expert on small mammals. He became an army surgeon (1868), serving in India from that year and in the Andaman Islands (1872–1876). He returned to England (1876), retiring (1888) as surgeon-major. He became Curator, Royal Victoria Museum, Netley Hospital (1878). He took pictures of local peoples as he was an early aficionado of photography. He published a number of papers, including one of medical hints for travellers. A bird and eight mammals are named after him.

Dodson

Dodson's Toad *Duttaphrynus dodsoni* **Boulenger**, 1895

Edward Dodson (1872–1948) was a taxidermist working at the British Museum. He undertook a number of collecting expeditions in the Middle East, Morocco, Libya and Somalia at the end of the nineteenth century. He accompanied Donaldson-Smith on an expedition (1894–1895) to Somaliland (Somalia) during which they succeeded in reaching Lake Rudolph. With the ornithologist, Arthur Blayney Percival, he took part in a Royal Society expedition to Arabia (1899). He led expeditions to Patagonia (1899) and Tripoli (Libya) (1901–1902). Thereafter his health broke down and he had to abandon further planned trips. He was also a skilled engineer and inventor of the 'Zenith' carburettor for engines and the special laminated wooden propeller for WW1 aircraft. Three birds are named after him.

Doflein

Doflein's (Mushroomtongue) Salamander *Bolitoglossa dofleini* **Werner**, 1903

Dr Franz Theodor Doflein (1873–1924) was a German ichthyologist and herpetologist. He studied natural sciences at the universities of Strasbourg and Munich (1893–1898). He joined the Bavarian State Collection of Zoology, Munich (1901) and became Professor, Department of Zoology, University of Freiburg (1912), and at the University of Breslau (Wroclaw, Poland) (1918). He made a number of collecting expeditions: to Central America and the USA (1898), China, Japan (where he made a notable collection of marine specimens in Sagami Bay) and Ceylon (Sri Lanka) (1904–1905), and Macedonia (1917–1918), when he was attached to the German Army. He wrote *Lehrbuch der Protozoenkunde* (1906). A bird is named after him.

Dollo

Werner's Snouted Treefrog *Scinax dolloi* **Werner**, 1903

Dr Louis Antoine Marie Joséph Dollo (1857–1931) was a palaeontologist. French by birth, he qualified as a civil engineer (1877). He supervised the excavation of Iguanodon fossils found at Bernissart, Belgium (1878) and moved to Belgium (1882) on becoming Assistant-Naturalist, Royal Museum of Natural History, Brussels. He retired (1925). He proposed Dollo's Law (1893), which states that 'complex organs, once lost, can never be regained in exactly the same form'.

Domergue

Domergue's Madagascar Frog *Blommersia domerguei* **Guibé**, 1974

Dr Charles Antoine Domergue (1914–2008) was an ornithologist, speleologist and herpetologist who spent much of his life in Madagascar. He had (1946) a laboratory in the Institut Pasteur, where he was still working when he died. He was Professor at the University, Tuléar, Madagascar, and a member of the Academy of Madagascar. He previously worked for the Muséum National d'Histoire Naturelle, Paris. He wrote *Notes sur les Chamaleo de Madagascar* (1973). Two reptiles are named after him.

Don Simon

Don Simon Rocket Frog *Hyloxalus patitae* **Lötters**, Morales & Proy, 2003

Don Simon Camp, Cordillera El Sira, Departamento Pasco, Peru, is the type locality.

D'Orbigny

D'Orbigny's Toad *Rhinella dorbignyi* **Duméril** & **Bibron**, 1841

Alcide Dessalines d'Orbigny (1802–1857) was a traveller, collector, illustrator and naturalist. His father, Charles-Marie Dessalines d'Orbigny (1770–1856), was a ship's surgeon. He and Alcide studied shells. Alcide went to the Academy of Science, Paris, to pursue his methodical paintings and classification of natural history specimens. The Paris Natural History Museum sent him to South America (1826). There, the Spanish briefly imprisoned him, mistaking his compass and barometer for 'instruments of espionage'. After prison, he lived for a year with the Guarani Indians, learning their language. He spent five years in Argentina and then travelled north along the Chilean and Peruvian coasts, before moving into Bolivia; returning to France (1834). Once home he donated thousands of specimens of all kinds to the Muséum National d'Histoire Naturelle, Paris. His fossil collection led him to determine that there were many geological layers, revealing that they must have been laid down over millions of years. This was the first time such an idea had been put forward. He wrote *Dictionnaire Universel d'Histoire Naturelle*. Ten birds, five reptiles and two mammals are named after him.

Doria

Doria's Frog *Limnonectes doriae* **Boulenger**, 1887

Doria's Callulops Frog *Callulops doriae* Boulenger, 1888

Doria's Bush Frog *Chiromantis doriae* Boulenger, 1893

[Alt. Doria's Asian Treefrog]

Marchese Giacomo Doria (1840–1913) was a zoologist who collected in Persia (now Iran) with de Filippi (1862–1863), and in Borneo with Beccari (1865–1866). He founded the Museo Civico di Storia Naturale, Turin (1867–1913), becoming its first Director. Six mammals, eight reptiles and a bird are named after him.

Doris

Doris' Robber Frog *Eleutherodactylus glandulifer* **Cochran**, 1935

Doris' Lime Treefrog *Sphaenorhynchus dorisae* **Goin**, 1957

Dr Doris Mable Cochran (see **Cochran**).

Douglas, AM

Douglas' Toadlet *Pseudophryne douglasi* **Main**, 1964

Athol M. Douglas is a zoologist who worked at the Western Australian Museum, Perth (1950s–1980s). He wrote *Tigers in Western Australia?* (1986) about the possible survival of the thylacine or 'Tasmanian Tiger' on the Australian mainland. A mammal and two reptiles are named after him.

Douglas (Lynch)

Robber Frog sp. *Pristimantis douglasi* **Lynch**, 1996

Douglas Stewart Lynch is the author's son. His etymology reads: 'A patronym for my son, Douglas Stewart Lynch, who helped me collect during the summer of 1986 and who found many of this species.'

Douliot

Douliot's Treefrog *Boophis doulioti* **Angel**, 1934

Dr Henri Louis Douliot (1859–1892) was a botanist who worked at the Muséum National d'Histoire Naturelle, Paris. He was sent by the French government to explore Madagascar (1891–1892). He died of fever on the small island of Nossi-Bé, but his journals survived.

Drayton

California Red-legged Frog *Rana draytonii* **Baird** & **Girard**, 1852

Joseph Drayton (1795–1856) was an engraver and artist who worked in Philadelphia (1818–1838) and was employed as draftsman for the US Exploring Expedition (1838–1842), during which he collected specimens at San Francisco and on the Columbia River.

Drewes

Drewes' Moss Frog *Arthroleptella drewesii* **Channing**, Hendricks & Dawood, 1994

Dr Robert Clifton Drewes (b.1942) is Curator and Chairman, Department of Herpetology, California Academy of Sciences. He took his bachelor's degree at San Francisco State University (1969) and his doctorate (1981) at the University of California, Los Angeles. He worked at Nairobi Snake Park, Kenya (1969–1970). He taught at a number of institutions, including Harvard, the University of Kansas and, as Affiliate Professor, the University of Idaho, before joining the California Academy of Sciences (1998). His major area of interest is Africa; he has visited 30 different countries there. He co-wrote *Pocket guide to the reptiles and amphibians of East Africa* (2006). A reptile is named after him.

Dring

Dring's Borneo Frog *Leptobrachella brevicrus* Dring, 1983

Dring's Asian Toad *Leptolalax dringi* **Dubois**, 1987

Dring's Horned Frog *Xenophrys dringi* **Inger**, Stuebing & Tan, 1995

Julian Dring's Bush Frog *Philautus juliandringi* Dehling, 2010

Dr Julian Christopher Mark Dring (b.1951) of the Natural History Museum, London, is a herpetologist who collected in Sarawak (late 1970s). He wrote *Collection of amphibians and reptiles from Borneo*. Three reptiles are named after him.

Duarte

Duarte's Snouted Treefrog *Scinax duartei* **Lutz**, 1951

Dr Wanderbilt Duarte de Barros (1916–1997) graduated as an agronomist (1937) and worked at the Rio de Janeiro Botanical Gardens, Brazil (1940–1941). He worked at the Itatiaia National Park (1941–1942) and again (1943–1975) as Director. In retirement he became a professor at the University of Rio de Janeiro, and returned to the Botanic Gardens as Superintendent (1994–1997).

Dubois

Madagascar Frog subgenus *Duboimantis* **Glaw** & Vences, 2006

[Subgenus of *Gephyromantis*]

Dubois' Frog *Nanorana rostandi* Dubois, 1974

Dubois' Whipping Frog *Rhacophorus duboisi* Ohler et al., 2000

Koadaikanal Bush Frog *Raorchestes dubois* **Biju** & Bossuyt, 2006

[Syn. *Philautus dubois*]

Professor Alain Dubois is Director, Department of Reptiles and Amphibians, Muséum National d'Histoire Naturelle, Paris. His speciality is the study of Himalayan species of anurans. He wrote *The genus in zoology, a contribution to the theory of evolutionary systematics* (1988).

Duellman

Brook Frog genus *Duellmanohyla* **Campbell** & Smith, 1992

Duellman's Robber Frog *Pristimantis duellmani* **Lynch,** 1980

Purple-and-orange Leaf Frog *Phyllomedusa duellmani* **Cannatella**, 1982

Duellman's Poison Frog *Ranitomeya duellmani* **Schulte**, 1999

Strabomantid frog sp. *Phyllonastes duellmani* **Lehr, Aguilar** & **Lundberg**, 2004

[Syn. *Noblella duellmani*]

Ceuthomantid frog sp. *Ceuthomantis duellmani* Barros-Amoros, 2010

Treefrog sp. *Osteocephalus duellmani* Jungfer, 2011

Dr William Edward Duellman (b.1930) is a herpetologist. He became Curator of Herpetology, University of Kansas (1959) and retired (1997) as Curator Emeritus and Professor Emeritus, Department of Ecology and Biological Evolution. His wife, Linda Trueb (q.v.), is also a herpetologist and succeeded her husband as Curator of Herpetology, University of Kansas. They co-wrote *Biology of Amphibians* (1986). Eight reptiles are named after him.

Dugrite

Dugrite's Whipping Frog *Rhacophorus dugritei* **David**, 1872

Monsieur Dugrite was Rector of the College, established by Paris Foreign Missions, in Mouping, western China, where he had served as a missionary for many years before he was visited by Père David, who named this frog after him in gratitude for all that Dugrite had done for him.

Duméril, AMC & AHA

Duméril's Striped Frog *Leptodactylus gracilis* Duméril & **Bibron**, 1840
Duméril's Bright-eyed Frog *Boophis tephraeomystax* Duméril, 1853
Duméril's Madagascar Frog *Mantidactylus lugubris* Duméril, 1853
Duméril's Wrinkled Ground Frog *Platymantis dorsalis* Duméril, 1853
Eastern Banjo Frog *Limnodynastes dumerilii* **Peters**, 1863
[Alt. Pobblebonk, Bonking Frog, Four-Bob Frog]
Lake Pátzcuaro Salamander *Ambystoma dumerilii* **Dugès**, 1870

Dr André Marie Constant Duméril (1774–1860) was a zoologist and herpetologist. He qualified as a physician (1793). He was Professor of Anatomy, Muséum National d'Histoire Naturelle, Paris (1801–1812), changing to Professor of Herpetology and Ichthyology (1813–1857). He built up one of the largest collections of the time. Towards the end of his career his son, Auguste Henri André (1812–1870), also a distinguished zoologist, assisted him and later took over his father's professorship (1857). Twenty-one reptiles are named after one or other of them. Two birds are named after André. (See **Auguste**)

Dundee

Robber Frog sp. *Pristimantis dundeei* **Heyer** & **Muñoz**, 1999

Dr Harold A. Dundee (b.1924) is a retired Professor of Biology and Zoology at Tulane University, Plaquemines, Louisiana, where he acts as Curator of Amphibians and Reptiles in the University's museum. The University of Oklahoma awarded his bachelor's degree and his master's and doctorate (1958) were from the University of Michigan. He co-wrote *The Amphibians and Reptiles of Louisiana* (1989).

Dunn

Dunn's Oriental Salamander *Hynobius dunni* **Tago**, 1931
Dunn's Mushroomtongue Salamander *Bolitoglossa dunni* **Schmidt**, 1933
Dunn's Salamander *Plethodon dunni* **Bishop**, 1934
Dunn's Caecilian *Caecilia dunni* Hershkovitz, 1938
Dunn's Robber Frog *Craugastor monnichorum* Dunn, 1940
Dunn's Cave Salamander *Plethodon dixi* Pope & J Fowler, 1949 NCR
[JrSYN *Plethodon wehrlei*]
Patzcuaro Frog *Lithobates dunni* **Zweifel**, 1957
[Syn. *Rana dunni*]
Dunn's Rocket Frog *Prostherapis dunni* **Rivero**, 1961
Dunn's Marsupial Frog *Gastrotheca dunni* **Lutz**, 1977
Dunn's Red Salamander *Pseudotriton rubber nitidus* Dunn, 1920
[Alt. Blue Ridge Red Salamander]
Dunn's Cuban Long-nosed Toad *Peltophryne longinasus dunni* **Barbour**, 1926
Dunn's Spring Salamander *Gyrinophilus porphyriticus dunni* Mittleman & Jopson, 1941
[Alt. Carolina Spring Salamander]

Dr Emmett Reid Dunn (1894–1956) was a leading American herpetologist of his time. He took both his bachelor's degree (1915) and his master's (1916) at Haverford College, Pennsylvania, where he later became Professor of Biology (1934). He took his doctorate (1921) at Harvard, where he worked at the Museum of Comparative Zoology. He was a Zoology Assistant, Smith College (1916–1928). He visited London, Paris and Berlin to study their museum collections (1928). He was Secretary of the *Journal of American Society of Ichthyologists and Herpetologists* (1924–1929) and was President of the American Society of Ichthyologists and Herpetologists (1930–1931). From 1937 he was closely associated with the Philadelphia Academy of Natural Sciences, becoming Curator of Herpetology (1944). He tried to become an army officer (WW1) but was rejected, as it was considered that his week-end pursuit of salamanders and snakes was unbecoming in an officer and a gentleman! Instead he was an ensign in the US Navy (1917–1918). Among his publications is *The American Caecilians* (1942). Eighteen reptiles are named after him.

Durant

Durant's Glass Frog *Hyalinobatrachium duranti* **Rivero**, 1985

Durant's Rocket Frog *Aromobates duranti* **Péfaur**, 1985

Professor Dr Pedro Durant is a biologist and zoologist at the Department of Animal Ecology, Faculty of Sciences, University of the Andes, Mérida, Venezuela. He graduated in chemistry and biology (1958), and later took a master's degree in animal ecology. Péfaur, who described the rocket frog, also works at the University of the Andes.

Durrell

Tzunantza Giant Glass Frog *Centrolene durrellorum* Cisneros-Heredia, 2007

Gerald 'Gerry' Malcolm Durrell (1925–1995) and Lee McGeorge Durrell née Wilson (b.1949) are best known for setting up the Durrell Wildlife Preservation Trust, based in Jersey, Channel Islands. Gerald was born in India and first went to England (1928) upon his father's death. The family lived a Bohemian existence on Corfu (1935–1939), before returning to England at the outbreak of the Second World War. Gerry's call-up for war service came (1943), but he was exempted from military duty on medical grounds. His first expedition, to British Cameroons (Cameroon) (1947), was financed by his inheritance from his father. He sold the animals he brought back and so financed a further expedition to Cameroon, and one (1950) to British Guiana (Guyana). He founded his own zoo on Jersey (1958) with the help of his first wife, Jacqueline Sonia Wolfenden. After their divorce (1979), he married (1979) Lee McGeorge Wilson, a naturalist, zoo-keeper and author from Tennessee. She has carried on the work that Gerald started. He wrote *My Family and Other Animals* (1956), which was a financial success and provided funding for further expeditions. A reptile and a mammal are named after him.

Dury

Kentucky Spring Salamander *Gyrinophilus porphyriticus duryi* **Weller**, 1930

Ralph Dury (1899–1984) was Director, Cincinnati Museum of Natural History (1918–1975). He founded the Junior Society of Natural History (1928). The University of Cincinnati awarded him an honorary master's degree (1958) and an honorary doctorate (1976).

Dusén

Ypiranga Button Frog *Cycloramphus duseni* **Andersson**, 1914

Dr Per Dusén (1855–1926) was a Swedish naturalist, botanist, cartographer, explorer and bryologist. His first overseas collecting expedition was to Cameroon (1890). He was in Argentina on the Princeton expeditions to Patagonia (1896–1899). He was on board the 'Antarctic', responsible for the cartography, as a member of Nathorst's expedition to Spitzbergen (1899). Princeton awarded him an honorary doctorate (1904). A reptile is named after him.

Du Toit

Du Toit's Torrent Frog *Petropedetes dutoiti* **Loveridge**, 1935

[Alt. Mount Elgon Torrent Frog]

Dr C. A. du Toit was a member of Cape Town University, a founder member of the Zoological Society of South Africa, and Professor of Zoology at Stellenbosch University. He wrote *Zoological research in South Africa* (1961). He collected the frog holotype (1934).

Dutra

Treefrog sp. *Dendropsophus dutrai* Gomes & **Peixoto**, 1996

Professor José Alfredo Pinheiro Dutra is a Brazilian entomologist who established the Entomology Laboratory at the Department of Zoology, Institute of Biology, Universidade Federal do Rio de Janeiro, where his personal collection is held.

Dutta

Toad genus *Duttaphrynus* **Frost** et al., 2006

Dr Sushil Kumar Dutta is an Indian biologist, specialising in cell and molecular biology. He is Professor of Zoology, North Orissa University, Baripada.

Dybowski

Dybowski's Salamander *Salamandrella keyserlingii* Dybowski, 1870

[Alt. Siberian Salamander]

Dybowski's Frog *Rana dybowskii* **Günther**, 1876

Dr Benedykt Dybowski (1833–1930) was a Polish biologist who was born in what is now Belarus. He was an ardent proponent of Darwin's theory of evolution. He became (1862) Adjunct Professor of Zoology, Warsaw, but after the failure of the (1863) revolt he was banished to Siberia. There, support from the Zoological Cabinet, Warsaw, allowed him to investigate the natural history of Lake Baikal and other parts of the Russian Far East. He was pardoned (1877) and went to Kamchatka as a physician (1878). He was Professor of Zoology (1883–1906), University of Lemburg, Poland (Lwow, Ukraine). Three birds and a mammal are named after him.

E

Ebenau

Ebenau's Mantella *Mantella ebenaui* **Boettger**, 1880

Karl Ebenau, a zoologist who was German Consul in Madagascar (1880–1890), collected specimens with Anton Stumpff (q.v.) and Boettger. Two reptiles are named after him.

Edelca

Ayarzagüena's Treefrog *Tepuihyla edelcae* **Ayarzagüena, Señaris** & **Gorzula**, 1993

'Edelca' stands for Electrificación **del Car**oní. It is a Venezuelan company involved in hydroelectric power generation and the promotion of energy saving.

Edelmo

Heart-tongued Frog sp. *Phyllodytes edelmoi* **Peixoto, Caramaschi** & Freire, 2003

Edelmo de Melo Gonçalves is a biologist and herpetologist at the Department of Zoology, Universidade Federal de Alagoas, Brazil.

Edwardine

Edwardine's Horned Toad *Borneophrys edwardinae* **Inger**, 1989

[Alt. Rough Horned Frog; Syn. *Megophrys edwardinae*]

Edwardine Nodzenski is a biochemist who (1988) was at the Rheumatology Section, Department of Medicine, University of Chicago. She worked as a volunteer in the Zoology Department, Field Museum (1983). She co-wrote with Inger *Uncoupling of related structural changes in metamorphosing torrent-dwelling tadpoles* (1990).

Edwards

Edwards' Rocket Frog *Hyloxalus edwardsi* **Lynch,** 1982

Dr Stephen R. Edwards was Lynch's colleague at the Museum of Natural History, University of Kansas. Edwards was Executive Director, Association of Systematics Collections (Natural Science Collections Alliance) (1978). He started a programme to develop high-quality taxonomic catalogues to meet the needs of CITES (Convention on Trade in Endangered Species of Fauna and Flora). The second of these projects was *Amphibian Species of the World*.

Ehrenberg

Arabian Skittering Frog *Euphlyctis ehrenbergii* **Peters,** 1863

Dr Christian Gottfried Ehrenberg (1795–1876) was a natural scientist. He started studying theology at Leipzig (1815), but changed direction (1817) and went to Berlin to study medicine. He worked on fungi (1820) and lectured at the University of Berlin, where he became Professor of Medicine (1827). He travelled, mainly with his friend Hemprich (1820–1825), in northeast Africa and the Middle East. He travelled with Humboldt to Asia (1829). He met Darwin at Oxford (1847). He is regarded as the founder of micro-palaeontology. On his death his collection of specimens was deposited at the Museum für Naturkunde, Berlin. He was the first person to establish that phosphorescence in the sea is caused by the presence of plankton-like micro-organisms. A mammal and two birds are named after him.

Ehrhardt

Ehrhardt's Snouted Treefrog *Aplastodiscus ehrhardti* **Müller**, 1924

Wilhelm Ehrhardt (1860–1936) was a German collector and taxidermist who was born in Guyana. He collected professionally in Brazil (1897–1935), with a gap (1920–1927) when he lived in Hamburg, kept a shop with 'zoological supplies' and traded in natural history specimens. His herpetological collection is in the Museum für Naturkunde, Berlin, one of his clients. The place and date of his death are not known but are conjectured as being Brazil (1936).

Eichwald

Eichwald's Toad *Bufo eichwaldi* Litvinchuk et al., 2008

Professor Dr Karl Eduard Eichwald (1795–1876) was a Baltic German from Lithuania (in his day under Russian control). He took a degree in medicine (1819) at Vilnius University after studying in Berlin (1814–1817) and Paris (1818). He was a member of various expeditions to the Caspian Sea, Azerbaijan and the Caucasus. He was Professor of Zoology and Comparative Anatomy at Vilnius University (1827–1838), then moved to St Petersburg, where he was a Member of the Russian Academy of Science. He produced the 3-volume *Zoologia specialis, quam expositis animalibus tum vivis, tum fossilibus potissimuni rossiae in universum, et poloniae in specie, in usum lectionum publicarum in Universitate Caesarea Vilnensi. Zawadski, Vilnae* (1829–1831).

Eiffinger

Eiffinger's Tree Frog *Kurixalus eiffingeri* **Boettger**, 1895

Georg Eiffinger (1838–1920) was a German entomologist who lived in Frankfurt. His day job was as an official of the local Wiesbaden railway company.

Eileen

Eileeen's Robber Frog *Eleutherodactylus eileenae* **Dunn**, 1926

[Alt. Cuban Free-fingered Frog]

Dunn collected the holotype (1925) at Mina Carlota, Cuba, but put no clue in his description as to who Eileen was.

Eiselt

Treefrog sp. *Habrahyla eiselti* **Goin**, 1961 NCR

[JrSYN *Leptopelis notatus*]

Eiselt's Caecilian *Atretochoana eiselti* **Taylor**, 1968

Eiselt's Madagascar Frog *Gephyromantis eiselti* **Guibé**, 1975

Dr Josef Eiselt (1912–2001) was an Austrian herpetologist. He joined the Vienna Natural History Museum as a volunteer (1939). He served in the German armed forces for the duration of the Second World War. He then returned to Vienna to find that his job had been filled so worked as a labourer for the British occupation forces (1946–1949). He taught and worked as a scientific assistant (1950–1952), and re-joined the staff of the Natural History Museum, Vienna (1952). He started to make trips abroad (1962) to collect and research the herpetology of Turkey, Iraq and Afghanistan. He was Director, Vertebrate Collections (1972–1977). In retirement he travelled even more. Four reptiles are named after him.

Elbert

Lesser Sundas Frog *Hylarana elberti* **Roux**, 1911

Elbert's River Frog *Phrynobatrachus elberti* **Ahl**, 1925

Dr Johannes Eugen Wilhelm Elbert (1878–1915) was a German naturalist. An important collection was acquired by Johannes Elbert, his wife and his assistant C. Gründler (1910) during a geographic-botanical expedition to Sulawesi that he led on behalf of the Frankfurt Society for Geography and Statistics. The main purpose of the expedition was to explore the geographical relationship between the Asian and Australian areas. He explored in the Nusa Tenggara islands in East Indonesia e.g. Bali, Lombok, Sumbawa and Flores, as well as islands off Sulawesi such as Muna and Buton. He was also in the Cameroons (1914). Three birds are named after him.

Elena (Gavetti)

Elena's Treefrog *Boophis elenae* **Andreone**, 1993

[Alt. Vohiparara Bright-eyed Frog]

Dr Elena Gavetti is an Italian herpetologist and malacologist who is Curator at the Museo Regionale di Scienze Naturali, Torino and Head of the Zoological Section (2007). She has collaborated with Andreone on many projects and publications, including *Revised catalogue of the herpetological collection in Turin University I. Amphibia.* (1993).

Elena (Heyer)

Marbled White-lipped Frog *Leptodactylus elenae* **Heyer**, 1978

Elena Heyer is the describer's daughter.

Eliane

Treefrog sp. *Dendropsophus elianeae* **Napoli** & **Caramaschi**, 2000

Eliane de Freitas Napoli is the senior author's wife.

Elke

Treefrog sp. *Litoria elkeae* **Günther** & **Richards**, 2000

Elke Günther is the senior author's wife.

Elke Junginger

Henle's Slender-legged Treefrog *Osteocephalus elkejungingerae* **Henle**, 1981 NCR

[JrSyn *Osteocephalus mimeticus*]

Miss Elke M. Junginger helped the describer, Klaus Henle, on his expedition and in the preparation of his manuscript.

Emcela

Rio Claro Robber Frog *Craugastor emcelae* **Lynch**, 1985

Mrs Marsha C. Lynch. The best way to explain this is to quote Lynch's etymology: 'The name *emcelae* is a non-arbitrary combination of letters and is dedicated to my wife, Marsha C. Lynch, who has endured patiently my frequent absences in quest of eleutherodactyline frogs these past 20 years.'

Emile

Climbing Frog sp. *Anodonthyla emilei* Vences et al., 2010

Emile Rajeriarison is a Malagasy experimental naturalist and nature guide working at Ranomafana National Park. His brother is Theo Rajeriarison. (See **Theo**)

Emilia

Emilia's Robber Frog *Eleutherodactylus emiliae* **Dunn**, 1926

Dunn collected the holotype (1925) at Mina Carlota, Cuba, but put no clue in his description as to who Emilia was.

Emilio

Emilio's Ground Frog *Eupsophus emiliopugini* Formas, 1989

Professor Dr Emilio Pugin-Rios (d.1982) was a Chilean zoologist and herpetologist. He co-wrote *Comparative studies of the digestive tract development between* Rhinoderma darwinii *and* R. rufum (1982). The Faculty of Sciences building at the Universidad Austral de Chile is named after him. He died in Calfuco beach, Chile while doing scientific work.

Emlen

Robber Frog sp. *Craugastor emleni* **Dunn**, 1932

John Thompson Emlen Jr (1909–1997) was an American zoologist and ornithologist who worked with Dunn and wrote several papers with him, such as *Reptiles and amphibians from Honduras* (1932), referring to their collecting together in Honduras (1930). Primarily an ornithologist, Emlen's particular area of interest was ecological studies of peninsular effects and the structure of island avifaunas. He received (1973) the Elliott Coues Award in honour of his conservation work. He served as President of both the Wilson Ornithological Society (1956–1957) and the Wisconsin Society for Ornithology (1955–1956). Two birds are named after him.

Eneida

Eneida's Coqui *Eleutherodactylus eneidae* **Rivero**, 1959

[Alt. Villalba Robber Frog]

Mrs Eneida B. Rivero was the wife of the describer, who wrote that she 'helped him collect in the field.' She was at Boston University (1951), where she was awarded a master's degree in sociology. She wrote *Family structure and values: A methodological study for prediction of health* (1970).

Engelhardt

Engelhardt's Mushroomtongue Salamander *Bolitoglossa engelhardti* **Schmidt**, 1936

Teodoro Engelhardt was the owner of a coffee plantation, Olas de Moca, Guatemala. He and his wife repeatedly entertained Schmidt and other expedition members. They had a long history of welcoming visits from scientific parties.

Engels

Thin-toed Frog sp. *Adenomera engelsi* Kwet, **Steiner** & Zillikens, 2009

[Syn. *Leptodactylus engelsi*]

Dr Wolf Engels is a German biologist and entomologist who is a professor in the Department of Developmental Biology, Zoological Institute, Eberhard Karls Universität Tübingen. He has made extensive studies of honey bees. He is involved with the Pontifícia Universidade Católica do Rio Grande do Sul, Brazil, in a study project of the South Atlantic Forest.

Engländer

Englaender's Bright-eyed Frog *Boophis englaenderi* **Glaw** & Vences, 1994

Dr Hans Engländer (1914–2011) was a professor in the Zoological Institute, Universität zu Köln, Germany. The describers named the frog as a salute for his eightieth birthday.

Enki

Ambatolahy Madagascar Frog *Gephyromantis enki* **Glaw** & Vences, 2002

Enki was an ancient Sumerian god whose name is commonly translated as 'Lord of the Soil'. He was regarded as a god of wisdom and magic.

Ercepe

R.C.P.'s Paa Frog *Nanorana ercepeae* **Dubois**, 1974

[Alt. Torrent Paa Frog; Syn. *Paa ercepeae*]

'Ercepe' is phonetic for RCP. The original etymology reads as follows: 'L'espèce est dédiée aux membres de la R.C.P. 253 du C.N.R.S. (Ecologie et Géologie de l'Himalaya Central, responsable Corneille JEST) qui l'ont récoltée: M. le Professeur Paul OZENDA et MM. Jean-François DOBREMEZ, Corneille JEST et Alain MAIRE.' [C.N.R.S. = Le Centre National de la Recherche Scientifique, the largest governmental research organisation in France]

Erica

Bale Mountains Frog genus *Ericabatrachus* **Largen**, 1991

The original description carries no etymology for the genus, but notes that the vegetation at the type locality (in the Bale Mountains, Ethiopia) is dominated by *Erica arborea* (Tree Heath). Thus *Ericabatrachus* can be loosely translated as 'Heath Frog'.

Erica (Caramaschi)

Treefrog sp. *Hypsiboas ericae* **Caramaschi** & **Cruz**, 2000

Dr Erica Maria Pellegrini Caramaschi is an ichthyologist and a professor at the Departamento de

Ecologia, Universidade Federal do Rio de Janeiro, Brazil. Her bachelor's degree was awarded (1975) by the Faculdade de Ciências Médicas e Biológicas de Botucatu and both her master's (1980) and doctorate (1986) by the Universidade Federal de São Carlos. We think it likely that she and the senior author are husband and wife.

Erickson

Rocket Frog sp. *Aromobates ericksonae* Barrio-Amorós & Santos, 2012

Ronna B. Erickson is a research engineer at the Department of Astronomy, University of Massachusetts, Amherst. The University of Massachusetts awarded her bachelor's degree (1980). The etymology states: 'She has greatly supported the field research of the senior author.'

Erika

Dwarf Frog sp. *Physalaemus erikae* **Cruz** & Pimenta, 2004

Erika Costa Elias is married to the junior author. The etymology states: 'The specific epithet 'erikae' honors the second author's wife, Erika Costa Elias, in gratitude for her friendship and assistance.'

Erlanger

Erlanger's Grassland Frog *Ptychadena erlangeri* **Ahl**, 1924

Baron Carlo von Erlanger (1872–1904) was a German collector. He travelled in the Tunisian Sahara (1893 & 1897) and visited Abyssinia (Ethiopia) and Somaliland (1900–1901). He died in a car accident in Salzburg, so must have been one of the very first such casualties. Among other taxa named after him are a reptile, two mammals and thirty-seven birds.

Ermi Zhao

Warty Newt sp. *Paramesotriton ermizhaoi* Wu et al., 2009

Clawed Salamander sp. *Onychodactylus zhaoermii* Che, Poyarkov & Yan, 2012

Dr Zhao Er-mi (b.1930) is a zoologist and China's foremost herpetologist. He graduated from the University of Huaxi (1951). He has been Research Professor at Cornell and Visiting Professor at the University of California, Berkeley. He is an academician of the Chinese Academy of Sciences and is retired, albeit still active, from the Chengdu Institute of Biology. He wrote *Snakes of China* (2006).

Ernest

Ernest's Robber Frog *Pristimantis ernesti* **Flores**, 1987

Dr Ernest Edward Williams (1914–1998) was a herpetologist, Museum of Comparative Zoology, Harvard, retiring as Professor Emeritus of both Biology and Zoology. He was a renowned expert on Anole lizards and described many species. Among his publications is *The ecology of colonization as seen in the zoogeography of anoline lizards on small islands* (1969). Flores wrote: 'Named as an expression of deep gratitude to Dr. Ernest E. Williams, a most inspirational zoologist, teacher, advisor, and friend.' Twelve reptiles are named after him.

Ernesto

Marsupial Frog sp. *Gastrotheca ernestoi* **Miranda-Ribeiro**, 1920

Ernest W. Garbe. (See **Garbe**)

Eschscholtz

Eschscholtz's Lizard [archaic*] *Dicamptodon ensatus* Eschscholtz, 1833

[Alt. California Giant Salamander]

Eschscholtz's Salamander *Ensatina eschscholtzii* **Gray**, 1850

[Alt. Ensatina]

Dr Johann Friedrich von Eschscholtz (1793–1831) was a Russian physician of German descent. He qualified as a physician at the University of Dorpat (Tartu) (1815). He was on Kotzebue's 'Rurik' circumnavigation of the globe (1815–1818). He was Professor and Director, Zoological Cabinet, University of Dorpat (1819–1822). He then went on Kotzebue's 'Predpriaetie' expedition as Chief Naturalist, circumnavigating the world again (1823–1826). He was Professor of Zoology and Professor Extraordinary of Medicine at Dorpat (1828–1831). He began the report on the 'Predpriaetie' expedition, *Zoologischer Atlas* (1829–1833), which was completed after his death. One of the Marshall Islands and a bay in Alaska are named after him, as is a reptile.

* The name 'Eschscholtz's Lizard' was applied by Yarrow (1882), and is clearly unsuitable for an amphibian!

Espada

Espada's Robber Frog *Pristimantis galdi* Jiménez de la Espada, 1870

Espada's Marsupial Frog *Gastrotheca testudinea* Jiménez de la Espada, 1870

Inquisivi Water Frog *Telmatobius espadai* **De la Riva**, 2005

Dr Marcos Jiménez de la Espada (1831–1898) was a Spanish explorer, zoologist, geographer and author. He was a member of the Spanish Pacific Scientific Commission that visited the Americas (1862–1865). He graduated from the University of Madrid (1855) and became Assistant, Natural History Department, University of Madrid and additionally (1857) Assistant, National Museum of the Court (Natural Science Museum, Madrid). He co-founded the Spanish Society of Natural History (1871), becoming President (1895). He wrote *Batracians: Vertebrates from the Pacific trip* (1875). He turned his attention to geography (1875) and (1876) founded the Geographic Society, Madrid. He went to the History Academy (1883), where he edited the works of great mediaeval and modern travellers. Curiously, he only got round to presenting his doctoral thesis three months before his death. (See **Jimenez**)

Espinosa

Turquoise-bellied Poison Frog *Epipedobates espinosai* Funkhouser, 1956

This amphibian takes its binomial from the type locality, Hacienda Espinosa in Ecuador.

Esteban

Redbelly Toad sp. *Melanophryniscus estebani* Céspedez, 2008

Dr Esteban Orlando Lavilla is a herpetologist at the Fondaci6n Miguel Lillo, Tucumán, Argentina, where he is Director of the department dealing with systematics, taxonomy, distribution and conservation of neotropical amphibians. He co-wrote *Amphibians of Argentina: A second update, 1987–2000* (2001). The Superior Universitario de Grado awarded all his degrees: bachelor's, master's and doctorate.

Esteves

Treefrog sp. *Hyloscirtus estevesi* **Rivero**, 1968

Andrés Eloy Esteves is a Venezuelan photographer and field collector. With Rivero he co-wrote *Observations on the agonistic and breeding behavior of* Leptodactylus pentadactylus *and other amphibian species in Venezuela* (1969).

Estrada

Estrada's Robber Frog *Eleutherodactylus melacara* **Hedges**, Estrada & **Thomas**, 1992

Alberto R. Estrada Acosta is a Cuban zoologist and herpetologist working in Puerto Rico. His bachelor's degree in biology and zoology was awarded by the University of Havana (1981). He was Researcher, Zoology Institute, Academy of Sciences of Cuba, Havana (1982–1984), and (1984–1989) at the Institute of Forestry Research, Department of Agriculture, with a gap when he was Research Assistant, Department of Biology, Pennsylvania State University (1987). He left Cuba (1990s) for Puerto Rico. He was Research Assistant, Department of Biology, Universidad de Puerto Rico (1998) and Assistant Professor (1999). He was part-time Professor, Universidad Metropolitana, San Juan, Puerto Rico (2000–2004) and became Professor of Physics and Biology, Academia Maria Reina, Guaynabo, Puerto Rico (2004). He wrote *Data on a clutch of* Anolis evermani *(Sauria: Iguanidae) from Puerto Rico* (2002).

Etheridge

Etheridge's Robber Frog *Eleutherodactylus etheridgei* **Schwartz**, 1958

Professor Dr Richard Emmett Etheridge (b.1929) is a biologist and herpetologist. He took both his bachelor's (1951) and master's (1952) degrees at Tulane University. He served in the US Navy (1952–1956) as a sonar operator. He was a teaching assistant, University of Michigan (1958–1959), which awarded his doctorate (1959). He worked at the University of Southern California and Los Angeles County Museum (1959–1960), and at San Diego State University (1961–1997) as Professor of Biology and Curator of Herpetology at the University's museum, retiring as Professor Emeritus. Nine reptiles are named after him.

Eugenia

Robber Frog sp. *Pristimantis eugeniae* **Lynch** & **Duellman**, 1997

Dr Eugenia María del Pino Veintimilla (b.1945) is a developmental biologist who works at the Pontificia Universidad Catolica del Ecuador, where she was Head, Biological Sciences (1973–1975), as well as being Professor (1973). Her master's degree (1969) was awarded by Vassar College and her doctorate (1972) by Emory University, Georgia, USA. She is distinguished in being the first Ecuadorian woman to be elected to the Latin American Academy of Sciences, and the first Ecuadorian to be elected to the US National Academy of Sciences. Lynch and Duellman stated in their etymology: 'For more than a decade she extended many courtesies to us while we were working in Ecuador, and she accompanied WED (Duellman) to the type locality in 1977.'

Eugenio

Treefrog sp. *Xenohyla eugenioi* **Caramaschi,** 1998

Treefrog sp. *Aplastodiscus eugenioi* Carvalho-e-Silva & Carvalho-e-Silva, 2005

Eugenio Izecksohn. (See **Izecksohn**)

Eurydice

Maracas Snouted Treefrog *Scinax eurydice* **Bokermann**, 1968

In Greek mythology, Eurydice was the wife of the great musician Orpheus. He famously failed in his attempt to bring her back from the underworld. (Bokermann seems to have been fond of using names from mythology, e.g. see also **Ariadne**, and to have applied them without any particular reason).

Eusebio Diaz

Harlequin Frog sp. *Atelopus eusebiodiazi* Venegas et al., 2008

(See **Diaz**)

Euskirchen

Euskirchen's Water Frog *Petropedetes euskircheni* Barej et al., 2010

Oliver Euskirchen is a German biologist and herpetologist who is head of the environmental office in Koblenz. He has worked with Böhme and Schmitz and co-wrote with them *Zur Herpetofauna einer montanen Regenwaldregion in SW-Kamerun (Mt. Kupe und Bakossi-Bergland), 1.: Einleitung Bufonidae und Hyperoliidae.* (1999).

Evangelista

Evangelista's Dwarf Frog *Physalaemus evangelistai* **Bokermann**, 1967

This frog is named after a place – São João Evangelista, Minas Gerais, Brazil.

Evans

Evans' Stefania *Stefania evansi* **Boulenger**, 1904
[Alt. Groete Creek Treefrog]

Dr Richard Evans was appointed (1902) Curator, British Guiana Museum, Georgetown, British Guiana (Guyana). He held both a master's degree and a doctorate in science from Oxford. Evans collected the holotype (1902) and presented it to the Natural History Museum, London.

Everett

Zamboanga Frog *Hydrophylax everetti* **Boulenger**, 1882
[Syn. *Hylarana everetti*]

Everett's Flying Frog *Rhacophorus everetti* Boulenger, 1894

Everett's Asian Tree Toad *Pedostibes everetti* Boulenger, 1896

Everett's Timor Treefrog *Litoria everetti* Boulenger, 1897

Alfred Hart Everett (1848–1898) was a British colonial administrator in Borneo. He collected widely in the East Indies and Philippines. It is believed that a jawbone of an orang-utan, which he found in a cave, may have been used in the famous 'Piltdown Man' hoax. He was interested in all aspects of natural history and anthropology. His death made the front page of the *Sarawak Gazette*. Five mammals, three reptiles and twenty-nine birds and other taxa are named after him. (See **Alfred**)

Ewing

Ewing's Treefrog *Litoria ewingii* **Duméril** & **Bibron**, 1841
[Alt. Whistling Tree Frog]

The Reverend Thomas James Ewing (c.1813–1882) was born in England but moved to Tasmania (1833) and was admitted to holy orders (1838). He became Headmaster of the Queen's Orphan Schools and was a keen amateur naturalist and collector. When John Gould (q.v.) stayed in Tasmania, Revd Ewing helped him with his researches into the local avifauna. Two birds are named after him.

Eyre

Moaning Frog *Heleioporus eyrei* **Gray**, 1845

Edward John Eyre (1815–1901) was a British explorer and colonial administrator. He went to Australia to avoid going to university. He bought a flock of sheep (1833) and organized one of the first sheep drives from New South Wales to Adelaide, discovering Lake Eyre en route. He was the first European to traverse the coastline of the Great Australian Bight and cross Nullarbor Plain (1840–1841). Two native Australians murdered his companion, John Baxter, so Eyre finished his 2,000-mile journey accompanied by just one other heroic native Australian called Wylie. Eyre became Lieutenant-Governor of New Munster Province, New Zealand (1848–1853) and Governor of a number of Caribbean islands (1854), most notably of Jamaica where he put down a rebellion (1865). His actions involved considerable loss of life and the judicial execution of at least one person. This caused such uproar that a committee, including John Stuart Mill, Charles Darwin, John Bright and Thomas Huxley, was formed to press for Eyre to be tried for murder. A counter-committee, including Charles Dickens, John Ruskin, Charles Kingsley and Alfred Lord Tennyson, defended his actions. Although he was twice charged with murder, the cases were never prosecuted.

Ezra

Cross Frog sp. *Oreophryne ezra* Kraus & **Allison**, 2009

Ezra Kraus is the wife of Fred Kraus, the senior describer.

F

Fafnir

Mid-montane Rainforest Frog *Albericus fafniri* **Menzies**, 1999

Fafnir appears in *Der Ring des Nibelungen* as a giant who transforms himself into a dragon, in order to guard a gold hoard – see *Siegfried*, third opera in the Ring cycle.

Faivovich

Snouted Treefrog sp. *Scinax faivovichi* Brasileiro, Oyamaguchi & **Haddad**, 2007

Dr Julián Faivovich is a herpetologist at the Divisi6n Herpetología, Museo Argentino de Ciencias Naturales, Buenos Aires. The Universidad de Buenos Aires awarded his bachelor's degree (1999) and his master's (2002). He was a graduate student at the AMNH, New York, whilst completing a doctorate at Columbia University (2005). He wrote *The phylogenetic relationships of* Scinax *and other hylid frogs (Anura: Hylidae)* (2005).

Farc

Forest Stubfoot Toad *Atelopus farci* **Lynch**, 1993

This amphibian is named after FARC (*Fuerzas Armadas Revolucionarias de Colombia*), a Colombian guerilla army, which used to be active in the toad's habitat. It has been claimed that if this group had not used the area as a refuge, it would probably have been devastated, and the toad would not have become known to science.

Fausto

Fausto's Button Frog *Cycloramphus faustoi* Brasiliero et al., 2007

Dr Fausto Pires de Campos is a Brazilian biologist and conservationist.

Fay

Bright-eyed Frog sp. *Boophis fayi* **Köhler** et al., 2011

Andreas Norbert Fay is a Swiss supporter of BIOPAT.

Fazila

Lycian Salamander sp. *Lyciasalamandra fazilae* Basoglu & Atatür, 1974

Professor Dr Fazila Şevket Giz (1903–1981) was the first female professor at a Turkish university. She was educated in the US at the American College for Girls and graduated at the University of America (1927). She taught biology (1927–1933) at Arnavutkoy Girls College. She then taught at Istanbul University, becoming Associate Professor (1941) at the Faculty of Science and Nature. She became Director of the Institute of General Zoology (1955) and later Director of the Research Institute of Hydrobiology (1964) until retiring (1972). She was honoured as she was the teacher of the senior author at Istanbul University.

Fea

Fea's Short-legged Toad *Brachytarsophrys feae* **Boulenger**, 1887

[Alt. Fea's Horned Frog, Kakhien Hills Frog]

Ocellated Spiny Frog *Nanorana feae* Boulenger, 1887

Fea's Treefrog *Rhacophorus feae* Boulenger, 1893

Gaboon Dwarf Clawed Frog *Hymenochirus feae* Boulenger, 1906

Leonardo Fea (1852–1903) was an Italian explorer, painter and naturalist. He became an assistant at the Natural History Museum, Genoa. He made several collecting trips, including Burma (Myanmar) (1885), the Cape Verde Islands (1898), and the Gulf of Guinea area. Two mammals, five reptiles and two birds are named after him.

Fei

Rhacophorid frog genus *Feihyla* **Frost** et al., 2006

Chinese Newt sp. *Pachytriton feii* Nishikawa, Jiang & **Matsui**, 2011

Fei Liang (b.1936) is a zoologist at the Department of Herpetology, Chengdu Institute of Biology, Chinese Academy of Sciences. He co-wrote *An illustrated key to Chinese Amphibians* (1991).

Feichtinger

Robber Frog sp. *Eleutherodactylus feichtingeri* **Díaz, Hedges** & **Schmid**, 2012

Dr Wolfgang Feichtinger is a biologist who works at the Institute for Human Genetics (Biocenter), University of Würzburg, Germany. He has contributed to Cuban herpetology during fieldwork there. His boss is the third author.

Feio

Treefrog sp. *Bokermannohyla feioi* **Napoli** & **Caramaschi**, 2004

Dwarf Frog sp. *Physalaemus feioi* Cassini, **Cruz** & Caramaschi, 2010

Renato Neves Feio is a zoologist and arachnologist at the Animal Biology Department, Zoological Museum, Universidade Federal de Viçosa, Brazil.

Fejérvary

Dicroglossid frog genus *Fejervarya* Bolkay, 1915

Baron Géza Gyula von Fejérvary (1894–1932) was a Hungarian nobleman and herpetologist, grandson of a former Hungarian Prime Minister. He spent his childhood in Switzerland and, just 15, published a 50-page work on the herpetofauna of the Rhone Valley. He attended Budapest University (1912), studying zoology and medical science. When he graduated (1913) he became volunteer keeper of the Collection at the Hungarian Natural History Museum. He left (1915) to become assistant to Méhely (q.v.), the Professor of Zoology at Budapest University. He became Herpetology Curator (1916) while studying for his doctorate (1917). He took over as Professor of Zoology and was later succeeded by his wife. He wrote a number of papers including *Ascaphidae, a new family of the tailless batrachians* (1923), written the year he processed the herpetological material collected by E. Csiki in the Balkans (1916–1918).

Fenoulhet

Fenoulhet's (Pygmy) Toad *Poyntonophrynus fenoulheti* **Hewitt** & Methuen, 1912

[Alt. Transvaal Dwarf/Pygmy Toad]

Dr James Peter Fenoulhet MRCS LRCP (d.1914) qualified as a physician (1886). He went to South Africa (1897) and practised medicine in Muizenberg, where he remained until he disposed of his practice and became Relieving Medical Officer under the Cape Colonial Office, serving at different Government institutions including Robben Island. He resigned (1911) and went to the Transvaal as Construction Medical Officer on the Selati Railway. On completion of this (1912) he went to England. He was for some years a captain in the Cape Medical Corps.

Ferguson

Ferguson's Toad *Bufo fergusonii* **Boulenger**, 1892 NCR

[JrSYN *Duttaphrynus scaber*]

Ferguson's Shrub Frog *Pseudophilautus fergusonianus* **Ahl**, 1927

Harold Stuart Ferguson (1852–1921) was a British zoologist who spent most of his life in Travancore, southern India. He was associated with the State Museum at Trivandrum, becoming its Director (1894–1904). When he left India he wrote *Travancore Batrachians* (1904). A reptile is named after him.

Fernández

Whistling Dwarf Frog *Physalaemus fernandezae* **Müller**, 1926

Mrs Fernández was a professor at the University of La Plata, Argentina.

Fernández, K

Bella Vista Toad *Rhinella fernandezae* **Gallardo**, 1957

K. Fernández collected the holotype, but we know nothing more about this person.

Fernando

Fernando's Giant Glass Frog *Centrolene fernandoi* **Duellman** & **Schulte**, 1993

Fernando M. Cuadros is a herpetologist at the Universidad Nacional Mayor de San Marcos, Lima, Peru. This species was named after him '...in recognition of his cheerful companionship and dedication to collecting frogs in northern Peru in 1989.'

Ferner

Ferner's Fanged Frog *Limnonectes ferneri* Siler et al., 2009

Dr John W. Ferner joined Thomas More College, Kentucky (1977) and is Professor of Biology there. The College of Wooster, Ohio awarded his bachelor's degree and his doctorate was from the University of Colorado. He was on three expeditions (1990s) to the Philippines. He wrote *A review of marking and individual recognition techniques for amphibians and reptiles* (1977).

Ferreira

Rio Luinha Reed Frog *Hyperolius ferreirai* **Noble**, 1924

J. de Bethencourt Ferreira was a Portuguese naturalist working at the Bocage Museum, Lisbon, in the late nineteenth and early twentieth centuries. He wrote an article, *Algumas especies novas ou pouco conhecidas de amphibios e reptis de Angola* (1906), in which he described a frog species as *Rappia bivittata*. Because this name was invalid under taxonomic rules, Noble re-named it *Hyperolius ferreirai*, honouring the earlier describer.

Festa

Shy Robber Frog *Pristimantis festae* **Peracca**, 1904

Slender-legged Treefrog sp. *Osteocephalus festae* Peracca, 1904

Valle Santiago Beaked Toad *Rhinella festae* Peracca, 1904

Dr Enrico Luigi Festa (1868–1939) was an Italian naturalist who worked for the Zoological Museum, University of Turin, from where he graduated (1891). He visited Egypt, Palestine, Jordan, Lebanon and Syria (1893). He collected in Panama and Ecuador (1895–1898). He was at the Zoological Museum, Turin (1899–1923) as Deputy Assistant Professor (1899), retiring as Honorary Vice-Director. He also undertook an expedition to Cyrenaica, Libya (1921). Six reptiles and two birds are named after him.

Finch

Finch's Wart Frog *Limnonectes finchi* **Inger**, 1966

O. C. Finch worked for North Borneo Timbers in Sabah, Malaysia, and provided Inger with considerable logistical support and companionship in the field.

Firschein

Firschein's False Brook Salamander *Pseudoeurycea firscheini* Shannon & **Werler**, 1955

Dr Irwin Lester Firschein was a biologist, geneticist and anthropologist. His bachelor's degree was awarded (1949) by the University of Kansas. He worked at the Museum of Natural History, University of Illinois, which awarded his master's (1951), and was at Columbia University, New York (1959). He co-wrote *The ringed salamander,* Ambystoma annulatum, *in Oklahoma* (1951).

Firth

Firth's Leaf-litter Toad *Leptolalax firthi* Rowley et al., 2012

Denys Firth took a bachelor's degree at Oxford and then worked in the City of London, becoming experienced in the money, bond and derivatives markets. He has lived in Asia since 1996 and is Chief Operating Officer of ADM Capital and a member of the ADM Capital Foundation, set up (2006) to support social and ecological projects in Asia. This has included the senior author's amphibian research and conservation activities in Southeast Asia.

Fischer, E

Fischer's African Caecilian *Boulengerula fischeri* **Nussbaum** & Hinkel, 1994

Dr Eberhard Fischer (b.1961) is a botanist and lichenologist at Universität Koblenz-Landau, Germany. He was honoured in the binomial of the caecilian '...in recognition of his contributions to our knowledge of the fauna and flora of Rwanda' (where the caecilian is found).

Fischer, J

Fischer's Clawed Salamander *Onychodactylus fischeri* **Boulenger**, 1886

[Alt. Long-tailed Clawed Salamander]

Fischer's Dwarf Frog *Physalaemus fischeri* Boulenger, 1890

Dr Johann Gustav Fischer (1819–1889) was a vertebrate zoologist. He earned his first degree in Leipzig and his doctorate in Berlin (1843) with a dissertation on the cranial nerves in amphibians and reptiles. He went on to teach in various secondary schools, later establishing one himself. In his later years he was a volunteer Warden in charge of the fish, amphibian and reptile collections at the Museum für Naturkunde, Berlin. Seven reptiles are named after him.

Fisher

Vegas Valley Leopard Frog *Lithobates fisheri* **Stejneger**, 1893

[Syn. *Rana fisheri*]

Dr Albert Kenrick Fisher (1856–1948) was an American ornithologist and general naturalist. He was the ornithologist on the Death Valley Expedition (1891), during which Nelson collected the frog paratype. He was also on biological surveys in California, Nevada, Arizona, New Mexico and Utah (1892) and was a member of the Harriman Alaskan Expedition (1899). Three birds are named after him.

Fitzgerald

Mount Tucutche Treefrog *Flectonotus fitzgeraldi* **Parker**, 1934

Leslie Desmond Edward Foster Vesey-Fitzgerald (1909–1974) was an Irish zoologist, herpetologist environmentalist and naturalist. He was in Trinidad with Parker (1933–1934) collecting for the BMNH, London. He was in the Seychelles (1938) and led an expedition to the Comoro Islands (1940). He undertook several wildlife surveys in the Abu Dhabi desert and also collected many insects, amphibians and reptiles. He was involved in locust control in Northern Rhodesia (Zambia) (1949–1964). The Tanganyika (Tanzania) Government employed him as National Parks Officer (1965). He wrote *East African Grasslands* (1973).

Fitzinger

Fitzinger's Robber Frog *Craugastor fitzingeri* **Schmidt**, 1857

Leopold Joseph Franz Johann Fitzinger (1802–1884) was an Austrian zoologist who had a considerable influence on herpetology; he created seventy of the names of genera in use today. He read botany at the University of Vienna and worked at the Natural History Museum, Vienna (1817–1861). He became Director of the Munich and Budapest zoos. He wrote *Neue Classification der Reptilien* (1826). Six reptiles are named after him.

Fleay

Fleay's Barred Frog *Mixophyes fleayi* **Corben** & Ingram, 1987
[Alt. Queensland Barred Frog]

David Howells Fleay (1907–1993) was an Australian naturalist and a pioneer in the captive breeding of endangered species. He was the first person to breed platypus in captivity. He was employed in his father's chemist shop, then as a teacher, but went on to study for a degree in botany, zoology and education at Melbourne University (1927–1931). He was soon passionately involved in conservation and was the last person to photograph a thylacine in captivity in Hobart Zoo; it bit him on the buttock and he proudly carried the scar for life. He was asked (1934) to design the Australian animals section for Melbourne Zoo and he worked there for four years developing captive breeding programmes. He co-founded the Wildlife Preservation Society of Queensland (1962). He created a conservation park and, although it was sold to the Queensland government, he worked there for the rest of his life. A bird is named after him.

Fleischmann

Fleischmann's Robber Frog *Craugaster fleischmanni* **Boettger**, 1892
Fleischmann's Glass Frog *Hyalinobatrachium fleischmanni* Boettger, 1893

Carl Fleischmann was a collector in Costa Rica in the early 1890s.

Fletcher

Long-thumbed Frog *Limnodynastes fletcheri* **Boulenger**, 1888
[Alt. Barking Marsh Frog]
Fletcher's Forest Frog *Lechriodus fletcheri* Boulenger, 1890
[Alt. Black-soled Frog, Sandpaper Frog]
Fletcher's Ground Froglet *Geocrinia leai* Fletcher, 1898
[Alt. Lea's Frog]

Joseph James Fletcher (1850–1926) was a New Zealand-born Australian biologist. He went to London (1876), studying at the Royal School of Mines and University College, University of London, where he took his BSc (1879). He studied for a time at Cambridge before returning to Australia (1881). Initially he taught at the school where his father was principal, then took a post with the Linnaean Society of New South Wales (1886–1918). During this period he became an expert on amphibians, but later devoted more time to botany. Invalided in an accident (1922), he spent much of the rest of his life cataloguing his own collections, which he presented to the Australian Museum.

Flores

Robber Frog sp. *Pristimantis floridus* **Lynch** & **Duellman**, 1997
[Syn. *Eleutherodactylus floridus*]

Glenn Flores worked at the Harvard Museum of Comparative Zoology. The authors wrote that the specific name '...is a Latin adjective meaning 'abounding with flowers'. The name is used in allusion to Glenn Flores in recognition of his contributions to the study of Eleutherodactylus and whose surname is Spanish for 'flowers'.'

Flower

Flower's Ridged Frog *Ptychadena floweri* **Boulenger**, 1917
[Perhaps a junior synonym of *Ptychadena schillukorum*]

Captain Stanley Smyth Flower (1871–1946) was Director, Cairo Zoological Gardens, Giza, Egypt (1898–1924). He had previously spent two years as Scientific Adviser to the Siamese government. Flower visited the zoo at Madras (Chennai) as an adviser (1913) and described many zoos of the time. He wrote *Zoological Gardens of the World* (1908–1914). Three birds, three mammals and three reptiles are named after him.

Folkerts

Dwarf Black-bellied Salamander *Desmognathus folkertsi* **Camp** et al., 2002

Dr George W. Folkerts (1938–2007) was a professor in the Department of Biological Sciences, Auburn University, Alabama, where he taught for 38 years. His bachelor's degree in zoology and master's in botany were both awarded by Southern Illinois University, and his doctorate in herpetology by Auburn University.

Foote

Robber Frog sp. *Eleutherodactylus footei* **Stejneger**, 1913 NCR
[JrSYN *Gastrotheca marsupiata*]

Dr Harry Ward Foote (1875–1942) was Professor of Chemistry at the Sheffield Scientific School, Yale, which awarded his bachelor's degree (1895) and his doctorate in chemistry (1898). He was a member of the Yale Peruvian Expedition (1911), during which the frog holotype was collected, and in which he served as the expedition's naturalist.

Ford

Ford's Robber Frog *Craugastor daryi* Ford & **Savage**, 1984

Dr Linda S. Ford is an American herpetologist who is Director, Collections Operations, Museum of Comparative Zoology, Harvard University. She has worked at the Department of Herpetology, AMNH, New York and the Museum of Natural History, University of Kansas. She co-described this frog species in the article *A new frog of the genus Eleutherodactylus (Leptodactylidae) from Guatemala* (1984).

Forester

Forester's Cabin Robber Frog *Eleutherodactylus brittoni* **Schmidt**, 1920

The frog's vernacular name is derived from the type locality thus: 'El Yunque, near the Forester's Cabin (about 1300 feet)', Puerto Rico - not after a person called Forester.

Fornasini

Fornasini's Spiny Reed Frog *Afrixalus fornasini* Bianconi, 1849

[Alt. Fornasini's Banana Frog, Greater Leaf-folding Frog]

Cavaliere Carlo Antonio Fornasini (b.1805) was a collector who operated in the area around Inhambane, Mozambique (1839 on). He left Italy for unknown reasons (probably political) for Portugal, proceeding thence to Mozambique. He mainly collected spiders and botanical specimens, including the first example of the cycad that Bertolini named *Encephalartos ferox* (1851). A reptile is named after him.

Forrer

Forrer's Grass Frog *Lithobates forreri* **Boulenger**, 1883

Alphonse Forrer (1836–1899) was born in England but emigrated to the USA. He joined the Union Army at the outbreak of the America Civil War (1861). He was employed (1865) by the BMNH to collect zoological specimens on the west coasts of Mexico and the USA, and was still collecting in Mexico in the 1880s. He also supplied specimens to other museums and in that connection made four trips to Europe. Other taxa, including a reptile and three birds, are named after him.

Fouquette

Cajun Chorus Frog *Pseudacris fouquettei* Lemmon et al., 2008

Dr Martin J. 'Jack' Fouquette Jr is a biologist and herpetologist who studied this frog genus (1960s–1970s). His bachelor's and master's degrees and his doctorate (1959) were all awarded by the University of Texas, Austin. He was Interim Assistant Professor, University of Florida, and was at Arizona State University (1965–2005), retiring as Professor Emeritus. He wrote *Effect of environmental temperatures on body temperatures of aquatic-calling anurans* (1980).

Fowler, D

Fowler's Robber Frog *Eleutherodactylus fowleri* **Schwartz**, 1973

Danny C. Fowler often worked with Schwartz and they co-wrote *The Anura of Jamaica: A progress report* in *Studies on the Fauna of Curaçao and other Caribbean Islands* (1973). Three reptiles are named after him.

Fowler, HW

Fowler's Snouted Treefrog *Scinax quinquefasciatus* Fowler, 1913

Fowler's Rice Frog *Microhyla fowleri* **Taylor**, 1934

Henry Weed Fowler (1878–1965) was an American zoologist who studied at Stanford University, California. He joined the Academy of Natural Sciences, Philadelphia, working as an Assistant (1903–1922), Associate Curator of Vertebrates (1922–1934), Curator of Fishes and Reptiles (1934–1940) and Curator of Fishes (1940–1965). He was a co-founder and first treasurer of the American Society of Ichthyologists and Herpetologists (1927).

Fowler, S

Fowler's Toad *Anaxyrus fowleri* Hinckley, 1882

[Syn. *Bufo fowleri*]

Samuel Page Fowler (1800–1888) was a Massachusetts naturalist and historian. He developed an early interest in natural history, writing articles for the *New England Farmer*, and also studied American history. He is particularly noted for writing about the Salem witch trials (1692–1693).

France

France's Squeaker *Arthroleptis francei* **Loveridge**, 1953

[Alt. Ruo River Screeching Frog]

F. H. 'Freddie' France (d.1949) was the Forestry Officer, Mulanje, Nyasaland (Malawi). He was drowned whilst attempting to cross the Ruo River. The etymology says: 'They are called after France that his name may be linked with the forests he sought to preserve on the mountain he loved so well.' One of the last people to visit him, and a witness to his death, was Sir Laurens van der Post, who recorded it in his book *Venture to the Interior* (1951).

Francis

River Frog sp. *Phrynobatrachus francisci* **Boulenger**, 1912

A. C. Francis (d.1928), who presented the frog holotype, was a colonial administrator in Nigeria, where he was the Resident of Bauchi province at the time of his death.

Franciscus

Central Coast Stubfoot Toad *Atelopus franciscus* **Lescure**, 1974

The original description does not give an etymology. Our guess is that *franciscus* may be a Latinized reference to French Guiana, where the species is endemic.

Frankenberg

Frankenberg's Shrub Frog *Pseudophilautus frankenbergi* Meegaskumbara & Manamendra-Arachchi, 2005

Regina Bauer Frankenberg (1908–1991) established through her will a foundation for animal welfare. This foundation has given considerable support to the Global Amphibian Assessment.

Franklin

Franklin's Mushroomtongue Salamander *Bolitoglossa franklini* **Schmidt**, 1936

Franklin James White Schmidt (1901–1935) was the describer's brother. He was one of the collectors on an expedition to Guatemala. He was prominent in the field of wildlife management, a novel idea at that time. He and his mother died in a fire, which also destroyed all his notes, photographs and manuscripts.

Fraser

Fraser's Clawed Frog *Xenopus fraseri* **Boulenger**, 1905

Louis Fraser (1810–1866) was a zoologist, collector, curator, explorer, zookeeper, consul, author, dealer and taxidermist. He was employed by the Museum of the Zoological Society of London (1832–1841 & 1842–1846); first as an office boy, then clerk, Assistant Curator and finally Curator. He collected in West Africa (1841–1842), being official naturalist on the Niger expedition. He took charge of Lord Derby's zoological collections (1848–1850). He was appointed (1850) Consul at Whydah (Ouidah) (Benin). Later he collected natural history specimens in Ecuador (1857–1859) and California (1860). He tried to establish himself as a natural history dealer, opening a shop in London to sell exotic birds. This venture appears not to have been a success as he returned to the USA and died there. He wrote *Zoologica Typica* (1845–1849). A mammal and three reptiles are named after him.

Fred

Angra Tree Toad *Hylodes fredi* Canedo & **Pombal**, 2007

Dr Carlos Frederico Duarte da Rocha is Professor, Department of Ecology, Institute of Biology Roberto Alcantara Gomes, Universidade Estadual de Rio de Janeiro, Brazil. This university originally awarded his bachelor's degree (1984), and he has taught and researched there since 1988.

Freycinet

Freycinet's Frog *Litoria freycineti* **Tschudi**, 1838
[Alt. Wallum Rocket Frog]

Captain Louis-Claude de Saulces de Freycinet (1779–1841) was a French navigator who was involved, with Baudin (q.v.), in mapping the southern Australian coast. He commanded the *Uranie* (1817) and explored in the Pacific (1817–1820). Two birds are named after him.

Frith

Jessore Wart Frog *'Fejervarya' frithii* **Theobald**, 1868
[Unclear as to which genus this taxon belongs in]

Robert W. G. Frith owned an indigo factory at Khulna, Bangladesh, and was also a natural history collector. He was a great friend of Edward Blyth, who mentions him in his *Catalogue of Mammals and Birds of Burma* (1875) and who named two mammals after him.

Frontier

Frontier Forest Toad *Nectophrynoides frontierei* Menegon, Salvidio & Loader, 2004

This species is named after Frontier-Tanzania, an environmental research organisation whose members collected the type series.

Frost, C

Baw Baw Frog *Philoria frosti* **Spencer**, 1901

Charles Frost (1853–1915) was an Australian naturalist and a Fellow of the Linnaean Society. He was one of the earliest members (1885) of the Field Naturalists' Club of Victoria. He was on three of the club's expeditions (1887–1890). A reptile is named after him.

Frost, D

Toad genus *Frostius* **Cannatella**, 1986
Frost's Toad *Frostius pernambucensis* **Bokermann**, 1962
Treefrog sp. *Dendropsophus frosti* Motta et al., 2012

Dr Darrel Richmond Frost (b.1951) is an American herpetologist who works in the Division of Vertebrate Zoology, AMNH, New York. He was President of the Society for the Study of Amphibians and Reptiles (1998). Two reptiles are named after him.

Fry, A

Fry's Smooth Horned Frog *Proceratophrys fryi* **Günther**, 1873 NCR
[JrSYN *Proceratophrys boiei*]

Alexander Fry (1821–1905) went from England to Rio de Janeiro (1838) to enter his father's mercantile business, becoming a partner (1843). He returned to London (1854), still occasionally visiting Rio. His main interest was the study of Coleoptera, and he both collected beetles himself and expanded his collection with specimens bought from others. He bequeathed his 200,000 specimens, and library of 611 books, to the BMNH. Günther did not give the most detailed of etymologies, writing: 'A female of this fine species, from the Serra de Mantiqueira, Minas Geraes, has been presented to the British Museum by A. Fry, Esq., a gentleman who has contributed numerous valuable specimens from Brazil to the National Collection.'

Fry, DB

Fry's Frog *Austrochaperina fryi* **Zweifel**, 1962
[Replacement name for *Austrochaperina brevipes* Fry, 1915]

Dene Barrett Fry (1893–1917) worked at New South Wales University as a taxonomist and collector (1912–1915). He died in France (WW1). He wrote *On a collection of reptiles and batrachians from Western Australia* (1914). A reptile is named after him.

Fuente

Fuente's Powakka Treefrog *Hypsiboas fuentei* **Goin** & Goin, 1968

Murray H. de la Fuente collected the holotype and worked at the Carnegie Museum, Pittsburgh (1966–1977).

Fuhrmann

Fuhrmann's Backpack Frog *Cryptobatrachus fuhrmanni* **Peracca**, 1914

Professor Dr Otto Fuhrmann (1871–1945) was a Swiss zoologist and helminthologist (a studier of worms, particularly parasitic ones) who taught comparative anatomy and zoology at the University of Neuchatel (1896–1941). He wrote *Die Tänien der Amphibien* (1895).

Fülleborn

Fülleborn's Stream Frog *Strongylopus fuelleborni* **Nieden**, 1911

Dr Friederich Fülleborn (1866–1933) was a physician who worked in Tanganyika (Tanzania) (1896–1900). He became a Professor, Hamburg University, and was an expert on tropical diseases. Among other taxa named after him are ten birds and a reptile.

Fuller

Fuller's Caecilian *Gegeneophis fulleri* Alcock, 1904
[Alt. Kuttal Caecilian]

Sir Joseph Bampfylde Fuller (1854–1935) was a colonial administrator (1885–1906) who was Chief Commissioner of Assam (1902–1905) and Lieutenant-Governor of Eastern Bengal and Assam (1905–1906). He wrote *The Empire of India* (1913).

G

Gabriel

San Gabriel Slender Salamander *Batrachoseps gabrieli* **Wake**, 1996

This salamander is named after the San Gabriel Mountains, California.

Gadgil

Gadgil's Torrent Frog *Micrixalus gadgili* **Pillai** & Pattabitaman, 1990

Dr Madhav Gadgil (b.1942) is a biologist whose first degrees are from the Universities of Poona and Bombay, plus a doctorate at Harvard in mathematical ecology. He worked (1973–2004) at the Indian Institute of Science, Bangalore and was Visiting Professor at Stanford. He wrote *Diversity: The cornerstone of life* (1997).

Gadow

Gadow's False Brook Salamander *Pseudoeurycea gadovii* **Dunn**, 1926

Dr Hans Friedrich Gadow (1855–1928) was a German zoologist. His main contribution to zoology was to devise a method of taxonomy based on comparisons of 40 characteristics in birds; he was encouraged by Günther of the BMNH, London. Gadow became Curator, Strickland Collection, Cambridge (1884) and also lectured on the morphology of vertebrates. He co-wrote *A Dictionary of Birds* (1893–1896). A bird and three reptiles are named after him.

Gagliardo

Gagliardo's Robber Frog *Pristimantis gagliardoi* **Bustamente** & **Mendelson**, 2008

[Alt. Deviled Rain-Peeper]

Ron Gagliardo is an American herpetologist who is Training Officer with the Amphibian Ark project, seeking to breed endangered amphibians in captivity. His early years were spent near the Everglades National Park, Florida, where he developed an interest in reptiles and amphibians. His bachelor's (1987) and master's (1992) degrees were awarded by North Carolina State University. He was Curator and Manager, Plant Tissue Culture Laboratory, Atlanta Botanical Gardens (1993–2008).

Gaige

Fort Randolph Robber Frog *Pristimantis gaigei* **Dunn**, 1931

Gaige's Rapids Frog *Paratelmatobius gaigeae* **Cochran**, 1938

Mrs Helen Beulah Thompson Gaige (1890–1976) was a herpetologist who specialized in neo-tropical frogs. She studied at the University of Michigan, where she was Assistant Curator, Museum of Zoology (1910), Curator, Amphibians (1919), and Curator, Herpetology (1944). She became Editor-in-Chief of *Copeia* (1937). She married Frederick McMahon Gaige, an entomologist and one time Director of the Museum. In their honour the American Society of Ichthyologists and Herpetologists makes an annual award to a graduate student of herpetology. She co-wrote *The Herpetology of Michigan* (1928). Ten reptiles are named after her.

Galán

Señaris Treefrog *Tepuihyla galani* **Ayarzagüena**, **Señaris** & **Gorzula**, 1993

Dr Pedro Galán Regalado is a herpetologist who is Professor of Zoology, University of A Coruña, Spain. He wrote *Anfibios e Reptiles de Galicia* (1993). A reptile is named after him.

Gald

Espada's Robber Frog *Pristimantis galdi* **Jiménez de la Espada**, 1870

The original description is very brief and has no etymology at all. We have not been able to discover to whom / what *galdi* refers.

Galgano

West Iberian Painted Frog *Discoglossus galganoi* Capula et al., 1985

Mario Galgano (1907–1985) was a biologist who specialized in the study of the reproductive biology of amphibians. He was Professor of Histology, University of Naples, and was associated with the Italian Museum of Anthropology.

Gallardo

Gallardo's Toad *Rhinella gallardoi* **Carrizo**, 1992

José María Alfonso Félix Gallardo (1925–1994) was Director, Argentine Museum of Natural Sciences – like his grandfather Angel Gallardo (1867–1934) before him. He wrote *Anfibios y reptiles* (1994). Two reptiles are named after him.

Gámez

Monteverde Moss Salamander *Nototriton gamezi* Garcia-Paris & **Wake**, 2000

Dr Rodrigo Gámez Lobo (b.1936) is a Costa Rican scientist and public servant who was founding

Director, Instituto Nacional de Biodiversidad de Costa Rica (INBio) (1989). The University of Illinois awarded his doctorate in plant virology and botany (1966). He was Research Professor at the University of Costa Rica (1958–1990).

Gans

Grassland Frog sp. *Ptychadena gansi* **Laurent**, 1965
Gans' Caecilian *Uraeotyphlus gansi* Gower et al., 2008

Dr Carl Gans (1923–2009) was a zoologist who was born in Germany and emigrated to the USA (1939). He worked as an engineer (1947–1955). He was a Fellow in Biology (1957–1958), University of Florida. He taught biology at State University of New York, Buffalo (1958–1971), then became Professor of Biology, University of Michigan, retiring (1998). Six reptiles are named after him.

Garbe

Eirunepé Snouted Treefrog *Scinax garbei* **Miranda-Ribeiro**, 1926

P. Ernst Wilhelm Garbe (1853–1925) was a naturalist who collected in Brazil in the early twentieth century. He was associated with the Paulista Museum. A bird is named after him. (See **Ernesto**)

García

Glass Frog sp. *Nymphargus garciae* **Ruiz-Carranza** & **Lynch**, 1995

Dr Evaristo García Piedrahita (1845–1921) is described as being an illustrious physician and naturalist. He was also a surgeon, a nutritionist and a botanist who graduated in medicine at the Universidad Nacional, Bogotá, Colombia (1872). He wrote *Los Ofidios Venenos del Cauca* (1896). The hospital in Santiago de Cali, Colombia, is named Hospital Universitario del Valle Evaristo García after him.

Gardiner

Gardiner's Seychelles Frog *Sechellophryne gardineri* **Boulenger**, 1911

John Stanley Gardiner (1872–1946) was a zoologist and oceanographer. He was Professor of Zoology and Comparative Anatomy, Cambridge University (1909–1937). He travelled in the Indo-Pacific, visiting the Maldives (1899) and Fiji some time earlier. He wrote *The Natives of Rotuma* (1898). Two reptiles are named after him.

Garman

Garman's Toad *Amietophrynus garmani* Meek, 1897
[Alt. Eastern Olive Toad; Syn. *Bufo garmani*]

Dr Samuel Walton Garman (1843–1927) graduated in Illinois (1870), became a schoolteacher and was Professor of Natural Science at a seminary in Illinois (1871–1872). He became Louis Agassiz's special student (1872) and from 1873 onwards worked in the Herpetology and Ichthyology Section, Museum of Comparative Zoology, Harvard. He was in South America with Alexander Agassiz (1874) and surveyed Lake Titicaca. He wrote *Reptiles of Easter Island* (1908). Four reptiles and a bird are named after him.

Gascon

Rocket Frog sp. *Allobates gasconi* Morales, 2002
[Syn. *Colostethus gasconi*]
Treefrog sp. *Litoria gasconi* **Richards** et al., 2009

Dr Claude Gascon is a Canadian ecologist. He took his bachelor's degree at the University of Quebec, Montreal, and his doctorate at Florida State University (1990). He is a senior member in the management of Conservation International. He is Visiting Professor, Department of Ecology, National Amazon Research Institute and a research associate at the Smithsonian. He co-wrote *Lessons from Amazonia: The Ecology and Conservation of a Fragmented Forest* (2001). A reptile is also named after him.

Gaston

Buta River Frog *Phrynobatrachus gastoni* **Barbour** & **Loveridge**, 1928

Dr Gaston-François de Witte (see **De Witte**).

Gaucher

Treefrog sp. *Dendropsophus gaucheri* **Lescure** & **Marty**, 2000

Philippe Gaucher is Scientific Director of the Centre National de la Recherche Scientifique and Mission Parc de la Guyane, Cayenne, French Guiana. He has worked on a number of occasions with Marty, with whom he co-wrote *The Tailless Amphibians of French Guiana* (2000).

Gaudichaud

Brazilian Spinythumb Frog *Crossodactylus gaudichaudii* **Duméril** & **Bibron**, 1841

Charles Gaudichaud-Beaupré (1789–1854) studied pharmacy and became a dispenser in the French navy (1810). Later he took part in several major expeditions as a naturalist; first around the world aboard the 'Uranie' and 'Physicienne' (1817–1820), to South America with the 'Herminie' (1831–1833), and with a circumnavigation on 'Bonite' (1836–1837). After these travels, Gaudichaud was appointed professor in pharmacy, attached to the

Paris Museum and worked on the botanical collections from his expeditions. A bird and two reptiles are named after him.

Gaun

Inger's Flying Frog *Rhacophorus gauni* **Inger**, 1966

Gaun Sureng, who was a collector for the Sarawak Museum, accompanied Inger on three collecting expeditions in Borneo, including the one on which they took the frog holotype (1962).

Gay

Gay's Frog *Calyptocephalella gayi* **Duméril** & **Bibron**, 1841

[Alt. Helmeted Water Toad]

Claude Gay (1800–1873) was a French naturalist. He was in Chile (1828–1832 & 1833–1842) and is noted for making the first detailed studies of the flora and fauna of that country. He returned to France (1843) from Rio de Janeiro. He travelled in Russia and Tartary (1856–1858), and was in the USA studying American mining techniques (1859–1860). He wrote *Historia Física y Política de Chile* (24 volumes, 1843–1851). Three birds are named after him.

Geay

Brown Egg Frog *Ctenophryne geayi* **Mocquard**, 1904

Martin François Geay (1859–1910) was a pharmacist, natural history collector and traveller. He led an expedition for the Ichthyology and Herpetology Laboratory, Muséum National d'Histoire Naturelle, Paris (1889–1891). He collected plant and animal specimens in Madagascar (1907) and during an expedition to French Guiana and Venezuela (1908). A snake and a fish from Madagascar are also named after him.

Gehrt

Robber Frog sp. *Ischnocnema gehrti* **Miranda-Ribeiro**, 1926

Augusto Gehrt (b.1897) collected (mainly botanical) specimens in Brazil (1920–1940). An orchid was named after him as recently as 2009. He worked extensively in the area of Alto da Serra, where the frog holotype was collected.

Geijskes

Wilhelmina Cochran Frog *Cochranella geijskesi* **Goin**, 1966

Dirk Cornelis Geijskes (1907–1985) was a biologist, ethnologist and entomologist. He graduated from University of Basel, Switzerland (1935). He worked in Suriname (1938–1965), firstly in agricultural research and from 1954 as Chief Biologist and Director, Suriname Museum. After returning to the Netherlands, he went to work for the Leiden Museum, where he became Curator (1967). He wrote *Natuurwetenschappejik onderzoek van Suriname: 1945–1965* (1967).

Geisler

Madang Cross Frog *Oreophryne geislerorum* **Böttger**, 1892

Geisler's Treefrog *Dendropsophus geisleri* **Mertens**, 1950

Bruno Geisler (1857–1945) was a German taxidermist and curator at the Dresden Zoological Museum (1893–1923). His brother, Herbert G. Geisler, was also a taxidermist. The brothers collected together in New Guinea (1884–1892). The cross frog is named after both of them. We suspect the treefrog is named after Bruno but cannot be certain. Three birds are also named after them.

Gené

Gené's Cave Salamander *Atylodes genei* Temminck & **Schlegel,** 1838

[Alt. Sardinian Cave Salamander; Syn. *Hydromantes genei*]

Carlo Giuseppe Gené (1780–1847) was an Italian author, entomologist and naturalist who studied at the Università degli Studi di Pavia, where he became an assistant lecturer (1828). He travelled in Hungary (1829) and made four expeditions to Sardinia (1833–1838). He became Professor of Zoology at the Università degli Studi di Torino, and Director, Regio Museo Zoologico di Torino. He wrote *De quibusdam insectis Sardiniae novis aut minus cognitis* (1839). Two birds are named after him.

Gentry

Gentry's Robber Frog *Pristimantis gentryi* **Lynch** & **Duellman**, 1997

Alwyn 'Al' H. Gentry (1945–1993) was a US botanist who for some years was senior curator at Missouri Botanical Gardens. Over a 25-year research career he collected nearly 80,000 plant specimens. He wrote *Field Guide to the Families and Genera of Woody Plants of Northwest South America* (1993). He was killed in western Ecuador whilst doing fieldwork. A bird is also named after him.

George

Tuwa Flying Frog *Rhacophorus georgii* **Roux**, 1904

The original description has no specific etymology, but the text implies that Roux had George Albert **Boulenger** in mind.

Germán

Slender-legged Treefrog sp. *Osteocephalus germani* Ron et al., 2012

Germán Chávez is a Peruvian herpetologist at the Centro de Ornitología y Biodiversidad and a very good friend of Pablo J. Venegas, one of the describers. The etymology says that he collected 'the type series and tissues of this new species.'

Gert

Gert's Mountain Toad *Ophryophryne gerti* Ohler, 2003

Gertrud 'Gerti' Steiner is one of the author's sisters. The author told us that she and another sister (see **Hans**) '…did a lot for my research when I was the mother of a young boy, caring for my son when I was in the field.'

Ghesquière

Central Zaire Reed Frog *Hyperolius ghesquieri* **Laurent**, 1943

Jean Hector Paul Auguste Ghesquière (1888–1982) was a botanist who collected for the Belgian National Herbarium. He was in Angola, Belgian Congo (DRC), Niger, São Tomé and Principe and Uganda (1918–1938).

Ghosh

Ghosh's Frog *Euphlyctis ghoshi* Chanda, 1991

Professor Dr Asish Kumar Ghosh (b.1938) is President, Centre for Environment and Development, Kolkata, and a former Director General, Zoological Survey of India. Ghosh and Chanda have worked together and co-wrote *A new frog of the genus* Philautus *Gistel from the proposed Namdapha Biosphere Reserve, Arunachal Pradesh* (1989).

Giaretta

Swamp Frog sp. *Pseudopaludicola giarettai* Carvalho, 2012

Dr Ariovaldo Antonio Giaretta is a Brazilian zoologist and herpetologist who is Professor at the Federal University of Uberlândia, Minas Gerais, and curator of the university's zoological collection of frogs. He also works at the University of São Paulo. He has been working on the taxonomic and natural history aspects of neo-tropical anurans since the 1980s.

Gigas

Bougainville Sticky-toed Frog *Batrachylodes gigas* Brown & **Parker**, 1970

Stubfoot Toad sp. *Atelopus gigas* **Coloma** et al., 2010

Gigas was a giant in Greek mythology, the child of Uranus and Gaea. The name is applied in zoology to taxa which are 'giants' of their kind, including a bird, four mammals and six reptiles.

Giglioli

Giglioli's Fire Salamander *Salamandra salamandra giglioli* **Eiselt** & **Lanza**, 1956

[Alt. Italian Fire Salamander]

Professor Dr Enrico Hillyer Giglioli (1845–1909) was an Italian zoologist, anthropologist, photographer and ornithologist who graduated (1864) at the University of Pisa. He started teaching zoology at the University of Florence (1869) and was Director, Royal Zoological Museum, Florence. He succeeded de Fillipi on the 'Magenta' expedition after de Fillipi's death from cholera in Hong Kong (1867) and wrote up the expedition's results and report. Two birds and a mammal are named after him.

Gill

Gill's Clawed Toad *Xenopus gilli* **Rose** & **Hewitt**, 1927

[Alt. Cape Platanna]

Edwin Leonard Gill (1877–1956) worked in various museums in the United Kingdom before being appointed Director, South African Museum, Cape Town (1925). He implemented and carried through a programme of modernization and expansion. The Museum suffered from a lack of finance during the Second World War and Gill decided to ameliorate the situation by resigning (1942), so that his own salary would not be a burden. He was a skilled taxidermist but it appears to have been a hobby as he practised it at home with his sister's help. Two birds are named after him.

Gillen

Centralian Tree Frog *Litoria gilleni* **Spencer**, 1896

Francis James Gillen (1855–1912) was an Australian anthropologist. He started work as a postal messenger (1867). He moved to Adelaide (1871) to work as a telegraph operator and attend evening classes at the Mines and Industries School. He worked on the overland telegraph line (1875–1892) and was appointed telegraph and postmaster at Alice Springs where he helped the Horn Expedition (1894). He became very interested in ethnography and collected artefacts from the local tribes but, owing to gambling, lost nearly everything and was forced to sell his collection of ethnographic items to the National Museum of Victoria (1899). He became a special magistrate and protector of Aboriginals and was fearless in upholding the native peoples' rights, even charging a policeman with the murder of an Aboriginal. He met Sir Walter Baldwin Spencer in Alice Springs (1894 & 1896). They co-wrote *The Native Tribes of Central Australia* (1899) and set out to cross the continent; the Spencer-Gillen Expedition (1901–1902) aroused considerable public interest. His last years were spent in an invalid's chair. A reptile is also named after him.

Gilliard

Gilliard's Wrinkled Ground Frog *Platymantis gilliardi* **Zweifel**, 1960

Ernest Thomas Gilliard (1912–1965) was an American ornithologist who was closely associated with the AMNH, New York. He participated in several ornithological expeditions, especially to New Guinea where the birds-of-paradise were a particular fascination. He died of a sudden heart attack. A fruit bat is named after him and his wife, Margaret; three birds and a mammal are named after him.

Gimmel

Benavony Stump-toed Frog *Stumpffia gimmeli* **Glaw** & Vences, 1992

Günther Gimmel-Schmitt is a German whose generous sponsorship made possible the publication of the colour plates in the describers' *A field guide to the amphibians and reptiles of Madagascar* (1992), in which the original description can be found.

Ginés

Rangel Robber Frog *Pristimantis ginesi* **Rivero**, 1964
Tepuis Treefrog *Stefania ginesi* Rivero, 1968

Hermano Ginés or Brother Ginés (1912–2011) was born in Navarre, Spain. Before becoming a monk, he was called Pablo Mandazen Soto. He is regarded as a notable Venezuelan naturalist who founded the Sociedad de Ciencias Naturales La Salle. He studied theology at Lambecque, Belgium and Natural Sciences at the Central University of Venezuela. He led 28 scientific expeditions and published over 40 scientific papers. A research ship is named 'Hermano Ginés' after him. A bird subspecies is also named after him.

Giorgi

Yambata River Frog *Phrynobatrachus giorgii* **De Witte**, 1921

Stephano Oronzo Vicenzo de Giorgi collected botanical specimens in the Belgian Congo (now DRC) (1913–1923), and entomological specimens around Yambata River (DRC) just before the First World War.

Girard

Girard's Dwarf Frog *Physalaemus signifier* Girard, 1853
Girard's Robber Frog *Ischnocnema parva* Girard, 1853

Dr Charles Frederic Girard (1822–1895) was a French herpetologist who was Louis Agassiz's pupil and assistant at Neuchâtel, moving with Agassiz to the USA. He was in Cambridge, Massachusetts (1847–1850) and worked with Baird (1850–1857) at the Smithsonian. He became an American citizen (1854) and, while continuing his work at the Smithsonian, studied medicine and graduated from Georgetown College (1856). He briefly visited Europe (1860). During the American Civil War he sided with the Confederacy and supplied the Confederate Army with medical and surgical supplies. He left the USA (1864), returned to France and practised medicine, including serving as a physician during the siege of Paris (1870). He retired (1891) and died in France. Four reptiles are named after him.

Glauert

Australian Toadlet genus *Glauertia* **Loveridge**, 1933 [Now synonymized with *Uperoleia*]
Glauert's Froglet *Crinia glauerti* Loveridge, 1933 [Alt. Rattling Froglet]

Ludwig Glauert (1879–1963) was born in England and trained as a geologist. He emigrated to Western Australia (1908) and joined the geological survey in Perth as a palaeontologist. He worked for two years in the Western Australian Museum as a volunteer, joining the permanent staff (1910) as Scientific Assistant, and Keeper of Geology and Ethnology (1914). He worked on the Margaret River caves (1909–1915), studying remains from the Pleistocene. He served in the Australian Army (1917–1919) and then studied Australian material in the British Museum before returning to Perth (1920) as Keeper of the Western Australian Museum's biological collections, later becoming Curator (1927) and Director (1954). His interests were legion – he was the leading authority on Western Australian reptiles, used his own money to buy books for the Museum and helped with the taxidermy. He finally retired (1956) and went on working on reptiles and scorpions. A reptile and three birds are named after him.

Gnusta

Rio Grande Toad *Rhinella gnustae* **Gallardo**, 1967

The original description has no etymology and there is no hint in the text as to who Gallardo may have intended.

Godman

Godman's Treefrog *Tlalocohyla godmani* **Günther**, 1901

Dr Frederick duCane Godman (1834–1919) was a British naturalist who, with his friend Osbert

Salvin, compiled the massive *Biologia Centrali-Americana* (1879–1915). Godman and Salvin gradually presented their joint collection to the Natural History Museum (1885–1900) and sometimes travelled together. Godman was a lawyer, but wealthy enough to have no real need to work, so he devoted his life to natural history, particularly ornithology. He visited Norway, Russia, the Azores, Madeira, the Canary Islands, India, Egypt, South Africa, Guatemala, British Honduras (Belize) and Jamaica. He wrote *Natural History of the Azores, or Western Islands* (1870). His widow and daughters set up the 'Godman Memorial Exploration Fund'. Three mammals, five reptiles and four birds are named after him.

Goebel

Goebel's False Brook Salamander *Pseudoeurycea goebeli* **Schmidt**, 1936

Dr H. Goebel was director of the Central American Plantations Company, Guatemala City. This company owned a large coffee plantation called El Porvenir, situated at the foot of the volcano Tajumulco, and Schmidt and his colleagues visited it at Goebel's invitation.

Goeldi

Goeldi's Frog *Flectonotus goeldii* **Boulenger**, 1895
Rio Big-tooth Frog *Megaelosia goeldii* **Baumann**, 1912

Emil August Goeldi (1859–1917) was a Swiss zoologist. He went to Brazil (1880) and worked at the Museu Nacional. Later he reorganised the Pará Museum of Natural History and Ethnography: the institution today bears his name – Museu Paraense Emílio Goeldi. He became well known for his studies of Brazilian birds and mammals. He was reputed to be a racist who did not like Brazilians and worked in what was a Swiss enclave! He wrote *Aves do Brasil* (1894). He returned to Switzerland (1907) to teach biology and physical geography at the University of Bern until 1917. Two mammals and three birds are named after him.

Goin

Guaniguanico Yellow-mottled Frog *Eleutherodactylus goini* **Schwartz**, 1960
Tarauaca Snouted Treefrog *Scarthyla goinorum* **Bokermann**, 1962
Vegas Falls Treefrog *Stefania goini* **Rivero**, 1968

Dr Coleman Jett Goin (1911–1986) was an American herpetologist. After starting studying at University of Pittsburgh (1935), he moved to the University of Florida (1936) where he gained a bachelor's degree, a master's degree (1941), and a doctorate (1946). He then joined the faculty of the University of Florida, becoming Professor (1956) before retiring (1971) and moving to Arizona where he worked with the Museum of Northern Arizona, Flagstaff. Despite being badly affected by an attack of polio, he was a good field worker. He wrote a number of books, often with his wife, a fellow herpetologist, Olive Lynda Bown Goin (the plural form *goinorum* in the name of the snouted treefrog honours them both). He co-authored *Guide to Reptiles, Amphibians, and Fresh-Water Fishes of Florida* (1955).

Goldman

Robber Frog sp. *Eleutherodactylus goldmani* **Noble**, 1924 NCR
[JrSYN *Craugastor gollmeri*]

Major Edward Alphonso Goldman (1873–1946) was a field naturalist and mammalogist, born in Illinois. Edward Nelson (q.v.) hired him first (1892) to assist his biological investigations of California and Mexico, then as Field Naturalist and eventually Senior Biologist with the US Bureau of Biological Survey. Goldman collected in every region of Mexico (1892–1906) amassing an enormous fund of information on the natural history of the country. He was on the Biological Survey of Panama (1911–1912) during the construction of the Canal. He assisted the US Government in negotiating with Mexico to protect migratory birds (1936). He had a honorary position with the Smithsonian as an Associate in Zoology (1928–1946). He was President, Biological Society of Washington (1927–1929). He wrote *The Mammals of Panama* (1920). Many taxa including eight mammals, eleven birds and a reptile are named after him, as is Goldman Peak in Baja California.

Goliath

Goliath Frog *Conraua goliath* **Boulenger**, 1906
[Alt. Giant Slippery Frog]

Goliath of Gath (about 1,030 BC) was a Philistine warrior of giant size who was killed with a slingshot by David, later King of the Jews (see 1 Samuel XVII.iv, Old Testament of the Bible). His name is sometimes used as a binomial to denote the exceptional size of a species. Four birds, three mammals and a reptile are named after this character.

Gollmer

Evergreen Robber Frog *Craugastor gollmeri* **Peters**, 1863

Julius Gollmer (d.1861) was a pharmacist and amateur biologist who was also German Consul in Caracas, Venezuela. He sent a number of consignments of specimens of all kinds to Berlin's Zoological Museum, Botanical Gardens and Zoo (1857–1858). He sent a batch of colourful fishes to Germany, but these were given little attention until Peters started to look at Gollmer's collections (1859) – leading to the formal scientific description of one of the most popular fishes in home aquaria, the Guppy *Poecilia reticulata*.

Gollum

Gollum's Toad *Ingerophrynus gollum* **Grismer**, 2007

Grismer is obviously a fan of Tolkien's *The Lord of the Rings*, where the pitiful character Gollum plays a pivotal role.

Gómez

Gómez's Web-footed Salamander *Bolitoglossa gomezi* **Wake, Savage** & **Hanken,** 2007

Dr Luis Diego Gómez (1944–2009) was a Costa Rican botanist who was Director of the Museo Nacional de Costa Rica (1970–1985). After leaving the Museum, he served as Director of the Estación Biológica Las Cruces (1986–2005).

Gordon

Gordon's Warty Frog *Theloderma gordoni* **Taylor**, 1962

[Alt. Gordon's Bug-eyed Frog]

Oliver Gordon Young (b.1927). To quote Taylor: 'It is named for Mr. O. Gordon Young of Chiang Mai, Thailand, who discovered the specimen.' (See **Young**)

Gore (Al)

Spotted Nurse Frog *Allobates algorei* Barrio-Amoros & Santos, 2009

Albert Arnold 'Al' Gore Jr (b.1948) was Vice-President of the USA (1993–2001). He has received a Nobel Peace Prize for his work as a climate change activist. He is the founder of the Alliance for Climate Protection. [His life is well-documented and needs no lengthy biography here.]

Gorzula

Bolivar Giant Glass Frog *Vitreorana gorzulae* **Ayarzagüena**, 1992

Dr Stefan Jan Filip Gorzula is a herpetologist and environmental scientist. He is a US citizen whose scientific education was in England: the University of London awarded both his bachelor's degree in zoology (1970) and his doctorate in marine biology (1977). He has worked on major projects in Venezuela, Guyana, Nepal and Vietnam as well as having field experience in many other countries. He co-wrote with Ayarzagüena *Dos nuevas especies del género* Thamnodynastes *(Serpentes: Colubridae) de los tepuyes de la Guayana Venezolana* (1995).

Gosse

Spaldings Robber Frog *Eleutherodactylus gossei* **Dunn**, 1926

Philip Henry Gosse (1810–1888) was an English naturalist, Plymouth Brethren missionary and science populariser. He tried to reconcile biblical ideas on creation with the growing evidence of geological eras lasting millions of years in his work *Omphalos: An Attempt to Untie the Geological Knot* (1857). In this, he advanced the theory that fossil animals had never lived but that God had inserted them into rocks during the Creation as a way of making the world appear older than it was. His son Edmund, his only child, famously wrote *Father and Son*, his birth having been noted by Gosse in his diary thus: '*Received green swallow from Jamaica. E delivered of a son.*' He was clearly a man engrossed in his work! Gosse made a career of writing textbooks on a wide range of subjects, from Jamaican natural history to marine biology. Dunn spent ten days in Jamaica (1925) and in that time discovered seven new *Eleutherodactylus* species, all endemic to Jamaica, and named most of them after Jamaican naturalists or worthies. Three birds are named after him.

Gottlebe

Gottlebe's Narrow-mouthed Frog *Scaphiophryne gottlebei* Busse & **Böhme**, 1992

[Alt. Malagasy Rainbow Frog, Red Rain Frog]

Gunter Gottlebe is a German botanical collector who lives in Madagascar. He collected the frog holotype (1991). He owns a private reserve and palmerium, which are open to the public.

Goudot

Goudot's Bright-eyed Frog *Boophis goudotii* **Tschudi**, 1838

Jules Prosper Goudot was a French traveller and collector in Madagascar in the first half of the nineteenth century. His nickname was 'Bibikely', which is Malagasy for 'insect', emphasizing his interest in entomology. He learned Malagasy, adopted local customs, married a local woman and became so immersed in everything Malagasy that he resigned his position as collector for the Paris Museum. He found the remains of a number of huge eggs, which turned out to be those of the extinct Elephant Bird (*Aepyornis*). Two mammals and a reptile are named after him.

Gould

Turtle Frog *Myobatrachus gouldii* **Gray**, 1841

John Gould (1804–1881) was the son of a gardener at Windsor Castle who became an ornithologist, artist and taxidermist and was acknowledged around the world as 'The Bird Man'. The newly formed Zoological Society of London employed him as a taxidermist. He was certainly the most prolific publisher and original author of ornithological works in the world. He produced in excess of 46 volumes of reference works in colour (1830–1881). He travelled to see birds in their natural habitats: Gould and his wife, Elizabeth, arrived in Australia (1838) to spend 19 months studying and recording the natural history of the continent. By the time they left, Gould had not only recorded most of Australia's known birds but had also gathered data for a major contribution to the study of Australian mammals. He was fascinated by the hummingbirds, a group with which his name is particularly associated. His superb paintings and prints of these and other birds were greatly sought after; so much so that he probably had trouble keeping up with the demand. Forty-six birds, two reptiles and five mammals are named after him.

Gouvêa

Itatiaia Treefrog *Bokermannohyla gouveai* **Peixoto** & **Cruz**, 1992

[Syn. *Boana gouveai*]

Professor Élio Gouvêa (d.1996) was a biologist who worked and researched in the Parque Nacional do Itatiaia, Brazil, where the holotype was acquired and where he started a programme of bird-ringing. He wrote *Uma nova espécie de elosiineo da Serra do Itatiaia (Amphibia, Anura, Leptodactylidae)* (1979) and co-authored a number of papers with the senior author.

Grabham

Clarendon Parish Robber Frog *Eleutherodactylus grabhami* **Dunn**, 1926

Dr Michael Grabham (d.1938) was a physician and surgeon who was born and brought up in Madeira, where his father was a doctor and an influential figure. After graduating as a physician at Cambridge, he went to Jamaica (1867). In his will he left the Government of Jamaica a capital sum to provide an income to be used exclusively for the promotion and extension of maternity services at Victoria Jubilee Lying-in Hospital. He was a friend of E. A. Andrews (q.v.) with whom he went on local collecting expeditions. Dunn spent ten days in Jamaica (1925) and in that time he discovered seven new species of the genus *Eleutherodactylus*, all endemic to Jamaica, and named most of them after Jamaican naturalists or worthies.

Grace

Continental Divide Treefrog *Isthmohyla graceae* **Myers** & **Duellman**, 1982

Grace M. Tilger worked at the AMNH, New York, and collected the frog holotype in Panama. The etymology reads: 'The name honors Grace M. Tilger, first collector of the new species, in recognition of her many contributions to the collections of the American Museum of Natural History.'

Graf, J

Graf's Hybrid Frog *Pelophylax* kl. *grafi* Crochet et al., 1995

Jean-Daniel Graf is a herpetologist and was the first person to discover this frog, which arose from hybridization between *Pelophylax perezi* and *P. ridibundus* in southern France.

Graham, DC

Graham's Frog *Odorrana grahami* **Boulenger**, 1917

[Alt. Diskless-fingered Odorous Frog]

The Revd Dr David Crockett Graham (1884–1961) was a Baptist missionary in Szechuan, China (1911–1918). He returned to the USA for post-graduate study and started to correspond with the Smithsonian with a view to collecting natural history specimens for them. He was back in Szechuan (1920–1926), during which period he encountered the Ch'uan Miao, a Chinese aboriginal tribe. The University of Chicago awarded his doctorate (1927). He was again in Szechuan (1928–1930). He became a Fellow of the Royal Geographical Society, London (1929) and during an expedition to Moupin acquired a Giant Panda skin for the Smithsonian. He taught anthropology and archaeology at West China Union University and was Curator, West China Union University Museum of Archaeology, Art and Ethnology, Chengtu, Szechuan province (1932–1948). After retiring (1948) he lived in Colorado (1949–1961). A reptile is also named after him.

Graham, ED

Graham's Robber Frog *Eleutherodactylus grahami* **Schwartz**, 1979

Eugene D. Graham Jr was the co-discoverer of the frog with Thomas Thurmond and Schwartz. Along with William Sommer, they all collected extensively along the coast of Haiti (1978). He wrote *A new spe-*

*cies of lizard (*Sphaerodactylus*) from northwest Haiti* (1981). A lizard, *Anolis eugenegrahami*, is also named after him.

Grandidier

Grandidier's Madagascar Frog *Mantidactylus grandidieri* **Mocquard**, 1895

Alfred Grandidier (1836–1921) was a French explorer, geographer and ornithologist who collected in Madagascar (1865) and discovered bones of the huge extinct 'Elephant Bird' *Aepyornis maximus* (1866). He returned to France (1870) and began to write his great work *L'Histoire physique, naturelle et politique de Madagascar*, which ran to 40 volumes (the final ones being published posthumously by his son, Guillaume Grandidier). The Malagasy mineral Grandidierite is named after him, as are three mammals, eight reptiles and a bird.

Grandison

Caecilian genus *Grandisonia* **Taylor**, 1968
Grandison's Ridged Frog *Ptychadena grandisonae* **Laurent**, 1954
Mazaruni Treefrog *Dendropsophus grandisonae* **Goin**, 1966
Aleku Caecilian *Sylvacaecilia grandisonae* Taylor, 1970
Red-dotted Glass Frog *Nymphargus grandisonae* **Cochran** & Goin, 1970
Ambrana Madagascar Frog *Blommersia grandisonae* **Guibé**, 1974
Puerto Eden Toad *Atelognathus grandisonae* **Lynch**, 1975
Grandison's Wrinkled Frog *Nyctibatrachus aliciae* **Inger** et al., 1984
[Alt. Alice's Night Frog]
Grandison's Toad *Poyntonophrynus grandisonae* **Poynton** & Haacke, 1993

Alice Georgie Cruickshank Grandison (b.1927) was Curator of Herpetology at the BMNH, London (1960s–1970s). She described several new species and co-wrote *Snakes: A Natural History* (1977) among other works and papers. Three reptiles are named after her.

Grant, T

Poison Frog sp. *Allobates granti* Kok et al., 2006

Dr Taran Grant is a herpetologist, described in the etymology as an 'esteemed colleague'. He became Assistant Professor (2007) at the Faculty of Biosciences, Pontifícia Universidade Católica do Rio Grande do Sul, Brazil, whilst remaining a Research Associate at the AMNH, New York, where he had been a post-doctoral fellow. He co-wrote *A new species of* Colostethus *(Anura: Dendrobatidae) from the eastern slopes of the Cordillera Oriental of Colombia* (2002).

Grant, W

Grant's Big-eyed Treefrog *Litoria granti* **Boulenger**, 1914

William Robert Ogilvie-Grant (1863–1924) was a Scottish ornithologist who worked at the BMNH, London (1882–1918) becoming Curator of Birds (1909). He enlisted in the Army (WW1) and suffered a stroke while helping to build fortifications near London (1916). He collected in Somalia and Socotra Island, and organised the trip to New Guinea on which the treefrog holotype was collected. He wrote *A Hand-book to the Game Birds* (1895). Two mammals and twenty-one birds and a reptile are named after him.

Grauer

Grauer's Puddle Frog *Phrynobatrachus graueri* **Nieden**, 1911

Rudolf Grauer (1870–1927) was an Austrian explorer and zoologist who collected in the then Belgian Congo (1909 & 1910–1911) on an expedition paid for by the Austrian Imperial Museum. He suffered from actinomycosis contracted in Africa and eventually succumbed to this bacterial infection. Other taxa, including sixteen birds, two reptiles and two mammals are named after him.

Gray

Gray's Stream Frog *Strongylopus grayi* **Smith**, 1849
[Alt. Clicking Stream Frog, Gray's Grass Frog]

John Edward Gray (1800–1875) was a zoologist and entomologist. He started work at the British Museum (1824) with a temporary appointment at 15 shillings (75p) per day, and rose to become Keeper of Zoology. Gray was regarded as the leading authority on many reptiles, including turtles. He worked at the Museum with his brother George Robert Gray (1808–1872). John Gray suffered a severe stroke, which paralysed his right side, including his writing hand (1869). Nevertheless, he continued to publish to the end of his life by dictating to his wife, Maria Emma, who had always worked with him as an artist and occasional co-author. The brothers co-wrote *Catalogue of the Mammalia and Birds of New Guinea in the Collection of the British Museum* (1859). He has a large number of taxa, including nine mammals, twenty-three reptiles, and two birds named after him.

Green

Sri Lanka Paddy-field Frog *Fejervarya greenii* **Boulenger,** 1905

Edward Ernest Green (1861–1949) was a mycologist and entomologist who was born into an English family of tea and coffee planters in Ceylon (Sri Lanka). He managed his family's plantations there (1880), combining agriculture with being the Government Entomologist, Ceylon. He wrote the 5-volume *The Coccidae of Ceylon* (1896–1922) and more than 200 papers, mainly on entomology. He presented many specimens to the British Museum, where two of his collections are housed today.

Greening

Greening's Frog *Corythomantis greeningi* **Boulenger,** 1896

Linnaeus Greening (1855–1927) was an English businessman and naturalist. He worked in his family company – Greening's Wireworks, Warrington, Lancashire – and was the leading light for 40 years of the local museum, to which he donated his collections. Beyond his native county he was known nationally for his work on arachnids, reptiles and amphibians. Part 1 of *Check List of Fauna of Lancashire and Cheshire* (1930) was dedicated to him.

Gregg

Gregg's Stream Frog *Craugastor greggi* Bumzahem, 1955

Colonel Clifford C. Gregg (1895–1992) was Director (1937–1962), then President (1962–1964), Field Museum, Chicago. He was originally an upholstery and drapery buyer for the Marshall Field department store, Chicago. He was appointed Assistant Director of the Museum under Stanley Field, who was the Museum's President at the time and a nephew of its founder.

Gregory

Mud-dwelling Caecilian *Schistometopum gregorii* **Boulenger,** 1895

[Alt. Witu Caecilian]

Dr John Walter Gregory (1864–1932) was an explorer, stratigrapher, invertebrate palaeontologist and geomorphologist who undertook a number of expeditions to India, Spitzbergen, Australia, Africa and the Himalayas. He was recorded collecting fishes at Ngatana, East Africa (also the type locality of the caecilian). He wrote *The Great Rift Valley* (1896). He was Professor of Geology, University of Melbourne (1899) and was Director, Geological Survey of Victoria (1901–1904). He returned to Britain (1904) to take up a professorship at the University of Glasgow. His final expedition, to Peru, ended in disaster when his canoe overturned and he drowned in the Urubamba River. A mammal is named after him.

Greshoff

Greshoff's Wax Frog *Cryptothylax greshoffii* Schilthuis, 1889

Anton (sometimes Antoine) Greshoff (1856–1905) was a trader who arrived in the Congo (1877) and was there for more than 20 years, running the company's business at Boma. He was also Dutch Consul in Leopoldville. He was an active collector, particularly of entomological specimens. He supplied ethnographic material to the Ethnographical Museum, Leiden, where he presented a large percentage of the spears now on display.

Grey

Grey's Robber Frog *Eleutherodactylus greyi* **Dunn,** 1926

Robert M. Grey (fl.1936) was the first Superintendent of the Harvard Botanical Garden at Soledad, Cuba. A reptile is named after him.

Griet

Griet Bush Frog *Raorchestes griet* Bossuyt, 2002

[Syn. *Philautus griet*]

The etymology reads: 'this species is named after my wife Griet Decock, as a token of appreciation for her active involvement in fieldwork and many other aspects of my herpetological research'.

Griffiths

Ecuador Cochran Frog *Nymphargus griffithsi* **Goin,** 1961

Dr Ivor Griffiths was on the staff of the Department of Zoology, Birkbeck College, University of London. He established the amphibian genus *Euparkerella* (1959). In the mid-1970s he was at the Department of Veterinary Surgery, University of Glasgow.

Grindley

Grindley's Pygmy Toad *Poyntonophrynus fenoulheti grindleyi* **Poynton,** 1963

Dr John Richard Grindley (1935–2004) was a marine biologist and ecologist. He took a bachelor's degree in zoology at the University of Cape Town (1956); the same university awarded his doctorate (1965) following study at Southampton University, England (1959–1960). He joined (1960) the South African Museum, where his collection of crustaceans is now housed, as Deputy Director and Curator of Invertebrates. He became Director, Port Elisabeth Museum (1967). He was active in the study of plankton and a member of the South African Antarctic research programme. He returned to the University of Cape Town (1975) as Associate Professor, School of Environmental Studies. Ill health forced his resignation (1989).

Griswold

Griswold's Marsupial Frog *Gastrotheca griswoldi* **Shreve**, 1941

John Augustus Griswold Jr (1912–1991) was an aviculturist and ornithologist who took part in expeditions to Borneo (1936), Thailand (1937) and Peru (1939). He became Curator of Birds, Philadelphia Zoological Gardens (1947). A reptile and a bird are named after him.

Grobman

Grobman's Slimy Salamander *Plethodon grobmani* **Allen** & Neill, 1949

[Alt. Southeastern Slimy Salamander]

Dr Arnold Brams Grobman (b.1918) is an American zoologist. The University of Michigan awarded his bachelor's degree (1939); Rochester University, where he was instructor in zoology (1942–1944), awarded his master's and doctorate (1943). During the Second World War he was Research Associate on the Manhattan Project (developing the atom bomb). He was at the University of Florida (1946–1949), rising to Associate Professor. After two years at Oak Ridge Institute of Nuclear Studies, he was Director, Florida State Museum (1952–1959). He also became Dean, College of Arts and Sciences, Rutgers University (1965–1972), leaving to be Professor of Biological Science, University of Illinois, Chicago, which he left (1975) to become Chancellor, University of Missouri, St Louis (1976–1985). He wrote *The distribution of the races of* Desmognathus fuscus *in the southern states* (1960).

Gualter

Organ Mountains Robber Frog *Ischnocnema gualteri* **Lutz**, 1974

Gualter Adolpho Lutz (1903–1969) was a wildlife photographer and Professor of Medicine, Rio de Janeiro. The frog was described by Bertha Lutz, who we believe was his sister.

Gudrun

Kowat Rainforest Frog *Albericus gudrunae* **Menzies**, 1999

In the Volsunga saga Gudrun was the sister of Gunnar, King of the Nibelungs, and wife of Sigurd. This saga was important source material for Wagner's *Der Ring des Nibelungen*.

Guibé

Madagascar Frog genus *Guibemantis* **Dubois**, 1992

Guibé's Ridged Frog *Ptychadena guibei* **Laurent**, 1954

Mesilau Stream Toad *Ansonia guibei* **Inger**, 1966

Guibé's Bright-eyed Frog *Spinomantis microtis* Guibé, 1974

Guibé's Digging Frog *Rhombophryne serratopalpebrosa* Guibé, 1975

Guibé's Stump-toed Frog *Stumpffia tridactyla* Guibé, 1975

Warty Bright-eyed Frog *Boophis guibei* McCarthy, 1978

Guibé's Madagascar Frog *Spinomantis guibei* **Blommers-Schlösser**, 1991

[Syn. *Mantidactylus guibei*]

Dr Jean Guibé (1910–1999) was a French zoologist and herpetologist at the Muséum National d'Histoire Naturelle, Paris, where he was Professor of Zoology (Reptiles and Fish) (1957–1975). He wrote *Les batraciens de Madagascar* (1978). Three reptiles are named after him.

Guille

Guille's Andes Frog *Psychrophrynella guillei* **De la Riva**, 2007

Guillermo 'Guille' de la Riva (b.2002) is the describer's son.

Guinet

Guinet's Tomato Frog *Dyscophus guineti* **Grandidier**, 1875

[Alt. Sambava Tomato Frog]

Mr Guinet was a planter in the Sambava region of northeast Madagascar. He clearly had an interest in natural history, as he also discovered a lemur, the Silky Sifaka *Propithecus candidus*, in this area. Unfortunately, his first name(s) do not appear to be on record.

Gundlach

Gundlach's Robber Frog *Eleutherodactylus gundlachi* **Schmidt**, 1920

Gundlach's Caribbean Toad *Peltophryne gundlachi* Ruibal, 1959

[Alt. Cuban High-crested Toad]

Johannes Christoph (Juan Cristóbal) Gundlach, (1810–1896) was a German zoologist with a great interest in land snails. He was a curator at the University of Marburg, and later at the Senckenberg Museum, Frankfurt. He lived in Cuba (1839–1896) except during its Civil War (1868–1878) when he stayed in Puerto Rico. An event that nearly killed him allowed him to follow his chosen profession: during a hunting accident he discharged a small gun so close to his nose that he lost his sense of smell. After that he could calmly dissect, macerate and clean skeletons without difficulty. He wrote *Ornitología Cubana* (1893). Eight birds, a reptile and two mammals are named after him.

Gunnar

Gunnar's Rainforest Frog *Albericus gunnari* **Menzies**, 1999.

Gunnar, King of the Nibelungs, is a character in the Volsunga saga, which was Wagner's main inspiration for his opera cycle *Der Ring des Nibelungen*.

Günther, AC

Günther's Bubble-nest Frog *Pseudophilautus microtympanum* Günther, 1858
Günther's Shrub Frog *Pseudophilautus variabilis* Günther, 1858
[Alt. Variable Bush Frog]
Günther's Tropical Toad *Incilius intermedius* Günther, 1858
Günther's Whipping Frog *Polypedates eques* Günther, 1858
Günther's White-lipped Frog *Leptodactylus albilabris* Günther, 1859
[Alt. Hispaniolan Ditch Frog]
Guenther's Frog *Ischnocnema guentheri* **Steindachner**, 1864
[Alt. Steindachner's Robber Frog]
Günther's Ground Froglet *Geocrinia laevis* Günther, 1864
[Alt. Southern Smooth Froglet]
Günther's Sharp-nosed Reed Frog *Hyperolius nasutus* Günther, 1864
[Alt. Long-nosed Reed Frog]
Günther's Frog *Litoria latopalmata* Günther, 1867
[Alt. Broad-palmed Frog]
Günther's Borneo Frog *Meristogenys jerboa* Günther, 1872
Günther's Horned Frog *Proceratophrys appendiculata* Günther, 1873
Günther's Bush Frog *Raorchestes chalazodes* Günther, 1876
Günther's Toad *Duttaphrynus hololius* Günther, 1876
Günther's Flathead Toad *Pelophryne guentheri* **Boulenger**, 1882
Günther's Marsupial Frog *Gastrotheca guentheri* Boulenger, 1882
Günther's Red-and-black Frog *Hylarana guentheri* Boulenger, 1882
Günther's Streamlined Frog *Nannophrys guentheri* Boulenger, 1882
Günther's Toadlet *Pseudophryne guentheri* Boulenger, 1882
Günther's Wrinkled Ground Frog *Platymantis guentheri* Boulenger, 1882
Günther's Triangle Frog *Ceratobatrachus guentheri* Boulenger, 1884
[Alt. Solomon Islands Eyelash Frog]
Günther's Brazilian Treefrog *Hypsiboas guentheri* Boulenger, 1886
Günther's Papua Frog *Barygenys atra* Günther, 1896
Günther's Robber Frog *Craugastor omiltemanus* Günther, 1900
Günther's Costa Rican Treefrog *Isthmohyla pseudopuma* Günther, 1901
Günther's Caribbean Toad *Peltophryne guentheri* **Cochran**, 1941
Günther's Caecilian *Caecilia guntheri* **Dunn**, 1942

Dr Albert Carl Ludwig Gotthilf Günther (1830–1914) was a German-born British zoologist. He is noted for recognizing (1867) that the Tuatara is not a lizard but belongs to an entirely separate order of reptiles. He was educated as a physician in Germany, joining the British Museum (1856). He was appointed Keeper, Zoological Department (1857). He became a naturalized British subject (1862) and changed his second two Christian names to Charles Lewis. He became President, Biological Section, British Association for the Advancement of Science (1880) and was President of the Linnean Society (1881–1901). He wrote *The Reptiles of British India* (1864) and the eight-volume *Catalogue of Fishes* (1859–1870). Three mammals, two birds and sixty-seven reptiles are named after him.

Günther, R

Guenther's Digging Frog *Plethodontohyla guentheri* **Glaw** & Vences, 2007
Huon Mawatta Frog *Hylophorbus rainerguentheri* **Richards** & Oliver, 2007
River Frog sp. *Phrynobatrachus rainerguentheri* Rödel et al., 2012

Dr Rainer Günther (b.1941) is a German herpetologist at the Museum für Naturkunde, Berlin. He is particularly noted for his work on the anurans of New Guinea. A reptile is named after him.

Guppy

Shortland Island Webbed Frog *Discodeles guppyi* **Boulenger**, 1884
Treasury Wrinkled Ground Frog *Platymantis guppyi* Boulenger, 1884

Dr Henry Brougham Guppy (1854–1926) was a traveller and collector who began working life as a naval surgeon in the China and Japan station on 'HMS Hornet' (1877–1880), visiting Korea

(1878), and on 'HMS Lark' (1881–1884), which was involved in survey work in the West Pacific, particularly the Solomon Islands. He was in Fiji (1896–1900) and Hawaii (1903–1904). He worked on botany in the West Indies (1907–1911) and the Azores (1913–1914). He died in Martinique. He wrote the two-volume *Observations of a Naturalist in the Pacific between 1896 and 1899* (1903–1906). A reptile is named after him.

Gyldenstolpe

Gyldenstolpe's Frog *Limnonectes gyldenstolpei* **Andersson**, 1916

[Alt. Capped Frog, Koontan Frog]

Count Nils Gyldenstolpe (1886–1961) was a zoologist and ornithologist attached to the Riksmuseum, Stockholm (1914–1961). He travelled in Siam (Thailand) (1911), Central Africa (1921) and New Guinea (1951). A bird and a reptile are named after him.

H

Haddad

Robber Frog genus *Haddadus* **Hedges, Duellman** & Heinecke, 2008

Treefrog sp. *Dendropsophus haddadi* Bastos & **Pombal**, 1996

Dr Célio Fernando Baptista Haddad (b.1959) is an ecologist and herpetologist who is a professor in the Zoology Department, Universidade Estadual Paulista, Brazil. The Universidade Estadual de Campinas awarded his bachelor's degree (1982), master's (1987) and doctorate (1991).

Hahnel

Hahnel's Poison Frog *Ameerega hahneli* **Boulenger**, 1884

Paul Hahnel (1843–1887) was a German collector, mainly of Lepidoptera and Coleoptera, in Venezuela (1877–1879). He made two expeditions to the Amazon (1879–1884 & 1885–1887); he died at Manicoré, Brazil. He collected the frog holotype in northern Peru.

Hall

Hall's Water Frog *Telmatobius halli* **Noble**, 1938

Dr Frank Gregory Hall (1896–1967) was a physiologist, pharmacologist and expert on the effects on the human body of high altitude. Milton College awarded his bachelor's degree (1917), after which he served in the Signals Corps of US Army in the First World War. He then returned to teach at Milton College (1918–1920), also attending the University of Wisconsin for a master's degree (1921) and doctorate (1923), and teaching again at Milton College (1923–1926). He joined the faculty of Duke University (1926–1962), rising to Professor of Zoology and (1946) Professor of Physiology and Pharmacology at Duke Medical School, where he was Departmental Chairman (1946–1962). He retired as Professor Emeritus. He was in England at Cambridge (1933), where he developed an interest in high altitude respiratory physiology, which resulted in his being recalled to the US Army Air Corps immediately after Pearl Harbour (1941), becoming Chief, Physiology Branch, Aero Medical Laboratory. He was on the International High Altitude Expedition to Chile (1935) when he collected the frog holotype. The FG Hall Hypobaric and Hyperbaric Center Research Laboratories at Duke University are named after him.

Halliday

Halliday's Shrub Frog *Pseudophilautus hallidayi* Meegaskumbura & Manamendra-Arachchi, 2005

Timothy Richard Halliday (b.1945) was International Director of the Declining Amphibian Populations Task Force (DAPTF) of the IUCN. He is co-editor of *The New Encyclopedia of Reptiles and Amphibians* (2008).

Hallowell

Hallowell's Sedge Frog *Hyperolius concolor* Hallowell, 1844

Hallowell's Toad *Amietophrynus maculatus* Hallowell, 1854

Hallowell's Treefrog *Hyla hallowellii* **Thompson**, 1912

Dr Edward Hallowell (1808–1860) was an American herpetologist and physician in Philadelphia. He is credited with describing 55 new species of reptile and four are named after him.

Haly

Haly's Bubble-nest Frog *Pseudophilautus halyi* **Boulenger**, 1904

Dr Amyrald Haly (1845–1921) was the first Director of the Ceylon Museum, Colombo (1877–1901). He wrote a Natural History of Ceylon.

Hamid

Jelak Asian Toad *Leptolalax hamidi* **Matsui**, 1997

Aban Abdul Hamid helped Matsui in survey work in Sarawak. He wrote, with Matsui and others, *Acoustic characteristics and systematics relationships of arboreal microhylid frogs of the genus* Metaphrynella *from Malaysia* (1996).

Hamilton

Hamilton's Frog *Leiopelma hamiltoni* McCulloch, 1919

Harold Hamilton (1885–1937) was a biologist who graduated from Otago University, New Zealand. He was employed by the New Zealand Geological Survey and was entomological collector for the Dominion Museum, Wellington. He was on board 'Aurora' in the Antarctic (1913–1914). He became the first Director of the School of Maori Arts, Rotorua.

Hamiota

Nono Robber Frog *Pristimantis hamiotae* **Flores**, 1994

Kenneth Ichiro Miyata (1951–1983) collected the holotype (1979). Interestingly the etymology states: 'The specific epithet is the Latin for 'of the angler,' to honor the collector, the late Dr. Ken I. Miyata, who, like the frogs of the *surdus* assembly, was a lover of streams. Dr. Miyata died tragically in 1983 at the age of 32. He already had made significant contributions to our understanding of tropical herpetofaunas, and his outstanding photography, popular writings, and enthusiasm reflected a great love and appreciation for the tropics and all of nature.'

Hammond

Hammond's Spadefoot Toad *Spea hammondii* **Baird**, 1859

[Alt. Western Spadefoot]

Dr William Alexander Hammond (1828–1900) was a physician, naturalist and soldier. He started collecting on the Pacific Railroad Survey for the Smithsonian (c.1847) before qualifying as a physician (1848) at the University of New York and joining the army (1849) as Assistant Surgeon. He served (and collected birds for Baird) at various frontier stations (1849–1859). He became Professor of Anatomy and Physiology, University of Maryland (1859) but rejoined the army (1861) and served in the American Civil War as Surgeon-General of the Union army. He clashed with Edward M. Stanton, Secretary of War, and was court-martialled (1864, verdict reversed 1878). He lectured on mental diseases at the College of Physicians and Surgeons, New York (1865–1867), and was Professor of Nervous and Mental Diseases, Bellevue Hospital Medical College, New York (1867–1873). He was a co-founder, New York Medical School (1882). He went into private practice (1888). A bird and a reptile are named after him.

Hanitsch

Kadamaian Stream Toad *Ansonia hanitschi* **Inger**, 1960

Dr Karl Richard Hanitsch (1860–1940) was a German biologist and museum curator. He was Demonstrator, Zoology, University College, Liverpool and then Director, Raffles Library and Museum, Singapore (1895–1919). He wrote *An expedition to Mt. Kinabalu, British North Borneo* (1900). An Australian 50-cent postage stamp, issued by Christmas Island (1977), bears his portrait. A reptile is named after him.

Hanken

Hanken's Shrub Frog *Pseudophilautus hankeni* Meegaskumbara & Manamendra-Arachchi, 2011

Dr James Hanken is Professor of Zoology, Curator of Herpetology and Director, Museum of Comparative Zoology, Harvard. The University of California, Berkeley, awarded both his bachelor's degree and his doctorate. After post-doctoral research at Dalhousie University he became a member of the faculty at the University of Colorado, Boulder, before moving to Harvard (1999). A reptile is named after him.

Hans

Hans' Mountain Toad *Ophryophryne hansi* Ohler, 2003

Johanna 'Hansi' Waldl is one of the author's sisters. The author told us that she and another sister (see **Gert**) '...did a lot for my research when I was the mother of a young boy, caring for my son when I was in the field.'

Hansen, D

Hansen's Bushfrog *Chiromantis hansenae* **Cochran**, 1927

Miss Dora Hansen was, like the describer, a member of the Biological Society of Washington and was Doris Cochran's friend.

Hansen, J

Hansen's Warty Frog *Callulina hanseni* Loader et al., 2010

Dr James E. Hansen (b.1941) is a climatologist and a supporter of the African Rainforest Conservancy Trust. The University of Iowa awarded all his degrees – bachelor's (1963), master's (1965) and doctorate (1967). He joined NASA Goddard Institute for Space Studies, New York (1967), becoming its Head (1981). He is also Adjunct Professor in the Department of Earth and Environmental Sciences, Columbia University.

Harald

Harald's Treefrog *Dendropsophus haraldschultzi* **Bokermann**, 1962

Harald Schultz (1909–1966) was a Brazilian ethnologist and anthropologist who studied native South American tribes. He worked at the Museu Paulista, São Paulo (1947–1966) and during much of that time Bokermann was at Fundação Parque Zoológico de São Paulo. He wrote *Hombu* (1962). As a fish hobbyist he also made a particular study of piranhas. He is commemorated in the names of other Brazilian taxa including at least two fishes.

Harald Meier

Harald Meier's Mantella *Mantella haraldmeieri* Busse, 1981

Harald Meier (1922–2007) was a German herpetologist who made several trips to Madagascar and the Mascarene Islands. He described several new species of the gecko genus *Phelsuma*. A reptile is named after him.

Hardy, DE

Sacramento Mountains Salamander *Aneides hardii* **Taylor**, 1941

Dr Dilbert Elmo Hardy (1914–2002) was an American entomologist. He graduated from Brigham Young University (1937) and worked as Research Assistant at Utah State University (1937–1938), where he received his master's degree (1938). The University of Kansas awarded his doctorate (1941). He worked for the Institute of Entomology, University of Kansas (1938–1942). He was a Medical Entomologist in the US Army (WW2). He was stationed in India, Burma (Myanmar) and China. He was Assistant Professor and Assistant State Entomologist, based at Iowa State University (1945–1948). He worked at the University of Hawaii (1948–1981) retiring as Professor Emeritus. He carried out extensive fieldwork in many parts of the world including Europe and Australia. He collected the salamander holotype (1940) in New Mexico.

Hardy, FH

Hardy's Caecilian *Siphonops hardyi* **Boulenger**, 1888

F. Hardy du Dréneuf was a Belgian collector and traveller who made several visits to Brazil. He supplied specimens to both the British Museum and the Leiden Museum.

Harlan

Harlan's Antilles Frog *Osteopilus crucialis* Harlan, 1826

Dr Richard Harlan (1796–1843) was an American physician, naturalist and writer. When still a medical student (1816–1817), he acted as Ship's Surgeon on an East Indiaman bound for Calcutta. He was a member of the Philadelphia Academy of Natural Sciences and an amateur palaeontologist, but not an infallible one: it is owing to Harlan that a fossil whale is called *Basilosaurus* (meaning 'king lizard') because he thought it was a reptile. He wrote *Fauna Americana* (1825). A bird and three mammals are named after him.

Harrisson

Harrisson's Flying Frog *Rhacophorus harrissoni* **Inger** & Haile, 1959

Major Thomas Harnett Harrisson (1911–1976) was a British polymath. In his career he was an anthropologist, ornithologist, ecologist, explorer, journalist, broadcaster, soldier, ethnologist, museum curator, mass observer, archaeologist, filmmaker, writer and conservationist. He spent much of his life in Sarawak and during the Second World War, after having been a radio critic (1942–1944), he was parachuted (1945) into Borneo to organize tribesmen against the Japanese. He was Curator, Sarawak Museum (1947–1966) and undertook pioneering excavations at Niah, Sarawak, discovering a 40,000-year old skull. After leaving Sarawak he lived in the USA, UK and France, before being killed in a motor accident in Thailand. He wrote *Savage Civilisation* (1937). A bird is named after him.

Hartweg

Hartweg's Spikethumb Frog *Plectrohyla hartwegi* **Duellman**, 1968

Hartweg's Mushroomtongue Salamander *Bolitoglossa hartwegi* **Wake** & **Brame**, 1969

Dr Norman Edouard 'Kibe' Hartweg (1904–1964) was an American herpetologist whose specialty was the distribution and taxonomy of turtles. He took his doctorate (1934) at the University of Michigan, and became Curator of Herpetology at its Museum of Zoology. Two reptiles and a bird are named after him.

Harvey

Harvey's Andes Frog *Psychrophrynella harveyi* **Muñoz, Aguayo** & **De la Riva**, 2007

Dr Michael B. Harvey is a herpetologist who has worked extensively in Bolivia with the frog's describers. He attended graduate school at the University of Texas at Arlington (1996) and is presently at the Department of Biological Sciences, Broward College, Florida. Harvey is currently (2012) the editor of *Herpetological Monographs* and the recipient of a NSF grant to study herpetology in Indonesia. He has published numerous papers on the systematics of reptiles and amphibians, including a substantial monograph on teiid lizards.

Hasche

Hasche's Frog *Limnonectes hascheanus* **Stoliczka**, 1870

[Alt. Stoliczka's Frog]

Alfred Hasche was appointed Prussian Consul in Penang (Malaysia) (1866) and was re-appointed (1868).

Hasselt

Hasselt's Litter Frog *Leptobrachium hasseltii* **Tschudi**, 1838

[Alt. Tschudi's Frog]

Johan Coenraad van Hasselt (1797–1823) was a Dutch biologist and physician who graduated at the University of Groningen (1820). He was more interested in natural history, like his friend Heinrich Kuhl (q.v.).

They collected together in Europe and (1820) were sent to Java to study its natural history. They studied pelagic fauna en route, as well as that of Madeira, the Cape of Good Hope and the Cocos Islands, before arriving in Java. Kuhl died after less than a year on Java. Van Hasselt continued collecting, until he himself died. A mammal is named after him.

Haswell

Haswell's Frog *Paracrinia haswelli* **Fletcher**, 1894

William Aitcheson Haswell (1854–1925) was a zoologist who was born in Scotland, took both his bachelor's and master's degrees at Edinburgh (1878) and immediately, for reasons of health, left for Australia where he remained for the rest of his life. He became Curator of Queensland Museum, Brisbane (1879) but moved to Sydney (1880), where he gave a course of public lectures on zoology. He became Acting Curator, Australian Museum (1882), and taught at the University of Sydney, where he became Professor of Biology (1889–1913). After the post was split between botany and zoology, Haswell became Professor of Zoology. He resigned (1917), becoming Professor Emeritus. He co-wrote *A Text-book of Zoology* (1898), which remained the standard textbook for zoology courses in Australia for many years. He was a Fellow of the Royal Society, London, and of the Royal Society of New South Wales, in addition to being for 33 years a trustee of the Australian Museum.

Hauthal

Calientes Water Frog *Telmatobius hauthali* Koslowsky, 1895

Dr Rudolph Hauthal (1854–1928) was a German-born geologist at La Plata Museum, Argentina. He was involved in investigations in Patagonia, which led him and others to conclude that early inhabitants there had kept captive ground sloths in caves, farming them presumably as a food source. On this subject he co-wrote *The mysterious mammal of Patagonia,* Grypotherium domesticum (1899).

Hay

Hay's Snouted Treefrog *Scinax hayii* **Barbour**, 1909

Clarence Leonard Hay (1885–1969) was a research associate of the Department of Anthropology, AMNH, New York, and was at one time Secretary of the Museum board. He was heir to a fortune and son of John Hay, President Lincoln's Private Secretary who became Secretary of Sate under McKinley. Always interested in natural history and anthropology he travelled widely in the Americas and was once rescued in Veracruz by American marines during the Mexican Revolution (1914). During the First World War he served in military intelligence. His anthropological work was in Mesoamerican archaeology. Barbour's etymology states: 'Named for Mr. Clarence L. Hay, an enthusiastic helper on many collecting excursions.' (Hay & Barbour collected the treefrog holotype in Petropolis, Brazil). With others he wrote *The Maya and their Neighbors* (1941), among other books and papers such as *Notes on the Indians of the Great Slave Lake Area* (1946). The Clarence Hay Library in Cape Cod is also named after him. He died in the Ritz Hotel in Paris while on vacation.

Haydée

El Vivero Rocket Frog *Aromobates haydeeae* **Rivero**, 1978

The original description has no etymology and no hint in the text as to who may have been in Rivero's mind. We assume that it refers to Professor Haydée Solano de Chacin, who was a herpetologist at the Tropical Institute of Zoology, Caracas, Venezuela. She wrote *Amphibios comunes del valle de Caracas* (1968).

Hazel

Hazel's Wrinkled Ground Frog *Platymantis hazelae* **Taylor**, 1920

Hazel's Treefrog *Plectrohyla hazelae* Taylor, 1940

Mrs Hazel Roberts, along with her husband H. Radclyffe Roberts, assisted in making collections on the Cerro San Felipe, Mexico.

Heatwole

Heatwole's Dwarf Frog *Physalaemus enesefae* Heatwole, Solano & Heatwole, 1965 NCR

[JrSYN *Physalaemus fischeri*]

Frog sp. *Rana heatwolei* Stuart & Bain, 2005 NCR

[JrSYN *Odorrana tiannanensis*]

Dr Harold Franklin Heatwole (b.1934) is an ecologist, herpetologist and botanist. He took his bachelor's degree in botany (1955) in Indiana and his master's in zoology at the University of Michigan (1958). He has doctorates in both zoology and botany. He taught at the University of Michigan (1959–1960) and then at the University of Puerto Rico (1960–1966), leaving as Associate Professor to join the University of New England, New South Wales (1966–1991). He was Head of the Zoology Department, North Carolina State University (1991–1996) and Professor (1991). He was President of the Australian Society of Herpetologists (1977–1978). He co-wrote *Ecology of Reptiles* (1987). A reptile is named after him and he has suffered snakebite!

Heckscher

Heckscher's Frog *Lithobates heckscheri* **Wright**, 1924
[Alt. River Frog]

August Heckscher (1848–1941) was a German-born American businessman and philanthropist. He emigrated to the USA (1868), working in a coal business (1869–1884). He became a director of many different companies and was a successful investor in Manhattan property. He set up Heckscher Foundation for the Advancement of Research, which provided much of the funding Wright needed for his work on this amphibian.

Hedges

Hedges' Robber Frog *Eleutherodactylus dolomedes* Hedges & **Thomas**, 1992
Canasi Frog *Eleutherodactylus blairhedgesi* **Estrada, Diaz** & **Rodriguez**, 1998

Dr S. Blair Hedges is Professor of Biology at Pennsylvania State University. His master's degree (1984) and his doctorate (1988) were awarded by the University of Maryland. He wrote *Paleogeography of the Antilles and Origin of West Indian Terrestrial Vertebrates* (2006). He might be described as a 'micro-herpetologist', having discovered the world's smallest snake, lizard and frog.

Hedrick

Hedrick's Robber Frog *Eleutherodactylus hedricki* **Rivero**, 1963

Hedrick J. Rivero (b.1953) is the describer's son. He was aged nine when this frog was discovered. He announced that he was his father's assistant and duly followed him, day and night, wherever he went collecting.

Heilprin

Los Bracitos Treefrog *Hypsiboas heilprini* **Noble**, 1923

Professor Angelo Heilprin (1853–1907) was a Hungarian-born American geographer, scientist, explorer, geologist, artist, philosopher and student of natural phenomena in the West Indian region. He also took part in the Peary expedition to Greenland (1891–1892).

Heim

Andes Frog sp. *Phrynopus heimorum* **Lehr**, 2001

Doris and Klaus Peter Heim of Gräfeling, Germany, were honoured in this frog's binomial in recognition of their support for taxonomic research and nature conservation in Peru.

Heinrich

Heinrich's Wart Frog *Limnonectes heinrichi* **Ahl**, 1933

Dr Gerd Herrmann Heinrich (1896–1986) was a German explorer, entomologist and ornithologist, taxonomist, collector and philanderer! He explored, particularly in Sulawesi, but also in Burma, Mexico, North America, Europe and Africa, collecting mammal and bird skins to sell to museums. He collected most taxa but particularly ichneumon wasps. He is the subject of a biographical work by his son, Bernd Heinrich, *The Snoring Bird: My Family's Journey through a Century of Biology*. Thirteen birds and two mammals are named after him.

Helen (Gaige)

Helena's Treefrog *'Hyla' helenae* **Ruthven**, 1919
Helena's Marsupial Frog *Gastrotheca helenae* **Dunn**, 1944

Helen Beulah Thompson Gaige (1890–1976) – a reptile is named after her. (see **Gaige**)

Helen (Rowley)

Helen's Flying Frog *Rhacophorus helenae* Rowley et al., 2012

Helen M Rowley is the mother of the senior describer, Australian herpetologist Dr. Jodi Rowley.

Helena (Bigler)

Helena's Stump-toed Frog *Stumpffia helenae* **Vallan**, 2000
[Alt. Ambohitantely Stump-toed Frog]

Helena Bigler is the describer's wife.

Helena (Fonseca)

Helena's Glass Frog *Vitreorana helenae* **Ayarzagüena,** 1992

Helena Fonseca collected the holotype in 1984.

Helianne

Oval Frog sp. *Elachistocleis helianneae* **Caramaschi**, 2010

Helianne de Niemeyer is a herpetologist at the Department of Vertebrates, Museu Nacional, Rio de Janeiro, Brazil. She has often worked with Caramaschi – e.g. they are co-describers of *Bufo scitulus* (2003).

Helio

Cuzco Horned Treefrog *Hemiphractus helioi* Sheil & **Mendelson**, 2001

Dr Helio Ricardo da Silva is Adjunct Professor of Biology, Universidade Federal Rural do Rio de Janeiro, which he joined (2002) and where his master's degree was awarded (1991). His doctorate (1998) is from the University of Kansas, Lawrence. He wrote *Two Character States New for Hylines and the Taxonomy of the Genus* Pseudacris (1997).

Hellmich

Hellmich's Rocket Frog *Hyloxalus vergeli* Hellmich, 1940

Professor Dr Walter Hellmich (1906–1974) was Director, Natural History Museum, Munich. He was very interested in Himalayan fauna and flora and was Director, Nepal Research Centre, Katmandu (1972–1974). Two reptiles are named after him.

Helmrich

Cobán Mushroomtongue Salamander *Bolitoglossa helmrichi* **Schmidt**, 1936

Gustav Helmrich was a planter at Finca Samac in Guatemala. Finca Samac is a few kilometers away from Cobán, which gives the animal its vernacular name. He and his wife provided a great deal of help and hospitality for Schmidt and his expedition.

Helversen

Helversen's Salamander *Lyciasalamandra helverseni* Pieper, 1963

[Alt. Karpathos Lycian Salamander]

Dr Otto von Helversen (1943–2009) was a German biologist and orthopterist who studied acoustic communication in bats and insects. He studied at Johannes Gutenberg-Universität Mainz, Eberhard Karls Universität Tübingen and Albert-Ludwigs-Universität Freiburg, which awarded his doctorate (1970). He was Professor of Zoology, Friedrich-Alexander-Universität Erlangen (1979–2008). He co-wrote *Handbuch der Fledermäuse Europas und Nordwestafrikas* (2007).

Hendrickson

Thai Spadefoot Toad *Leptobrachium hendricksoni* **Taylor**, 1962

[Alt. Spotted Litter Frog]

Professor Dr John Roscoe Hendrickson (1921–2002) was a zoologist who was an expert on sea turtles and fought tenaciously for their conservation. His bachelor's degree was awarded (1944) by the University of Arizona, after which he spent the rest of the Second World War in the US Navy and was discharged (1946). He went to the University of California, Berkeley, and gained a master's degree (1949) and doctorate (1951). He moved (1959) to Kuala Lumpur as the first Professor of Zoology of the newly created University of Malaya. He was Vice-Chancellor of the University of Hawaii (1963–1969). He was Professor and Director, Marine Biology Programme, University of Arizona (1969). He wanted to be remembered for something worthwhile and arranged that a fund, the John R Hendrickson Scholarship Fund, be set up for the benefit of outstanding young naturalists in Malaysia. He also worked on amphibians and wrote *Ecology and systematics of the salamanders of the genus* Batrachoseps (1954).

Henle

Henle's Slender-legged Treefrog *Osteocephalus elkejungingerae* Henle, 1981 NCR

[JrSYN *Osteocephalus mimeticus*]

Henle's Snouted Treefrog *Scinax pedromedinae* Henle, 1991

Dr Klaus Henle is a herpetologist who is Departmental Head, Conservation Biology, Helmholtz Centre for Environmental Research, Leipzig, Germany. He co-wrote *Population ecology and life history of a lizard community in arid Australia* (1988).

Hensel

Hensel's Swamp Frog *Pseudopaludicola falciceps* Hensel, 1867

Hensel's Dwarf Frog *Physalaemus henselii* **Peters**, 1872

Hensel's Toad *Rhinella henseli* **Lutz**, 1934

Reinhold Friedrich Hensel (1826–1881) was a zoologist, ichthyologist, palaeontologist and naturalist. He taught natural history in Berlin (1850–1860). He was in the southern Brazilian province of Rio Grande do Sul (1863–1866), making a study of fish on behalf of the Berliner Akademie. He became Professor of Zoology (1867). He wrote *Beiträge zur Kenntnis der Wirbelthiere Süd-Brasiliens* (1868). A reptile is also named after him.

Hermine

Hermina's Poison Frog *Mannophryne herminae* **Boettger**, 1893

Mrs Hermine Boettger was the describer's wife. A reptile is named after her.

Hermógenes

Saddleback Toad sp. *Brachycephalus hermogenesi* **Giaretta** & Sawaya, 1998

Professor Dr Hermógenes de Freitas Leitão Filho (1944–1996) trained as an agronomist (1966) but later obtained a doctorate in botany (1973) and became a professor (1978). He wrote more than 100 articles and 8 books, such as *Invasive Plants of Cultures in the State of São Paulo*, as well as innumerable papers and contributions to other works.

The etymology says the toad's binomial '...honors the memory of Hermógenes F. Leitão Filho for his great contribution to the knowledge of the Atlantic Forest flora.' He died whilst on a collecting expedition. There is an ecological park named after him.

Hernández

Hernández's Robber Frog *Pristimantis hernandezi* **Lynch** & **Ruiz-Carranza**, 1983

Jorge I. Hernández-Camacho (1935–2001) was a Colombian biologist, primarily a mammalogist and conservationist, and a good friend of the senior author. He was particularly noted for his efforts to protect Colombia's biodiversity via a network of protected areas. Three mammals are named after him. (See **Mono Hernandez**)

Hewitt

Hewitt's Reed Frog *Hyperolius semidiscus* Hewitt, 1927
[Alt. Yellow-striped Reed Frog]
Hewitt's Bush Squeaker *Arthroleptis xenodactyloides* Hewitt, 1933
[Alt. Dwarf Squeaker; Syn. *Schoutedenella xenodactyloides*]
Hewitt's Long-toed Frog *Strongylopus rhodesianus* Hewitt, 1933
Hewitt's Moss Frog *Anhydrophryne hewitti* **FitzSimons**, 1947
Hewitt's Ghost Frog *Heleophryne hewitti* Boycott, 1988

Dr John Hewitt (1880–1961) was the British-born Director, Albany Museum in Grahamstown, South Africa (1910–1958). He had been Curator, Sarawak Museum (1905–1908) when he collected entomological specimens in Borneo. He was Assistant, Lower Vertebrates, Transvaal Museum (1909–1910), studying vertebrate zoology and archaeology. There he commenced his systematic work on South African arachnids. A mammal, three birds and two reptiles are named after him.

Heyer

Heyer's Button Frog *Cycloramphus rhyakonastes* Heyer, 1983
Heyer's Leaf Frog *Noblella heyeri* **Lynch**, 1986
Heyer's Snouted Treefrog *Scinax heyeri* **Peixoto** & **Weygoldt**, 1986
Eldorado Tree Toad *Hylodes heyeri* **Haddad, Pombal** & Bastos, 1996
Slender-legged Treefrog sp. *Osteocephalus heyeri* Lynch, 2002
Leptodactylid frog sp. *Adenomera heyeri* Boistel, Massary & Angulo, 2006
[Syn. *Leptodactylus heyeri*]

Dr William Ronald Heyer (b.1941) is an American herpetologist whose doctorate was awarded by the University of Southern California. He was at the Field Museum, Chicago (1969) and is now at the Smithsonian's Department of Vertebrate Zoology. A reptile is named after him and his wife, Miriam.

Heymons

Heymons' Narrow-mouthed Frog *Microhyla heymonsi* **Vogt**, 1911
[Alt. Black-sided Narrow-mouthed Frog]

Dr Richard Heymons (1867–1943) was a German zoologist who coined the name *Chelicerata* for arachnids. He studied in Berlin (1886–1891) and was Professor of Zoology, Institute of Zoology, Museum für Naturkunde, Berlin (1915–1935).

Hildebrandt

African Burrowing Frog genus *Hildebrandtia* **Nieden**, 1907

Johannes Maria Hildebrandt (1847–1881) was a German botanist and explorer who collected and travelled in Arabia, East Africa, Madagascar and the Comoro Islands (1872–1881). He was also interested in local languages and customs. He died of yellow fever in Madagascar. Two birds, three reptiles and two mammals are named after him.

Hill

Treefrog sp. *Litoria hilli* Hiaso & Richards, 2006

Dr Lance Hill is a biologist who taught at the University of Papua New Guinea (1969–2008), where he was Professor, School of Natural Sciences. He is now (2011) a professor at James Cook University in Cairns, Australia.

Hillen

Perinet's Bright-eyed Frog *Boophis hillenii* **Blommers-Schlösser**, 1979
[Perhaps a synonym of *B. idae*]
Betic Midwife Toad *Alytes dickhilleni* **Arntzen** & Garcia-Paris, 1995

Dick Hillenius (1927–1987) was a Dutch biologist, writer and teacher whose speciality was the taxonomy of anurans and chameleons. He became (1957) Curator of Reptiles and Amphibians, Zoological Museum, Amsterdam. His writing is regarded as more light-hearted (and therefore readable!) than the heavy scientific thought of many academics; he wrote *De vreemde eilandbewoner* (1967). He is probably better known in Holland as a writer-poet than a herpetologist. Two reptiles are named after him.

Hillman

Bale Mountains Treefrog *Balebreviceps hillmani* **Largen** & **Drewes**, 1989

Dr Stanley S. Hillman is Professor of Biology, Portland State University, Oregon, where he was Chairman, Biology Department (1997–2006). His doctorate was awarded (1977) by the University of California, Los Angeles. He co-wrote *Ecological and Environmental Physiology of Amphibians* (2008).

Hinton

Water Frog sp. *Telmatobius hintoni* **Parker**, 1940

Howard Everest Hinton (1912–1977) was a British zoologist. He was born in Mexico, where his father was working as a metallurgist and mining engineer. California University awarded his BSc (1934) and his PhD was awarded by Cambridge (1939). He became Junior Curator and Assistant Keeper at the University Museum of Zoology in Cambridge (1937), then Assistant Keeper at the BMNH (1939–1949), before taking a lectureship at Bristol University for the rest of his career. Here he became Reader in Entomology (1951), Professor of Entomology (1964) and finally Professor of Zoology and Head of Department (1970). His ill health forced him to resign as head of department shortly before he died. He wrote a great number of scientific papers on entomology such as *The Lathridiidae of economic importance* (1941) and several books. He collected several specimens of the water frog in Bolivia whilst on the Percy Sladen Expedition to Lake Titicaca (1937).

Hmong

Odorous Frog sp. *Odorrana hmongorum* Bain et al., 2003 NCR

[JrSYN *Odorrana jingdongensis*]

The Hmong are an ethnic minority group in Vietnam.

Hobbs

Hobbs' Treefrog *Hypsiboas hobbsi* **Cochran** & **Goin**, 1970

Dr Horton Holcombe Hobbs Jr (1914–1994) was an expert on crayfish. His doctorate was from the University of Florida, where he taught before going to the University of Virginia (1946–1962). He then went to the Smithsonian (1962–1984) as Head Curator, Department of Zoology and (1964) Senior Scientist, Invertebrate Zoology Department. He is commemorated in other taxa such as crustaceans.

Hochstetter

Hochstetter's Frog *Leiopelma hochstetteri* **Fitzinger**, 1861

Dr Christian Gottlieb Ferdinand Ritter von Hochstetter (1829–1884) was a German geologist. He visited Australia and New Zealand during the 'Novara' expedition (1857–1859). His doctorate was awarded in geology (1852) by Tübingen University, after he had studied theology (1847–1851). He worked for the Austrian Geological Survey (1853–1856) and lectured at the University of Vienna (1856). He was appointed official Geologist to the 'Novara' expedition, which allowed him to travel the world and expand his geological knowledge. He returned to Austria and was Professor, Mineralogy and Geology, Royal and Imperial Polytechnic Institute, Vienna (1860–1874), resigning because of ill health. He had become tutor in natural history to Crown Prince Rudolph (1872). He was First Superintendent, Imperial Natural History Museum in Vienna (1876–1884). A bird is also named after him.

Hocking

Hocking's Water Frog *Telmatobius hockingi* Salas & Sinsch, 1996

Dr Peter 'Pedro' J. Hocking is a Peruvian biologist and ornithologist at the Peruvian Natural History Museum, Lima. He was born in Peru to American missionary parents and studied at Bible College in the USA before returning to Peru (1963), where he renounced his American citizenship in favour of Peruvian. He is interested in cryptozoology and wrote *Large Peruvian Mammals Unknown to Zoology* (1992).

Hödl

Rocket Frog sp. *Allobates hodli* Simões, **Lima** & Farias, 2010

Dr Walter Hödl is an Associate Professor at the Institut für Zoologie, Universität Wien, Austria. He co-wrote *The influence of internote-interval variation of the advertisement call on the phonotactic behaviour in male* Allobates femoralis *(Dendrobatidae)* (2007).

Hoehne

Hoehne's Robber Frog *Ischnocnema hoehnei* **Lutz**, 1958

Frederico Carlos Hoehne (1882–1959) was a self-taught Brazilian botanist and conservationist. He finished his formal education (1899) without the possibility of a university education. He was accomplished as a collector and made money by selling orchids. Despite a lack of scientific training, he was appointed (1907) head gardener at the Scientific Institution, Rio de Janeiro, and (1908) went on the first of many field trips for this and other organizations. He was appointed Botanist (1913) of the (Theodore) Roosevelt-Rondon Scientific Expedition to Mato Grosso region. He worked (1917–1952) for São Paulo State Botanical Institute, becoming Director. The botanical journal *Hoehnea* was named in his honour.

Hoesch

Hoesch's Toad *Poyntonophrynus hoeschi* **Ahl**, 1934

Dr Walter Hoesch (1896–1961) was a German zoologist badly wounded in the First World War and invalided out of the Army, who then trained as a lawyer. He emigrated to Southwest Africa (Namibia) (1930) and bought a farm, but was bankrupted (1932) by drought. He started collecting specimens for Professor Karl Jordan, an entomologist, and accompanied Jordan on his South-West African Tour (1933). He also made a private collection of small mammals and birds. He wrote *Die vögelwelt Deutsches-Südwestafrikas namentlich des Damara- und Namalandes* (1940). Five birds and a reptile are named after him.

Hoffman

Valley and Ridge Salamander *Plethodon hoffmani* Highton, 1971

Dr Richard Lawrence Hoffman (1927–2012) was an American zoologist who started by working as a chemist supervisor in a gunpowder factory (1950–1957). Cornell awarded his master's degree (1958) and Virginia Polytechnic Institute a Doctorate in Zoology (1960). He was Research Assistant, helping with study of arthropods, for the National Science Foundation at the United States National Museum (1959–1960). He joined the faculty, Radford University, Virginia, as Associate Professor (1960), becoming Professor of Biology (1962). He collected the salamander holotype (1954).

Hoffmann

Hoffmann's Shrub Frog *Pseudophilautus hoffmanni* Meegaskumbura & Manamendra-Arachchi, 2005

Dr Hans Lukas 'Luc' Hoffmann (b.1923) is a Swiss conservationist and philanthropist who (1947) acquired the Tour du Valat estate (France) and (1954) set up a biological station there. He was a co-founder of WWF International and its vice-president (1961–1988), and was (1966–1969) vice-president of IUCN (International Union for the Conservation of Nature). He has given much support to Global Amphibian Assessment.

Hoipollo

Hoipollo's Bubble-nest Frog *Pseudophilautus hoipolloi* Manamendra-Arachchi & Pethiyagoda, 2005

[Alt. Anthropogenic Shrub Frog]

This is an example of someone inventing a 'personalized' name based on a false expectation that the binominal was in the genitive. 'Hoipolloi' is more correctly written as 'Hoi Polloi' or, in the original 'όί πoλλoί' – it is Greek for 'The many', as the etymology makes clear. The name is described as '... an allusion to the exclusively synanthropic distribution of this species'.

Holbrook

Holbrook's Spadefoot *Scaphiopus holbrookii* **Harlan**, 1835

[Alt. Eastern Spadefoot]

Holbrook's Triton *Eurycea guttolineata* Holbrook, 1838

[Alt. Three-lined Salamander]

Dr John Edwards Holbrook (1794–1871) was a zoologist who has been described as the father of North American herpetology. He qualified as a physician (1818) and went to Edinburgh for post-graduate studies (1819). He visited Paris and became friendly with the great French naturalists of the day, Cuvier and Duméril *inter alios*. He returned to the USA (1822) and practised medicine in Charleston, South Carolina. He was Professor of Anatomy, Medical School of South Carolina (1824–1854). During the American Civil War he was, despite his age, a surgeon in the Confederate army. After Union troops captured Charleston and his manuscripts and collections were looted, he gave up all scientific research. He was elected to the National Academy of Sciences (1868). He wrote *North American Herpetology* (1836–1842). Three reptiles are named after him.

Holdridge

Holdridge's Toad *Incilius holdridgei* **Taylor**, 1952

Dr Leslie Rensselaer Holdridge (1907–1999) was an American botanist and ecologist with a particular interest in forests and forestry. His bachelor's degree in forestry (1931) was awarded by the University of Maine. He achieved both his master's and his doctorate in botany (1947), both from the University of Michigan. He worked in Puerto Rico (1935–1941), firstly for the Caribbean National Forest and subsequently as a forester for the US Forestry Service. He divided his time (1941–1949) between working in Haiti and in cinchona research in Guatemala and Colombia. He worked (1949–1960) as forester and ecologist for the Inter-American Institute of Agricultural Sciences, Turrialba, Costa Rica, and founded (1954) La Selva research station. He was a senior staff member of the Associate Colleges of the Midwest's Central American field programme (1963–1967) and was engaged (1967) as an ecologist by the Tropical Science Center, San José, Costa Rica. He was a part-time Professor of Dendrology and Ecology (1978–1982) at the Instituto Tecnológico de Costa Rica. He wrote *Forest Environments in Tropical Life Zones: A Pilot Study* (1971).

Holst

Holst's Frog *Babina holsti* **Boulenger**, 1892

P. Aug. Holst (d.1895) was a Swedish naturalist who was in Japan (1889) and Taiwan (1893–1894). He sent specimens to Boulenger (1880s), although Boulenger's etymology just mentions Holst's name with no details. A bird is named after him.

Holt

Holt's Robber Frog *Ischnocnema holti* **Cochran**, 1948

Ernest Golsan Holt (1889–1983) was an American ornithologist and explorer. He had a roving commission for the US Biological Survey as there are reports of his activities in Alabama, Alaska, North Dakota and Virginia. He visited South America a number of times (1921–1947). He was a biologist (1930s) with the US Soil Erosion Service. He co-wrote *Birds of Autauga and Montgomery Counties, Alabama* (1914). He collected the frog during his expedition to Brazil (1921–1922). Two birds are named after him.

Holtz

Taurus Frog *Rana holtzi* **Werner**, 1898
[Sometimes regarded as conspecific with *Rana macrocnemis*]

Martin Holtz collected the frog holotype (1897). He collected entomological specimens in Greece (1901). We believe him to have been employed as a collector by the Naturhistorisches Museum Wien, Austria.

Hoogland

New Guinea Land Frog *Austrochaperina hooglandi* **Zweifel**, 1967

Dr Ruurd D. Hoogland (1922–1994) was a Dutch botanist with New Guinea Group, CSIRO Division of Land Research and Regional Survey (1952–1968), and (1968–1979) Research Fellow, Research School of Biological Sciences, Australian National University, Canberra. He visited Norfolk Island several times.

Hoogmoed

Hoogmoed's Harlequin Toad *Atelopus hoogmoedi* **Lescure**, 1974
Hoogmoed's Treefrog *Hypsiboas roraima* **Duellman** & Hogmoed, 1992
Hoogmoed's Toad *Rhinella hoogmoedi* **Caramaschi** & **Pombal**, 2006

Dr Marinus Steven Hoogmoed (b.1942) is a Dutch herpetologist who was at one time at Leiden Museum, but now lives in Belém, Brazil, and works at the Museu Paraense Emílio Goeldi. Among his publications is *Snakes of the Guianan Region* (1982). Three reptiles are named after him.

Horst Paul

Andes Frog sp. *Phrynopus horstpauli* **Lehr, Köhler** & Ponce, 2000

Horst Paul is a German patron of research. The senior author told us that he '...named several frogs after persons who sponsored taxonomic research by donating money to the German society BIOPAT e.V. The authors of the species description received half of the donated money after the publication of the species from the society to the account of their research institution and could ask for the money to pay for research (in my case flights to Peru). With the other half, conservation projects are founded in the country of origin of the newly described species.'

Horst, R

Horst's Cross Frog *Oreophryne frontifasciata* Horst, 1883
Horst's Treefrog *Litoria amboinensis* Horst, 1883

Dr Rutger Horst was a zoologist and herpetologist at the Leiden Museum, Netherlands.

Horstock

Horstock's Reed Frog *Hyperolius horstockii* **Schlegel**, 1837
[Alt. Arum Frog, Arum Lily Frog]

Dr Hubertus B. van Horstock, who practised medicine, also collected in Cape Province, South Africa (1826–1834) for the Leiden Museum.

Hose

Hose's Frog *Odorrana hosii* **Boulenger**, 1891
Hose's Tree Toad *Pedostibes hosii* Boulenger, 1892
[Alt. Boulenger's Asian Tree Toad]
Borneo Bubble-nest Frog *Philautus hosii* Boulenger, 1895

Dr Charles Hose (1863–1929) was a British naturalist and colonial administrator, who lived in Sarawak and other parts of Malaysia (1884–1907). He successfully investigated the principal cause of the disease beriberi. He was also a good cartographer who produced the first reliable map of Sarawak. He wrote *Fifty Years of Romance and Research* (1927). Fort Hose, Sarawak, which is now a museum, was named after him, as are seven mammals and five birds; a mammal is named after him and his brother Ernest.

House

House's [sic] Tree Toad *Pedostibes hosii* **Boulenger**, 1892

'House' is a mistranscription of 'Hose' (see entry above).

Hosmer

Hosmer's Frog *Cophixalus hosmeri* **Zweifel**, 1985
[Alt. Clicking Rainforest Frog]

William Hosmer (1925–2002) was an Australian herpetologist who worked at the Field Museum, Chicago. He was on the Spalding-Peterson Expedition to Northern Territory of Australia (1959–1960). He was a Field Associate, Department of Herpetology, AMNH, New York (1962). Among his many publications was *A New Frog of the genus* Hyla *from Northern Territory, Australia* (1964). A reptile is named after him.

Houy

Ussagara Reed Frog *Hyperolius houyi* **Ahl**, 1931

Dr Reinhard Houy (d.1912) collected in Tanganyika (Tanzania) (1911–1912). He wrote *Beiträge zur Kenntnis der Haftscheibe von Echeneis* (1909). Two birds are named after him.

Howell

Morara Forest Toad *Mertensophryne howelli* **Poynton** & Clarke, 1999

Kim Monroe Howell (b.1945) is Professor of Zoology and Marine Biology, University of Dar Es Salaam, Tanzania. He co-wrote *A Field Guide to the Reptiles of East Africa* (2006). Both Clarke and Poynton have written scientific papers with Howell. Three reptiles, a birds and a mammal are named after him.

Htun Win

Htun Win's Treefrog *Rhacophorus htunwini* Wilkinson et al., 2005 NCR
[JrSYN *Rhacophorus bipunctatus*]

Htun Win (d.2004) joined the Chatthin Wildlife Sanctuary, Myanmar (1993), becoming a forester (1995) and starting work on herpetology (1997). He fell ill and died whilst surveying the herpetofauna of Kachin state. A reptile is named after him.

Hu

Zhaojue Flying Frog *Rhacophorus hui* **Liu**, 1945

Professor Dr Hsen Hsu Hu (1894–1968) was a botanist. His bachelor's degree was awarded by the University of California, Berkeley (1916). He then taught at Nanjing Higher Normal School (1916–1923), after which he went to Harvard, where he was awarded his doctorate in science (1925). He was the first President of National Chung Cheng University (1940–1944) and was Professor at and Director of the Fan Memorial Institute of Biology in Peking (Beijing) (1945).

Huang

Huang's Spadefoot Toad *Xenophrys pachyproctus* Huang, 1981

Huang Y-Z is a Chinese herpetologist, specializing in amphibians. He wrote *A new species of pelobatid toads (Amphibia: Pelobatididae) from Ningxia Hui Autonomous Region* (1985).

Huber

Bolivar Bush Toad *Oreophrynella huberi* Diego-Aransay & **Gorzula**, 1990

Dr Otto Huber is a botanist and herpetologist who is Professor of Phytogeography at Universidad Simón Bolívar, Venezuela. Previously he was Director of Scientific Research, Fundación Instituto Botánico de Venezuela. His botanical collection is held at the Herbarium, Universidad Central de Venezuela. He undertook a number of expeditions in Venezuela (1980s). He has worked with Gorzula and they co-wrote *Consideraciones finales. En: El Macizo del Chimantá, Escudo de Guayana, Venezuela* (1992).

Hubricht

Peaks of Otter Salamander *Plethodon hubrichti* Thurow, 1957
Red Hills Salamander *Phaeognathus hubrichti* Highton, 1961

Leslie Hubricht (1908–2005) was a self-taught expert on molluscs who described 81 of the c.520 species of land snail known to exist in the USA. His collection of 500,000 land snail specimens was larger than the combined collections of the USA's major museums. He had no university education and went to work at an early age to help support his family. He worked as Research Assistant at Missouri Botanical Garden (1936–1943). He worked for Remington Rand as a mechanic, servicing tabulating machines and early computers (1943–1973), retiring to devote all his time to continuing his study of land snails. He deposited his collection at the Field Museum, Chicago (1990).

Hudson

Hudson's Humming Frog *Chiasmocleis hudsoni* **Parker**, 1940

C. A. Hudson was a collector, mainly of entomological specimens, for the Natural History Museum, London. He collected the frog holotype in British Guiana (Guyana). A reptile is named after him.

Hugo

Hugo's Spiny-chest Frog *Alsodes hugoi* Cuevas & Formas, 2001

Dr Hugo Hermes Campos Cereceda (1935–1998) was a Chilean zoologist and limnologist who was Professor at the Instituto de Zoología, Universidad Austral de Chile, Valdivia. He wrote *Mamíferos terrestres de Chile* (1986). He died from a heart attack whilst engaged in fieldwork.

Hull

Hull's Leaf Frog *Agalychnis hulli* **Duellman** & **Mendelson**, 1995

[Syn. *Phyllomedusa hulli*]

Dr Clark Hull was Worldwide Environmental Manager at Occidental Oil and Gas Corp, on whose concession in Peru the frog holotype was collected. He is presently Regulatory Manager at URS Corporation, which is a provider of water resources and environmental engineering and consultancy services.

Humayun

Humayun's Wrinkled Frog *Nyctibatrachus humayuni* Bhaduri & Kripalani, 1955

[Alt. Abdulali's Wrinkled Frog, Bombay Night Frog]

Shri Humayun Abdulali (1914–2001) was an ornithologist and a cousin of the famous Indian naturalist Salim Ali. He became Honorary Secretary of Bombay Natural History Society (1950) and, whilst there, catalogued their collection of bird skins. He was also a prolific contributor to their journal, writing more than 300 papers. He led two expeditions to the Andaman and Nicobar Islands (1964–1966). Two birds are named after him.

Humboldt, FWH

Humboldt's Toad *Rhinella humboldti* **Gallardo**, 1965

Baron Friedrich Wilhelm Heinrich Alexander von Humboldt (1769–1859) was a Prussian naturalist, explorer and politician. After attending universities at Frankfurt an der Oder and Göttingen, he enrolled at the Freiberg Mining Academy (1791) to learn natural history and earth sciences to help him with his intended future travels. To complete his experience he then worked as an inspector of mines in Prussia for five years. After two years of disappointments and delays, he set off (1799) to explore in South America (1799–1804). He collected thousands of specimens, mapped, and studied natural phenomena. The trip took in parts of Venezuela, Peru, Ecuador and Colombia. He returned via the USA, where Thomas Jefferson entertained him at Monticello. He made a journey of similarly epic proportions ranging from the Urals east to Siberia (1829). Humboldt's *Personal Narrative* (1819–1829) was inspirational to later travellers in the tropics, notably Darwin and Wallace. His most famous writing was the five-volume work *The Cosmos* (1845–1862). Humboldt did research in many other fields, including astronomy, forestry and mineralogy. The Humboldt Current, running south to north off the Pacific coast of South America, is named after him. Five mammals and five birds are named after him.

Humboldt, FWH & FWC

Treefrog sp. *Litoria humboldtorum* **Günther**, 2006

Baron Friedrich Wilhelm Heinrich Alexander von Humboldt (see above) and his older brother, Friedrich Wilhelm Christian Carl Ferdinand Baron von Humboldt (1767–1835) are both commemorated in this frog's binomial. F. W. C. Humboldt was a philosopher, diplomat, linguist (one of his languages was Basque) and educationalist: he was Minister of Education in the Prussian government and the founder of Humboldt Universität, Berlin.

Humboldt

Humboldt's Glass Frog *Vitreorana uranoscopa* **Müller**, 1924

Humboldt's Tree Toad *Hylodes perplicatus* **Miranda-Ribeiro**, 1936

Both amphibians are named after where the holotypes were collected: the town of Humboldt, Santa Catarina, Brazil.

Humphrey

Humphrey's Caecilian *Ichthyophis humphreyi* **Taylor**, 1973

Dr Philip Humphrey was an ornithologist and herpetologist who was Director, Museum of Natural History, Kansas University (1967–1995). He collected in many countries, including amphibians in Brazil (1964) and birds in Haiti, Argentina and Chile. Taylor said he honoured Humphrey '...in recognition of the courtesies extended to me at the Museum.'

Hunt

Treefrog sp. *Litoria hunti* **Richards** et al., 2006

The binomial ought to be in the plural (*huntorum*) as a family is honoured, but it appears that attempts to amend it have been rejected. This frog was named after the Hunt family of Adelaide, 'in recognition of their generous support for research at the South Australian Museum.'

Hurter

Hurter's Spadefoot *Scaphiopus hurterii* **Strecker**, 1910

Julius Hurter Sr (1842–1917) was a Swiss-born American mechanical engineer. He became Curator of the St Louis Academy of Sciences (1900) and wrote *Herpetology of Missouri* (1911).

Husain

Husain's Caecilian *Ichthyophis husaini* **Pillai** & Ravichandran, 1999

Dr Akhlaq Husain of the Zoological Survey of India collected the holotype.

Hutchins

Hutchins' Treefrog *Hypsiboas hutchinsi* **Pyburn** & **Hall**, 1984

Dr Melvin G Hutchins Jr was honoured '….as an expression of friendship and for his years of help and encouragement to the senior author.' We are unable to find more about him.

Hutchison

Hutchison's Climbing Frog *Anodonthyla hutchisoni* Fenolio et al., 2007

Dr Victor Hobbs Hutchison (b.1931) is an American biologist. North Georgia College awarded his bachelor's degree (1952), and both his master's (1959) and doctorate (1959) were awarded by Duke University. He served as a lieutenant in the US Army on active service in Korea (1952–1954). He was at the University of Rhode Island (1959–1970), progressing to become Professor of Zoology. He moved to the University of Oklahoma as Professor of Zoology (1970–1979), becoming George Lynn Cross Research Professor (1979–2002), Emeritus since 2001.

Hutsebaut

Ibembo Reed Frog *Hyperolius hutsebauti* **Laurent**, 1956

Father Franz Joseph Hutsebaut (1886–1954) was a Catholic missionary in the Belgian Congo from the 1920s until his death in Katanga. He was based for part of his career at Ibembo, where the frog holotype was collected. He was clearly a remarkable man, as he once shot a leopard which crawled through the window of the mission late one night and was also known for rearing and shipping okapis to zoos outside Africa (1927–1941).

I

Ian

Ian's Andes Frog *Psychrophrynella iani* **De la Riva, Reichle** & **Cortez**, 2007

Ian Yure Domic (b.2004) is the son of the third author.

Iatamas

Andes Frog sp. *Psychrophrynella iatamasi* Aguayo-Vedia & **Harvey**, 2001

This is not an eponym, though it may look like one. Iatamas is a word in the Quechua language of the central Andes, loosely meaning 'the land where one was born'.

Iban

Tributary Wart Frog *Limnonectes ibanorum* **Inger**, 1964

The Iban are a branch of the Dayak peoples of Borneo. They were formerly renowned as headhunters.

Ibarra

Jalapa Toad *Incilius ibarrai* **Stuart**, 1954

Professor Jorge Alfonso Ibarra (1921–2000) was Director of the Museo Nacional de Historia Natural, Guatemala. He also founded the magazine *Pro Natura* (1985). A bird and a reptile are named after him.

Ibisch

Strabomantid frog sp. *Oreobates ibischi* **Reichle, Lötters** & **De la Riva**, 2001

Dr Pierre Leonhard Ibisch (b.1967) is a German biologist who is (2012) Professor for Nature Conservation at Eberswalde University for Sustainable Development, and Co-Director of the Centre for Econics and Ecosystem Management at the Faculty of Forest and Environment. He takes a keen interest in Bolivia and is honoured for 'his engaged promotion of Bolivian wildlife conservation, especially in the Department of Santa Cruz'. He is also honoured in the scientific names of two plant species.

Iboundji

Smalltongue Toad sp. *Werneria iboundji* Rödel et al., 2004

Iboundji is a mountain in Gabon and the type locality for this toad.

Ida

Ida's Bright-eyed Frog *Boophis idae* **Steindachner**, 1867

Ida Laura Pfeiffer (1797–1858) was a very remarkable Austrian lady. She was 45 before she started travelling, but then made up for lost time. She visited Egypt and Palestine (1842). She circumnavigated the globe (1846–1848), undertaking a second circumnavigation (1851–1854) taking in parts of the Americas she had missed first time round and spending 18 months in the Sunda Islands and Moluccas. She went to explore Madagascar (1856), where she became embroiled in a plot to overthrow the government for which she was expelled from the country. She wrote *Ein Frauenfahrt urn die Welt* (1850). A bird is named after her.

Ihering

Southern Walking Leaf Frog *Phyllomedusa iheringii* **Boulenger**, 1885

Dr Hermann Friedirich Ibrecht von Ihering (sometimes Jhering) (1850–1930) was a German-Brazilian zoologist, malacologist and geologist. He trained as a physician and served in the German army. He went to Rio Grande do Sul, Brazil (1880) and founded the São Paulo Museum (1894), spending 22 years as its first Director (1894–1916). He returned to Germany (1924) and died there. He co-wrote *Catálogos da Fauna Brasileira. As Aves do Brazil* (1907) with his son, Rudolpho Teodoro Gaspar Wilhelm von Ihering (1883–1939). Three mammals, three reptiles and six birds are named after him.

Ikakogi

Glass Frog genus *Ikakogi* Guayasamin et al., 2009

Named after the Ika and Kogi peoples who live in the Sierra Nevada de Santa Marta, Colombia.

Illiman

Illiman's Andes Frog *Psychrophrynella illimani* **De la Riva** & **Padial**, 2007

This frog is named after Nevado Illimani, the second highest mountain in Bolivia.

Indraneil

Indraneil's Night Frog *Nyctibatrachus indraneili* **Biju** et al., 2011

See **Das**.

Inger

Dicroglossid frog genus *Ingerana* **Dubois**, 1987

Toad genus *Ingerophrynus* **Frost** el al., 2006

Inger's Asian Tree Toad *Pedostibes rugosus* Inger, 1958

Inger's Robber Frog *Strabomantis ingeri* **Cochran** & **Goin**, 1961

Wrinkled Ground Frog sp. *Cornufer ingeri* Brown & **Alcala**, 1963 NCR

[JrSYN *Platymantis guentheri*]

Inger's Flying Frog *Rhacophorus gauni* Inger, 1966
Inger's Poison Frog *Ameerega ingeri* Cochran & Goin, 1970
[Alt (Brother) Niceforo's Poison-arrow Frog]
Inger's Wart Frog *Limnonectes ingeri* Kiew, 1978
Inger's Bubble-nest Frog *Philautus ingeri* **Dring**, 1987
Inger's Grassland Frog *Ptychadena ingeri* **Perret**, 1991
Bob Inger's Bush Frog *Raorchestes bobingeri* **Biju** & Bossuyt, 2005

Dr Robert 'Bob' Frederick Inger (b.1920) is an American herpetologist and ichthyologist who is Curator Emeritus of Amphibians and Reptiles, Field Museum, Chicago, having started as a University of Chicago student volunteer. Since the 1950s his special subject has been Southeast Asian (particularly Borneo) herpetology. He co-wrote *Living Reptiles of the World* (1957). Seven reptiles are named after him.

Iova

Rainforest Frog sp. *Cophixalus iovaorum* Kraus & **Allison**, 2009

Bulisa and Didibu Iova, from central Papua New Guinea, were of great assistance on the describers' collecting expedition.

Iquito

Snouted Treefrog sp. *Scinax iquitorum* Moravec et al., 2009

This frog is named after the Iquito native Peruvian tribe.

Irfan

Irfan's Lycian Salamander *Lyciasalamandra irfani* Göçmen, Arikan & Yalçinkaya, 2011

Irfan Salih Göçmen (1920–1994) was the senior author's father.

Iris

Western Highlands Treefrog *Litoria iris* **Tyler**, 1962

In Greek mythology, Iris was the goddess of the rainbow and messenger of the goddess Hera. In zoology, *iris* is usually applied to very colourful species (e.g. the Iris Lorikeet *Psitteuteles iris*). Six birds and a mammal are named after her.

Isern

Rio Perene Toad *Rhinella iserni* **Jiménez de la Espada**, 1875

Juan Isern Batlló y Carrera (1821–1866) was a Spanish botanist on the staff of the Madrid botanical gardens. He was on the Royal Spanish Pacific Expedition (1862–1865), which went to Ecuador and from there across South America, following the course of the Amazon, and thence home to Spain. The expedition made such a huge collection that up to 200 porters were needed to carry it and other equipment. Isern contracted an infection during the latter part of the expedition and died shortly after arriving home (Madrid).

Ishikawa

Ishikawa's Frog *Odorrana ishikawae* **Stejneger**, 1901

Professor Dr Chiyomatsu Ishikawa (1861–1935) was a Japanese zoologist at the College of Agriculture, Imperial University, Tokyo, and Curator of the Imperial Museum. He was an early member of Zoological Society of Japan and his lectures to it are legendary. A fish is also named after him.

Iskandar

Chianjur Wart Frog *Fejervarya iskandari* **Veith** et al., 2001
Indonesian Tree Frog sp. *Polypedates iskandari* Riyanto, Mumpuni & McGuire, 2011

Dr Djoko Tjahono Iskandar (b.1950) is an Indonesian zoologist and herpetologist. He has worked at the Institut Teknologi Bandung, Java (1978), where he is Professor of Biometrics and Ecology of Small Vertebrates. After he had obtained his master's degree he was a curator at the Museum Zoologicum Bogoriense. He spent time in France (1980–1984). He wrote *The Amphibians of Java and Bali* (1998). Two reptiles are named after him.

Izecksohn

Izecksohn's Brazilian Treefrog *Xenohyla truncata* Izecksohn, 1959
Izecksohn's Toad *Brachycephalus didactylus* Izecksohn, 1971
Izecksohn's Treefrog *Bokermannohyla izecksohni* **Jim** & **Caramaschi**, 1979
Izecksohn's Bromeliad Frog *Crossodactylodes izecksohni* **Peixoto**, 1983
Izecksohn's Button Frog *Cycloramphus izecksohni* **Heyer**, 1983
Izecksohn's Guanabara Frog *Euparkerella robusta Isecksohn, 1988*
Izecksohn's Robber Frog *Ischnocnema izecksohni* Caramaschi & Kisteumacher, 1989
Brazilian Toadlet sp. *Brachycephalus izecksohni* **Ribeiro** et al., 2005

Dr Eugenio Izecksohn is a Brazilian herpetologist, and a specialist in neo-tropical anurans. He worked for the Department of Animal Biology, Universidade Federal Rural do Rio de Janeiro, where he is now Emeritus Professor.

J

Jackie

Jackie's Reed Frog *Hyperolius jackie* Dehling, 2012

Jacqueline 'Jackie' Mary Buchan is the wife of Gordon Buchan, who donated 3,000 euros to the BIOPAT initiative (supporting biodiversity research). In return, he was given the opportunity to choose the name of the new species.

Jackson, HW

Salamander sp. *Plethodon jacksoni* Newman, 1954
NCR
[JrSyn. *Plethodon wehrlei*]

Dr Herbert W. Jackson (1911–2002) was Associate Professor of Biology at the Virginia Polytechnic Institute (1942–1950) and was Chief Biologist with the US Government Public Health Headquarters in Cincinnati (1950–1972).

Jackson, JL

Jackson's Mushroomtongue Salamander *Bolitoglossa jacksoni* Elias, 1984

Jeremy L. Jackson accompanied the describer on his field expedition in Guatemala and collected the holotype.

Jacobson

Jacobson's Bubble-nest Frog *Philautus jacobsoni* **Van Kampen,** 1912

Edward Richard Jacobson (1870–1944) was a Dutch businessman and skilled amateur naturalist. He was manager of a trading company in Java, but he also lived for some years in Sumatra. He made extensive collections for Dutch museums, leaving his business (1910) to devote himself to natural history. His main interest was entomology, but he collected other taxa types too. He died in an internment camp during the Japanese occupation. Two reptiles and two birds are named after him.

Jaeger

Jaeger's Bright-eyed Frog *Boophis jaegeri* **Glaw** & Vences, 1992

Dr Friedhelm Jaeger was a veterinary surgeon in Düsseldorf, Germany, and a friend of the describers. He now works in the veterinary service of the Ministry of Agriculture of North Rhine-Westphalia and became (1995) head of the unit dealing with animal disease control.

Jahn

Jahn's Treefrog *Hyloscirtus jahni* **Rivero**, 1961

Alfredo Jahn Hartmann (1867–1940) (his mother's maiden name is used as his second last name) was a famous man in Venezuela, known to many as *Papa Jahn*. The third largest cave in Venezuela, and the area surrounding it, was named after him. He collected plant specimens for the Smithsonian's Department of Botany. He also climbed the highest peak (Pico Bolivar) in the Venezuelan Sierra Nevada Mountains (1910). The etymology confirms that the frog is named after the man and not the locality.

Jaime

Jaime's Robber Frog *Pristimantis jaimei* **Lynch**, 1992

Jaime Ramírez Avila. The etymology reads: 'Named for my friend, Jaime Ramírez A, of the Sueros Section of the Instituto Nacional de Salud, Bogotá, who collected most of the material of the species'.

Jarujin

Jarujin's Flying Frog *Rhacophorus jarujini* **Matsui** & **Panha**, 2006

Jarujin's Stream Frog *Limnonectes jarujini* Matsui et al., 2010

Dr Jarujin Nabhitabhata is a Thai biologist who was Curator of Reptiles and Amphibians, Ecological Research Division, Thailand Institute of Scientific and Technological Research. He is currently Director, Natural History Museum, Bangkok. He co-wrote *A Photographic Guide to Snakes and Other Reptiles of Peninsular Malaysia, Singapore and Thailand* (1998). A reptile is named after him.

Jasper

Cayey Robber Frog *Eleutherodactylus jasperi* Drewry & Jones, 1976

Dr Jasper James Loftus-Hills (1946–1974) was an Australian herpetologist who was working in the USA at the Museum of Zoology, University of Michigan, when he was killed in a car crash. Before going to Michigan he was in the Section of Neurobiology and Behaviour, Cornell University, New York, and while there wrote *Analysis of an acoustic pacemaker in Strecker's chorus frog,* Pseudacris streckeri. (1974). He was one of the collectors of the frog holotype (1973).

Jaume

Orange Long-nosed Frog *Eleutherodactylus jaumei* **Estrada** & Alonso, 1997

Miguel Luis Jaume García (1905–1990) was a Cuban zoologist, entomologist and malacologist at the Museo y Biblioteca de Zoología de la Habana.

He co-wrote *Catálogo descriptivo de los anfibios y reptiles de Cuba* (1984). A gastropod is named after him.

Jaunsar

Jaunsar Stream Frog *Amolops jaunsari* Ray, 1992

The Jaunsari are an ethnic group in Uttar Pradesh, India, where this frog is found.

Jayaram

Jayaram's Bush Frog *Raorchestes jayarami* **Biju** & Bossuyt, 2009

K. Jayaram (b.1952) is an internationally acclaimed, Chennai-based wildlife photographer who has travelled extensively with the senior author. A spider is also named after him.

Jeanba

Climbing Frog sp. *Anodonthyla jeanbai* Vences et al., 2010

Jean Baptiste 'Jean-Ba' Ramanamanjato is a Malagasy naturalist and conservationist who provided logistical information for an expedition to the heights of Andohahela National Park. He is employed by a subsidiary of the Rio Tinto mining group in the company's environmental and conservation team. He co-wrote *Panther Chameleon: Color Variation, Natural History, Conservation and Captive Management* (2004).

Jeanne

Spanish Painted Frog *Discoglossus jeanneae* Busack, 1986

Jeanne A. Busack née Visnaw is the wife of the describer, and accompanied him during four months of fieldwork in Spain and Morocco (1982).

Jeet Sukumaran

Jeet Sukumaran's Slender Toad *Ansonia jeetsukumarani* Wood et al., 2008

Jeet Sukumaran is a herpetologist whose bachelor's degree was awarded (1998) by the University of Texas, Austin, and his master's (2005) by the University of Malaysia, after which he became a PhD student at the Department of Ecology and Evolutionary Biology, University of Kansas. The etymology states: 'the specific epithet honors Jeet Sukumaran for his past, present and future contributions to our knowledge of the biology of Malaysia's amphibians and for his untiring and invaluable contributions as a collaborator.'

Jefferson

Jefferson's Salamander *Ambystoma jeffersonianum* Green, 1827

The type specimen was collected close to Jefferson College, Pennsylvania.

Jefferson, T

Komodo Cross Frog *Oreophryne jeffersoniana* **Dunn**, 1928

Virginia Seal Salamander *Desmognathus monticola jeffersoni* **Hoffman**, 1951

Thomas Jefferson (1743–1826) was the third President of the United States (1801–1809) (so much is written elsewhere that a biography is unnecessary). The type specimen of the frog was collected on 4 July 1926 – the date inspired Dunn to name it thus: 'I have named this frog in honor of Thomas Jefferson, author of the Declaration of Independence of the United States of America, which was signed on July 4.' The etymology for the salamander reminds us that Jefferson was '... a naturalist of note in his day as well as statesman and educator.'

Jelski

Acancocha Water Frog *Telmatobius jelskii* **Peters**, 1873

Professor Constantin (Konstanty) Roman Jelski (1837–1896) was a Polish naturalist connected with the Zoological Museum of Warsaw. He collected widely, including in Peru as correspondent for the Museum. He was Curator, National Museum (1874–1878), where many of his specimens are still displayed. He returned to Poland and was Curator, Krakow Museum (1878–1896). One mammal and ten birds are named after him.

Jerdon

Jerdon's Bullfrog *Hoplobatrachus crassus* Jerdon, 1854

Jerdon's Narrow-mouthed Frog *Ramanella montana* Jerdon, 1854

Jerdon's Circular-flapped Frog *Ombrana sikimenses* Jerdon, 1870

[Alt. Sikkimese Frog]

Jerdon's Bush Frog *Philautus jerdonii* **Günther**, 1876

Thomas Claverhill Jerdon (1811–1872) was a British physician, zoologist and botanist. He studied medicine at Edinburgh and became an Assistant-Surgeon in the East India Company. He started collecting birds almost immediately on arrival in India (1836). He wrote *Birds of India* (1862–1864). Twenty-three birds, eight reptiles and a mammal are named after him.

Jeude

Garman New Guinea* Treefrog *Litoria jeudii* **Werner**, 1901

Dr Theodor Gerard van Lidth de Jeude (1788–1863) was a Dutch zoologist. He was Director (1821–

1826) of the Rijks Veeartsenijschool, the first school where veterinary medicine was taught in Holland. He assembled a large collection of natural history specimens, which became known as the Museum van Lidth de Jeude.

* The common name originates from someone who almost certainly meant to write *German* New Guinea Treefrog - as the species is known only from the imprecise type locality 'German New Guinea'. Their typo has misled other authors, thus promulgating a meaningless vernacular name.

Jim

Jim's White-lipped Frog *Leptodactylus viridis* Jim & Spirandeli-Cruz, 1973

Treefrog sp. *Dendropsophus jimi* **Napoli** & **Caramaschi**, 1999

Humaita Humming Frog *Chiasmocleis jimi* Caramaschi & **Cruz**, 2001

Toad sp. *Rhinella jimi* Stevaux, 2002 [Syn. *Chaunus jimi*]

Dr Jorge Jim is a herpetologist, professor and Head of the Department of Zoology, Universidade Estadual Paulista Júlio de Mesquita Filho, São Paulo, Brazil, where his private collections are housed. His bachelor's degree on agronomy was awarded (1966) by the Universidade Federal Rural do Rio de Janeiro, and both his master's (1970) and doctorate in zoology (1981) by the Universidade de São Paulo. He has been honoured 'for his contribution to the knowledge of Brazilian herpetology, especially anurans'.

Jiménez

Jiménez's Robber Frog *Pristimantis lacrimosus* Jiménez de la Espada, 1875

Marcos Jiménez de la Espada (see **Espada**).

Jiménez, J

Treefrog sp. *Hypsiboas jimenezi* **Señaris** & **Ayarzagüena**, 2006

Juan Jiménez runs Campamento Parakaupa, a guesthouse on Canaima Lagoon, Venezuela. His friendship and help were essential to the describers.

Joanna

Treefrog sp. *Dendropsophus joannae* **Köhler** & **Lötters**, 2001

Mrs Jo Ann Oxley-Foster of Prescott, Arizona, manages a blues-rock band called Muddbone and is honoured through BIOPAT 'in recognition of supporting taxonomic research and nature conservation'. She apparently loves nature but won't go anywhere 'where the spiders are bigger than the birds'.

Joaquin

Treefrog sp. *Hypsiboas joaquini* **Lutz**, 1968

This amphibian is named after São Joaquim, Brazil, where the holotype was collected.

Johannesde

Urrao Robber Frog *Pristimantis johannesdei* **Rivero** & **Serna**, 1988

John (hence Johannes) D. Lynch (see **Lynch**).

John C (Jacob)

Johncee's Bush Frog *Raorchestes johnceei* Zachariah et al., 2011

Professor John C. (Johnsi) Jacob (1936–2008) was born in Kerala, India and was one of the pioneers of the environmental movement there. He studied zoology at Madras Christian College, who awarded his bachelor's degree. He worked at Payanur College until retirement (1972–1992), eventually becoming Head of the Zoology Department. An ardent naturalist, he devoted his life to conserving the biodiversity of the Western Ghats. He wrote a number of books including *Urangunnavarude Thazhvaraka*. The etymology reads: 'The species is named after Late Prof. John C. Jacob (who was popularly known as 'Johncee'), in honor of his nature education movement initiatives in Kerala.' At least one plant species is named after him.

John (Moore)

Moore's Frog *Lithobates johni* **Blair**, 1965

Dr John Alexander Moore (see **Moore**). Blair originally (1947) named this frog *Rana moorei* but this proved invalid owing to an earlier usage of *moorei* (1942), so he had to find a replacement name. He took Moore's first name.

Johns

Johns' Frog *Rana johnsi* **Smith**, 1921

John Francis Johns (1885–fl.1967) was a student interpreter in Siam (Thailand) (1907); Vice-Consul (1915) and then Consul (1921) Senggora District, Thailand; Acting British Consul, Saigon (Ho Chi Minh City), Vietnam (1919); and Consul-General and Chargé d'Affairs, Bangkok (1924–1932), retiring on grounds of ill health. The frog's describer made a collecting trip to Annam (1917), and noted Johns' assistance 'in arranging many details of our expedition'. A bird is named after him.

Johnson

Johnson's Horned Treefrog *Hemiphractus johnsoni* **Noble**, 1917

R. D. O. Johnson was a mining engineer and amateur naturalist working in Colombia in the early twentieth century. He wrote *Notes on the habits of a climbing catfish (*Arges marmoratus*) from the Republic of Colombia* (1912). He collected the treefrog holotype.

Johnston

Johnston's Treefrog *Leptopelis flavomaculatus* **Günther**, 1864

[*Hylambates johnstoni* (Boulenger, 1897) is a junior synonym]

Johnston's Water Frog *Petropedetes johnstoni* **Boulenger**, 1888

Johnston's River Frog *Amietia johnstoni* Günther, 1894

Sir Harry Hamilton Johnston (1858–1927) was a British explorer, colonial administrator, linguist, painter, cartographer, naturalist and writer. He was a larger-than-life character and became known as the 'Tiny Giant', being just five feet tall. He started exploring tropical Africa (1882) and travelled over most of the continent south of the Sahara, including meeting Henry Morton Stanley in the Congo (1883). He joined the colonial service (1885), serving in all parts of Africa and established a British Protectorate in Nyasaland (Malawi). Johnston was Queen Victoria's first Commissioner and Consul-General to British Central Africa. He spoke over 30 African languages, as well as Arabic, Italian, Spanish, French and Portuguese. He made the first Edison cylinder recordings in Africa, which have preserved his squeaky voice for posterity. He stood for parliament for the Liberal Party (1903 & 1906), but was unsuccessful on both occasions. He discovered 100 new species, the Okapi being the most memorable. Six mammals, two reptiles and six birds are named after him.

Johnstone

Johnstone's Robber Frog *Eleutherodactylus johnstonei* **Barbour**, 1914

Sir Robert Stewart Johnstone (1855–1936) was a solicitor and colonial administrator who became Chief Justice, Grenada and the Grenadines (1909). Barbour wrote that he '…very kindly aided the members of the Museum's expedition while at Grenada.' He was Commissioner for Lagos Western District (late nineteenth century) and was an administrator in the Bahamas (early twentieth century). He donated ethnographic items from Nigeria and the West Indies to the British Museum (1923–1929). A bird is named after him.

Joly, AB

Joly's Frog *Leptodactylus jolyi* **Sazima** & **Bokermann**, 1978

Aylthon Brandao Joly (1924–1975) was a Brazilian botanist whose bachelor's degree (1945) and doctorate (1950) were awarded by the University of São Paulo. He joined the university's Department of Botany (1944), rising eventually to Professor (1973), being Head of the Botany Department and Vice-Director, Institute of Biosciences, University of São Paulo. He founded and was first Director, Department of Botany, Universidad Estadual de Campinas. He worked with Sazima and together they wrote *On the occurrence of Vaucheria longicaulis Hoppaugh (Vaucheriaceae, Chrysophyta) in the American South Atlantic* (1970). He died of cancer.

Joly, CA

Joly's Snouted Treefrog *Scinax jolyi* **Lescure** & **Marty**, 2000

Dr Carlos Alfredo Joly is a Brazilian botanist who followed in his father's footsteps (see above). The University of São Paulo awarded his bachelor's degree in biology (1976), the State University, Campinas his master's in plant biology (1979) and the University of St Andrews, Scotland his doctorate in botany (1982). He did post-doctoral study at the University of Bern, Switzerland (1993–1994). He became a full Professor of Plant Ecology at the State University, Campinas (1997).

Jordan

Jordan's Dwarf Frog *Physalaemus jordanensis* **Bokermann**, 1967

This frog is named after the type locality, Campos do Jordão, São Paulo, Brazil.

Jordan, DS

Jordan's Casque-headed Treefrog *Trachycephalus jordani* **Stejneger** & Test, 1891

Jordan's Salamander *Plethodon jordani* Blatchley, 1901 [Alt. Red-cheeked Salamander]

Dr David Starr Jordan (1851–1931) was a physician, biologist and ichthyologist who had studied under Agassiz. He graduated (1872) as the only undergraduate in Cornell's history to be awarded a master's degree *instead* of a bachelor's. He graduated as a physician at Indiana Medical College (1875). He was Professor of Biology, Butler University, Indiana (1875–1879) and Professor of Natural History, Indiana State University (1879–1885), being President of that university (1885–1891) and first President, Indiana Academy of Science (1886). He was President of Stanford, California (1900) and was the Commissioner representing the United States on the International Commission studying the fisheries of the US/Canadian Boundary (1908). He promulgated Jordan's Law: *'The species most closely related to another is found just beyond a barrier to distribution.'* He wrote *Manual of the Vertebrates of Northern US* (1876). He had the quaint notion of refusing to learn his students' names on the grounds that he would forget the name of a fish for every student's name he learned.

Jordan, HEK

Congulu Forest Treefrog *Leptopelis jordani* **Parker**, 1936

Toad sp. *Bufo jordani* Parker, 1936 NCR [JrSYN *Poyntonophrynus hoeschi*]

Dr Heinrich Ernst Karl Jordan (1861–1959) was a German entomologist. After graduating at Georg-August-Universität Göttingen (1886), he taught in a secondary school until he went to work at Rothschild's Museum, Tring. Thereafter, he lived in England (1893–1959). He led an expedition to South-West Africa (Namibia) and Angola in the early 1930s. He was President, Entomological Society of London (1929–1930). He described 2,575 new species, plus a further 851 as a collaborator with the Rothschilds. A reptile is named after him.

Jorge Velosa

Robber Frog sp. *Pristimantis jorgevelosai* **Lynch**, 1994

Jorge Luis Velosa Ruiz (b.1949) is a Colombian composer-musician, credited with being the co-initiator of the musical genre known as 'carranga'. He graduated as a veterinarian, but turned to acting, poetry and music.

Juan

Poison Frog sp. *Allobates juanii* Morales, 1994

Dr Juan Arturo Rivero-Quintero (see **Rivero** for biographical details).

Juancho

Robber Frog sp. *Pristimantis juanchoi* **Lynch**, 1996

Juan 'Juancho' Castro collected the holotype.

Juanita

Juanita's Earless Treefrog *Exerodonta juanitae* Snyder, 1972

Juanita Snyder is the wife of the describer, David Snyder. His etymology reads: 'The specific epithet is the genitive Latin form of Juanita, after my wife, who has been of invaluable assistance in the field on many expeditions.'

Juarez

Sierra Juarez False Brook Salamander *Pseudoeurycea juarezi* Regal, 1966

This salamander is named after the type locality, the Sierra de Juárez, Oaxaca, Mexico.

Julia Wurstner

African Torrent Frog sp. *Petropedetes juliawurstnerae* Barej et al., 2010

Julia A. Wurstner is a photographer who collected in Cameroon with Barej (2007).

Julian, F

Treefrog sp. *Dendropsophus juliani* Moravec, Aparicio & **Köhler**, 2006

(See entry for Dr Julian **Faivovich**)

Julian, L

Maya Mountain Frog *Lithobates juliani* Hillis & de Sá, 1998

Julian C. Lee is Professor of Biology at the University of Miami. The etymology says: 'This species is named in honor of the collector of the first specimens of the species, Julian C. Lee. This patronym is especially appropriate in view of the large toothy mouth of the tadpoles of this species.' Poor Julian must be noted for his big, toothy mouth! He collected the specimens in Belize, in 1974. They were not recognised at the time as representing a new species. His bachelor's degree is from the University of California (1966), his master's from San Diego (1973) and his doctorate from the University of Kansas (1977). Apart from numerous scientific articles, he wrote *The Amphibians & Reptiles of the Yucatán Peninsula* (1996) and *A Field Guide to the Amphibians and Reptiles of the Maya World: The Lowlands of Mexico, Northern Guatemala, and Belize* (2000). In addition to herpetology, he enjoys bird watching, and is attempting to master the guitar.

Junaluska

Junaluska Salamander *Eurycea junaluska* Sever, **Dundee** & Sullivan, 1976

The describers were inspired by Junaluska's statue and this salamander is named after it, rather than the subject who was a notable Cherokee chief (c.1775–1858). According to tradition, President Andrew Jackson is said to have told Junaluska: 'As long as the sun shines and the grass grows, there shall be friendship between us, and the feet of the Cherokee shall be toward the east.' But the chief later had cause to say: 'If I had known that Jackson would drive us from our homes, I would have killed him that day at the Horseshoe' (i.e. the Battle of Horseshoe Bend, 1814).

Juncá

Snouted Treefrog sp. *Scinax juncae* Nunes & **Pombal**, 2010

Dr Flora Acuña Juncá is a Brazilian herpetologist at the Department of Biological Sciences, Universidade Estadual de Feira de Santana, Bahia, Brazil, where she is a professor.

Junor

Spaldings Robber Frog *Eleutherodactylus junori* **Dunn,** 1926

Percy Sigismund Junor (b.1880) was a Jamaican planter, financier, businessman and Justice of the Peace, who owned an estate called Spaldings. He was born into relative poverty and his education was limited, but he became a trader in agricultural products, especially ginger, coffee, pimento, oranges and bananas. Dunn spent ten days in Jamaica (1925) and in that time he discovered seven new species of the genus *Eleutherodactylus*, all endemic to Jamaica, and named most of them after Jamaican naturalists or worthies, Junor being one of them. This was only fair as Dunn had stayed with him.

Justiniano

El Chape Toad *Rhinella justinianoi* **Harvey** & Smith, 1994

Hermes Justiniano is a conservationist and photographer who has worked for over 15 years in Bolivia's protected reserves. He is Director, Foundation of Friends of Nature, Santa Cruz de la Sierra, Bolivia, and of the Foundation for Conservation of Chiquitano Forest. He wrote *Biodiversity Portraits of Bolivia* (2006).

K

Kabisch

Asian Toad sp. *Bufo kabischi* Herrmann & Kühnel, 1997 NCR

[JrSYN *Bufo gargarizans andrewsi*]

Dr Klaus Kabisch (b.1933) is Professor at the University of Leipzig and Director, Interdisciplinary Institute for Nature and Environment Protection. He wrote *Wörterbuch der Herpetologie* (1990).

Kachowski

Sand Frog sp. *Tomopterna kachowskii* Nikolsky, 1900

A. W. Kachowski explored in Abyssinia (Ethiopia) and Somaliland (Somalia) (1898–1899).

Kadarsan

Nusa Tenggara Wart Frog *Limnonectes kadarsani* **Iskandar**, Boeadi, & Sancoyo, 1996

Dr Sampurno Kadarsan (b.1929) is a zoologist and a former Director, Museum Zoologicum Bogoriense, Indonesia (1960–1962, 1964–1968 & 1971–1977). He was Director of the Bogor Botanical Gardens (1987–1990).

Kaie

Kaie Rock Frog *Anomaloglossus kaiei* Kok et al., 2006

Kaie was a legendary great chieftain of the Patamona tribe in Guyana. The Kaieteur Falls are named after him as according to legend he canoed over them, sacrificing himself to the Great Spirit to save his people from the warlike Carib tribe.

Kaikatt

Kaikatt's Bush Frog *Raorchestes kaikatti* **Biju** & Bossuyt, 2009

[Syn. *Philautus kaikatti*]

This frog is named after the type locality, Kaikatti-Nelliyampathi, Kerala, India.

Kaiser

Kaiser's Spotted Newt *Neurergus kaiseri* **Schmidt**, 1952

[Alt. Luristan Newt, Emperor Spotted Newt]

E. W. Kaiser was a naturalist who collected in Iran for the Zoological Museum, University of Copenhagen, until the late 1940s. He is recorded collecting at Shah Bazan, the site where the newt holotype was taken. He also discovered the Iranian Cave Fish *Iranocypris typhlops*, and co-described it (1944).

Kalm

New Jersey Chorus Frog *Pseudacris kalmi* Harper, 1955

Pehr Kalm (1716–1779) was of Finnish descent but was born in Sweden, where his parents had gone in order to avoid Russian oppression. The family returned to Finland (1721) where Pehr graduated from the University of Åbo (1735) and then became a pupil of Linnaeus in Uppsala. He travelled with Baron Bielke, for whom he worked for seven years, to St Petersburg and Moscow. He became Professor of Ecology at the University of Åbo (1747), then went to America (1748), where he became friends with both Benjamin Franklin and James Bartram. He travelled in New York state (visiting Lake Champlain) and Canada (1749–1750). He returned to Åbo (1751) with a large collection of North American fauna and flora and founded its botanical gardens. He wrote *Travels in North America* (1753–1761). Harper named the frog 'In commemoration of Peter Kalm...the accomplished author...and a pioneer student of frogs in New Jersey.'

Kampen

Bander Baru Frog *Hylarana kampeni* **Boulenger**, 1920

Moroka Cross Frog *Oreophryne kampeni* **Parker**, 1934

Pieter Nicolaas van Kampen (1878–1937) was a Dutch herpetologist and ichthyologist who was Professor of Zoology, Leiden University, until his retirement (1931). He wrote *The Amphibia of the Indo-Australian Archipelago* (1923).

Kaplan

Kaplan's Garagoa Treefrog *Dendropsophus stingi* Kaplan, 1994

[Alt. Sting's Treefrog]

Melissa Kaplan describes herself as an educator, author and activist. Her master's degree was awarded by Sonoma State University. She runs a number of herpetological websites and organisations. She has been associated with the School of Biological Sciences, University of Nebraska, Lincoln. She described the treefrog in the article *A new species of frog of the genus* Hyla *from the Cordillera Oriental in northern Colombia with comments on the taxonomy of* Hyla minuta (1994).

Karelia

Robber Frog sp. *Pristimantis kareliae* **La Marca**, 2005

Karelia La Marca is the describer's daughter.

Karelin

Karelin's Newt *Triturus karelinii* **Strauch**, 1870
[Alt. Southern Crested Newt]

Grigori Silich Karelin (or Grigorij Silyč Karelin) (1801–1872) was a Russian naturalist and collector. He explored in Central Asia from his first expedition (1832) to when he made what is regarded as his great expedition (1840), involving a number of voyages of discovery on the Caspian Sea and penetrating further into Cossack territory than any Russian had done before. A bird is named after him.

Karen Anne

Vaupes Treefrog *Scinax karenanneae* **Pyburn**, 1993

Dr Karen Anne Pyburn is the describer's daughter. She is Professor of Anthropology, Indiana University, Bloomington. Her bachelor's degree is from Reed College, Portland, Oregon (1977) and her master's from the University of Arizona, Tucson (1984), as is her doctorate (1988). Most of her primary research has taken place in Belize, where she has directed major excavations and general mapping and reconnaissance in the remains of three ancient Mayan cities.

Karl

Karl's Robber Frog *Eleutherodactylus karlschmidti* **Grant**, 1931

(See under **Schmidt, KP**) Three reptiles are named after him.

Karunaratne

Karunaratne's Narrow-mouthed Frog *Microhyla karunaratnei* Fernando & Siriwarrdhane, 1996

G. Punchi Banda 'Karu' Karunaratne (1930–1996) was a Sri Lankan zoologist and entomologist who was Curator of Entomology at the Sri Lankan National Museum, Colombo. He wrote *Fauna of the Samanalawewa Area* (1992). A reptile is named after him.

Kassas

Nile Valley Toad *Amietophrynus kassasii* Baha El Din, 1993
[Alt. Damietta Toad; Syn. *Bufo kassasii*]

Dr Mohammed Abdel Fattah El-Kassas (1921–2012) was an Egyptian conservationist and botanist who was Professor of Applied Botany, Cairo University, where he had his early education, after which he studied for a doctorate in biology at Cambridge. He taught at the University of Khartoum, Sudan (1964–1968). It was through his actions that Egypt's first National Park was established (1983). Baha El Din, who collected the holotype (1987), regarded Professor Kassas highly for his work on soils and his understanding of landscape. He co-wrote *Studies on the ecology of the Red Sea coastal land, II: The district from El-Galala El-Qibliya* (1965).

Kathrin

Madagascar Frog sp. *Guibemantis kathrinae* **Glaw**, Vences & Gossmann, 2000

Kathrin Schmidt. Glaw wrote that the frog was: 'Dedicated to Kathrin Schmidt, Bonn, in recognition for her invaluable help with our research on Malagasy amphibians and reptiles; her call recordings from Andapa helped to clarify the relationships and distribution of the species of the *M. depressiceps* complex.'

Kato

Akaishi Salamander *Hynobius katoi* **Matsui** et al., 2004

Dr Makoto Kato (b.1932) is Professor of Human and Environmental Studies at Kyoto University, Japan. He was trained in botany and entomology. While still a student, he collected the first specimen of this salamander.

Kaulback

Kaulback's Torrent Frog *Amolops kaulbacki* **Smith**, 1940

Lieutenant-Colonel Ronald John Henry Kaulback (1909–1995) was a British botanist and explorer. He collected herpetofauna in Upper Burma (Myanmar) (1930s). He went to Tibet and China with Frank Kingdon-Ward (1933), and was again in Tibet, with J. Hanbury-Tracy (1935–1936). He wrote *Tibetan Trek* (1936). Two reptiles are named after him.

Kaune

Andes Frog sp. *Phrynopus kauneorum* **Lehr, Aguilar** & **Köhler**, 2002

Andreas, Meeta and Rebekka Kaune were honoured for their support of the BIOPAT programme.

Kaup

Kaup's Caecilian *Potamotyphlus kaupii* Berthold, 1859

Johann Jakob von Kaup (1803–1873) was a German zoologist and palaeontologist. He was Director of the grand-ducal natural history 'cabinet', Darmstadt. He wrote *Classification der säugethiere und vögel* (1844). A bird genus and bird species are named after him.

Kautsky

Brazilian Heart-tongued Frog *Phyllodytes kautskyi* **Peixoto** & **Cruz**, 1988

Kautsky's Snouted Treefrog *Scinax kautskyi* Carvalho-e-Silva & Peixoto, 1991

Dr Roberto Anselmo Kautsky (1924–2010) was a self-taught Brazilian botanist and orchid expert. His doctorate from the University of Rio de Janeiro was honorary. He and a group of friends founded the 'Kautsky Institute' (2003) to continue his work on the preservation and restoration of the flora of Espirito Santo state, southeast Brazil.

Keferstein

Keferstein's Toadlet *Pseudophryne coriacea* Keferstein, 1868

[Alt. Red-backed Toadlet]

Keferstein's Tree Frog *Litoria dentata* Keferstein, 1868

[Alt. Bleating Treefrog]

Wilhelm Moritz Keferstein (1833–1870) was a German naturalist. He was Professor of Zoology and Comparative Anatomy at Georg-August-Universität Göttingen. He wrote *Zoologische Beiträge gesammelt im Winter 1859/60 in Neapel und Messina* (1861). He described a number of reptiles as well as amphibians.

Keiling

Keiling's Ridged Frog *Ptychadena keilingi* Monard, 1937

R. P. Keiling was a missionary and amateur herpetologist at Cubango, Angola.

Keith

Keith's Toad *Amietophrynus kerinyagae* Keith, 1968

Keith's Striped Frog *Phlyctimantis keithae* **Schiøtz**, 1975

Ronalda Keith is a herpetologist who collected in Africa (1962) for the AMNH, New York. She co-wrote *Bufo of Africa* (1972).

Kelaart

Kelaart's Toad *Adenomus kelaartii* **Günther**, 1858

Lieutenant-Colonel Edward Frederick Kelaart (1819–1860) was a physician and zoologist born in Ceylon (Sri Lanka). He qualified at Edinburgh and also studied in Paris. He was in the Ceylon medical service but also served in Gibraltar (1843–1845). He was in Ceylon (1849–1854) then England (1854–1856). On his return he was appointed the Ceylon Government Naturalist, which paid £200 p/a (a lot of money then) plus expenses on top of his army pay! One task was to investigate why the Ceylon Pearl Fisheries were profitless. He investigated, and wrote four reports on the life history of the pearl oyster (1858–1863), the latter ones published posthumously. The Governor of Ceylon sailed for England (1860); his health was of such concern that Kelaart, with his wife and five children, accompanied him as personal medical attendant. The Governor died two days before 'Nubia' arrived at Southampton and Kelaart died the following day. He wrote *Prodromus fauna Zeylanica* (1852). Three birds, two reptiles and three mammals are named after him.

Keller

Keller's Foam-nest Frog *Chiromantis kelleri* **Böttger**, 1893

Professor Dr Conrad Keller (1848–1930) was a German herpetologist who studied zoology in Zurich, Jena and Lausanne. He went on Eugenio Ruspoli's expedition to Somaliland (1893). He taught in a high school in Zurich (1898–1905) and from 1905 in a Berlin high school. He wrote *Die Ostafrikanischen Insen* (1898). A reptile is also named after him.

Kelley

Kelley's Island Salamander *Ambystoma nothagenes* Kraus, 1985 NCR

[Hybrid population: *Ambystoma texanum* x *Ambystoma laterale*]

This salamander is named after the type locality, Kelley's Island, Erie County, Ohio.

Kellogg

Little Mexican Toad *Anaxyrus kelloggi* **Taylor**, 1938

Arthur Remington Kellogg (1892–1969) studied mammalogy at the Universities of Kansas and California, specializing in marine mammal evolution. He joined the Bureau of Biological Survey, US Department of Agriculture, as Assistant Biologist (1920). He had a research appointment at the Carnegie Museum (1921–1943). He joined the Smithsonian (1928) as Assistant Curator, Mammals, becoming Curator (1941). He was Director of the US National Museum (1948–1962) and was also an Assistant Secretary of the Smithsonian (1958–1962), holding after retirement an emeritus position with the Smithsonian in the Division of Vertebrate Palaeontology (1962–1969). He was involved in the international regulation of whaling (1930–1967), serving as delegate to many conferences including the League of Nations whaling conference (1930). He wrote *Mexican tailless amphibians in the United States National Museum* (1932). A mammal is named after him.

Kelly

Kelly's Salamander

Not an amphibian, but a make of bicycle!

Kemp

Kemp's Bush Frog *Philautus kempii* **Annandale**, 1912
Garo Hills Tree Toad *Pedostibes kempi* **Boulenger**, 1919
Tura Bubble-nest Frog *Philautus kempiae* Boulenger, 1919
[Alt. Kemp's Bush Frog*]

Dr Stanley Wells Kemp (1882–1945) (and his wife). Kemp was a zoologist and anthropologist. He joined the Fisheries Research Section, Department of Agriculture, Dublin as Assistant Naturalist (1903). He joined the Indian Museum, Calcutta (1911) as Superintendent, Zoological Section. There he worked very closely with Annandale; they became great friends. He later wrote Annandale's obituary (1925) in the *Records of the Indian Museum*. He was on the Abor Punitive Expedition (1911–1912) during which government scientists made extensive collections in the Garo Hills. Boulenger noted that both the species he described were 'obtained in the Garo Hills, by Mr. and Mrs. S. W. Kemp' so he named one after the doctor and one after his wife (*kempiae* being feminine). Kemp later joined the Colonial Office (1924) as Director of Research, Discovery Committee. He was Director of the Plymouth Marine Laboratory (1936–1945). He lost all his personal possessions, his library and his unpublished works as the result of a German air raid (1941). A bird is named after him.

* Two species of *Philautus* have at times been given the name Kemp's Bush Frog: they could, perhaps, be known as Mr Kemp's Bush Frog and Mrs Kemp's Bush Frog.

Kempff

Junin Andes Frog *Psychrophrynella kempffi* **De la Riva**, 1992

Professor Noel Kempff Mercado (1924–1986) was a Bolivian biologist and conservationist who, with three companions, was murdered by drug dealers during an expedition to the Huanchaca cliffs. The Natural History Museum, Santa Cruz, Bolivia was named after him and the National Park at Huanchaca renamed. A bird is named after him.

Kennedy

Kennedy's Snouted Treefrog *Scinax kennedyi* **Pyburn**, 1973

Dr J. P. Kennedy of the Biology Department, University of St Thomas, Houston, Texas worked with Pyburn. They co-wrote *Artificial hybridization on the grey treefrog*, Hyla versieolor. *(Hylidae)* (1960). Pyburn honoured Kennedy 'in recognition of his outstanding contributions to the Society for the Study of Amphibians and Reptiles.'

Keyserling

Dybowski's Salamander *Salamandrella keyserlingii* **Dybowski**, 1870
[Alt. Siberian Salamander]

Alexander Friedrich Michael Lebrecht Nikolaus Arthur, Graf von Keyserling (1815–1891) was a Russian geologist, zoologist, botanist and palaeontologist of Baltic-German descent. He travelled in Estonia, Northern Russia and the Urals (1839–1846) at the behest of Tsar Nicholas I. A reptile is named after him.

Kezer

Columbia Torrent Salamander *Rhyacotriton kezeri* Good & **Wake**, 1992

Dr James Kezer (1908–2002) was a biologist and herpetologist who worked in the Department of Biology, University of Oregon (1950s). The University of Iowa awarded his bachelor's (1930) and master's (1937) degrees. His doctorate (1948) was from Cornell.

Khare

Khare's Frog *Pterorana khare* Kiysetuo & Khare, 1986
[Alt. Indian Flying Frog, Winged Frog]

Dr M. K. Khare was a zoologist and herpetologist and a professor in the Department of Zoology, North-Eastern Hill University, India, which offers a memorial prize in his honour.

Khasi

Asian Toad sp. *Leptolalax khasiorum* **Das** et al., 2010

This frog is named after the East Khasi Hills, northeast India.

Kibara

Shaba Reed Frog *Hyperolius kibarae* **Laurent,** 1957

The Kibara Mountains, Katanga Province, Democratic Republic of Congo, lie partly in the Upemba National Park (where the type specimen of this frog was collected).

Kichwa

Ahuano Robber Frog *Pristimantis kichwarum* Elmer & **Cannatella**, 2008

Named after the indigenous Kichwa communities of the Oriente region, Ecuador.

Kimura

Hida Salamander *Hynobius kimurae* **Dunn**, 1923
[Alt. Hondo Salamander]

Dr Arika Kimura (1900–1996) collected the holotype with Wilder (1920). In the same year he also collected a spider that was named after him. He was Professor of Botany at Tohoku University, from which he retired (1963).

King

King's River Slender Salamander *Batrachoseps regius* Jockusch, **Wake** & Yanev, 1998

'King's' is a mistaken use of an apostrophe. In fact it is named after the Kings River, California.

Kinghorn

Kinghorn's Treefrog *Litoria kinghorni** **Loveridge**, 1950

James Roy Kinghorn (1891–1983) was an Australian ornithologist and herpetologist. He was on the staff of the Australian Museum, Sydney (1907–1956), becoming Assistant to the Director after service in the artillery of the Australian army (1915–1918). He became the resident radio and television naturalist for the Australian Broadcasting Corporation (1956). He wrote *The Snakes of Australia* (1929). Two reptiles are named after him.

* This species is often considered a junior synonym of *Litoria wilcoxii* (Günther, 1864) but such synonymy has been questioned.

Kingsbury

Kingsbury's Rocket Frog *Allobates kingsburyi* **Boulenger**, 1918

Isaac James Frederick Kingsbury (d.1918) was Boulenger's attendant at the Natural History Museum, London. He was killed in action in Palestine and Boulenger named this frog after him as a memorial.

Kirk

Kirk's Caecilian *Scolecomorphus kirkii* **Boulenger**, 1883

Dr Sir John Kirk (1832–1922) was a diplomat, explorer and naturalist. He was David Livingstone's chief assistant, physician and naturalist during his second Zambesi Expedition (1858–1863). He became Vice-consul, then Consul General, in Zanzibar (1866–1886). He obtained the local sultan's agreement to a treaty abolishing the slave trade (1873). Six birds, three mammals and a reptile are named after him.

Kirkland

Kirkland's Robber Frog *Pristimantis kirklandi* **Flores**, 1985

Benjamin E. Kirkland. Flores's etymology says: 'Named for Benjamin E. Kirkland, as an expression of gratitude to a most inspirational, devoted, caring high-school biology teacher.'

Kirtisinghe

Kirtisinghe's Frog *Fejervarja kirtisinghei* Manamendra-Arachchi & Gabadage, 1996

Dr Parakrama (originally Edmund Peter) Kirtisinghe (1903–1981) was a malacologist and herpetologist who was Professor of Zoology at the University of Ceylon. He wrote *The Amphibia of Ceylon* (1957).

Klappenbach

Klappenbach's Redbelly Toad *Melanophryniscus klappenbachi* Prigioni & **Langone**, 2000

Dr Miguel Angel Klappenbach Childe (1920–2000) was a Uruguayan zoologist and malacologist at the Museo Nacional de Historia Natural, Montevideo, where he was Director (1970–1984). He led the Uruguayan expedition (1957) to the Orinoco River. He co-wrote *Un Ilustrado Aragones en el Virreinato del Rio de la Plata: Felix de Azara, 1742–1821 Estudios Sobre su Vida, su Obra y su Pensamiento.*

Klauber

Klauber's Blotched Salamander *Ensatina eschscholtzii klauberi* **Dunn**, 1929

[Alt. Large-blotched Ensatina; Syn. *Ensatina klauberi*]

Dr Laurence Monroe Klauber (1883–1968) was an electrical engineer and inventor, who became overall chief of a California utility. He was a keen herpetologist, becoming the world expert on rattlesnakes. Reptiles were still just a hobby when the San Diego Zoo asked him about snakes that they had acquired (1923); he became Curator of Reptiles (1923–1958). He gave around 36,000 specimens and all his notes and his library to the San Diego Natural History Museum. He wrote *Rattlesnakes: Their Habits, Life Histories and Influence on Mankind* (1956). Nine reptiles are named after him.

Klemmer

Klemmer's Madagascar Frog *Gephyromantis klemmeri* **Guibé**, 1974

Dr Konrad Klemmer (b.1930) is a German herpetologist who was Curator of Herpetology at the Senckenberg Museum, Frankfurt-am-Main. Klemmer collected in Morocco and Western Sahara (1960s). Three reptiles are named after him.

Klinikowski

Klinikowski's Robber Frog *Eleutherodactylus klinikowskii* **Schwartz**, 1959

Ronald Francis Klinikowski (1939–2008) was an American herpetologist who accompanied Schwartz on a number of his collecting trips: to

Cuba (1959), Jamaica (1961) and the Dominican Republic (1962). A bird is also named after him. (See also **Ronald**)

Knudsen

Knudsen's Thin-toed Frog *Leptodactylus knudseni* **Heyer**, 1972

Dr Jens Werner Knudsen was a biologist who inspired Heyer to pursue a life of biological research rather than teach in a high school. He wrote *Biological techniques: Collecting, preserving and illustrating plants and animals* (1966).

Knysna

Knysna Spiny Reed Frog *Afrixalus knysnae* **Loveridge**, 1954

This frog is named after the type locality, Knysna, Cape Province, South Africa.

Koba

Koba's Frog *Rana kobai* **Matsui**, 2011

Dr Kazuo Koba (1904–1981) was Professor Emeritus of Kumamoto University, in Japan's Ryukyu Islands. He helped pioneer herpetological surveys of the Amami island group. His collection is in the Osaka Museum.

Kobayashi

Kobayashi's Horned Frog *Megophrys kobayashii* Malkmus & **Matsui**, 1997

Professor Teiichi Kobayashi (1901–1996) was a Japanese geologist and palaeontologist. He studied at the Imperial University in Tokyo (1924–1927). He remained at the university after he had graduated until his retirement (1962). In retirement he continued to conduct research and by 1989 had published nearly 800 books and papers. His works include *Sakawa Orogenic Cycle and its Bearing on the Genesis of the Japanese Island* (1941), which is regarded as a seminal publication. When well into 'retirement' (1970), jointly with H. Abe, K. Maeda, and T. Miyao, he wrote *Faunal Survey of the Mt. Ishizuchi Area, JIBP Main Area—II: Results of the Small Mammal Survey on the Mt. Ishizuchi Area.* A mammal is named after him.

Koechlin

Koechlin's Treefrog *Dendropsophus koechlini* **Duellman** & **Trueb**, 1989

José E. Koechlin von Stein is a businessman and entrepreneur with a number of private hotels and nature reserves, and a board member of the Peruvian National Park Service. He is the founder of Inkaterra, which promotes eco-friendly tourism in Peru. He sponsored Duellman's publication *Cusco Amazonico, the Lives of Amphibians and Reptiles* (2005) and is joint author with him in *Cusco Amazonico Reserva, Peru: Biological Investigations, Conservation and Ecotourism* (1991). The etymology for the treefrog says Koechlin '...has generously provided support for field parties undertaking biological surveys...'

Koehler, G

Worm Salamander sp. *Oedipina koehleri* Sunyer et al., 2011

Dr Gunther Köhler is Curator of Herpetology at the Senckenberg Museum, Frankfurt, Germany, and Head of the Department of Herpetology at the Research Institute there. He is also Director of the Conservation Project for the Utila Iguana, Honduras. He has a particular interest in Spiny-tailed Iguanas. Among his other books and papers he wrote *Reptiles of Central America* (2008). The etymology reads: 'The specific name koehleri is a patronym for our friend and colleague Gunther Köhler, in recognition of his many important contributions to the herpetology of Central America in general and Nicaragua in particular.'

Koehler, J

Robber Frog sp. *Pristimants koehleri* **Padial** & **De la Riva**, 2009

Dr Jörn Köhler is a zoologist and herpetologist with the Department of Natural History–Zoology, Hessisches Landesmuseum Darmstadt, Germany. He wrote *Amphibian diversity in Bolivia: A study with special reference to montane forest regions* (2000).

Koehler, M

Koehler's Green Frog *Chlorolius koehleri* **Mertens**, 1940

Max Köhler collected in West Africa with Dr H. Graf during expeditions for the Senckenberg Museum, Frankfurt (1930s). A reptile is named after him.

Koepcke

Quisto Cocha Caecilian *Oscaecilia koepckeorum* **Wake**, 1984

Professor Hans-Wilhelm Koepcke (1914–2000) and his wife Maria (1924–1971) were both honoured in the name of this amphibian. She was killed in an air crash in the Andes on Christmas Eve, 1971; he was not on board the aircraft but their daughter was and she miraculously survived. He was an ecologist at the Museum of Natural History, San Marcos, whilst Maria, the more famous partner, was known as the mother of Peruvian ornithology. She was born Maria Emilia Ana von Mikulicz-Radecki in Leipzig, Germany. She went to Peru (1950) and collaborated with Hans-Wilhelm on ornithological studies. Four birds and a mammal are named after her and a reptile is named after both of them.

Kohchang

Koh Chang Frog *Limnonectes kohchangae* **Smith**, 1922

Named after the type locality, Koh Chang island in the Gulf of Siam, Thailand.

Kokaci

Catanduanes Island Narrowmouth Toad *Kaloula kokacii* Ross & Gonzales, 1992

The original description contains no etymology and we are unaware of the meaning of this binomial.

Kopstein

Kopstein's Callulops Frog *Callulops kopsteini* **Mertens**, 1930

Dr P. Felix Kopstein (1893–1939) was an Austrian physician and naturalist. He collected in Albania (1914). He studied biology and medicine at the University of Vienna (1913–1920). He worked for the Dutch Government in the Dutch East Indies (Indonesia) (1921), initially in Amboina. He made field trips to New Guinea and acted as local agent for the Leiden Museum. He wrote *Een Zoölogische Reis Door de Tropen* (1930). Three reptiles are named after him.

Kotagama

Kotagama's Toad *Duttaphrynus kotogamai* Fernando & Dayawansa, 1994

Dr Sarath W. K. Kotogama is Professor of Environmental Science at Colombo University, Sri Lanka. His bachelor's degree was awarded in Sri Lanka (1974) and his doctorate by Aberdeen University (1982). He co-wrote *Preliminary studies on some biological and ecological aspects of the tadpole of the endemic frog* Limnonectes kirtisinghei *(Amphibia, Ranidae)* (2001).

Kotosh

Kotosh Andes Frog *Phrynopus kotosh* **Lehr**, 2007

Kotosh is an archaeological site near Huanuco, Peru, which gives its name to a pre-Columbian culture and people (c. 5,000 years ago).

Krauczuk

Krauczuk's Redbelly Toad *Melanophryniscus krauczuki* Baldo & Basso, 2004

Ernesto Rubén Krauczuk is a biologist, naturalist and ornithologist who collected the holotype (2000). He is head of the Department of Fauna and Flora of the Ministry of Ecology of Misiones Province, Argentina. He co-wrote *New record for* Phrynops vanderhaegei *in Argentina* (2000).

Krause

Krause's Tree Toad *Dendrophryniscus krausei* **Cruz** & Fusinatto, 2008

Dr Ligia Steyer Krause is a Brazilian herpetologist. Her bachelor's degree (1971) in biological sciences, her master's (1976) and her doctorate in geosciences (1983) were all awarded by the Universidade Federal do Rio Grande do Sul, where she is now a professor in the Faculty of Biological Sciences & Zoology. The etymology says she is honoured '... in gratitude for her friendship.' She wrote *Notes on biological aspects and reproductive behavior of* Dromicus poecilogyrus *in captivity (Serpentes, Colubridae)* (1988). A reptile is also named after her.

Krebs

Mebebque Frog *Arlequinus krebsi* **Mertens**, 1938

W. Krebs sent the holotype from Mubengue, Cameroon, but we know nothing more about him.

Krefft, JLG

Krefft's Frog *Litoria kreffti* **Günther**, 1863 NCR
[JrSYN *Litoria jervisiensis*]
Australian Frog sp. *Limnodynastes kreffti* Günther, 1863 NCR
[JrSYN *Limnodynastes peronii*]
San Cristobal Frog *Hylarana kreffti* **Boulenger**, 1882
[Syn. *Rana kreffti*]

Johann Ludwig (Louis) Gerhard Krefft (1830–1881) was a German adventurer, artist and naturalist who settled in Australia, where he is regarded as the father of Australian herpetology. He first went to the USA (1851) and worked as an artist in New York, but sailed for Australia (1852) to join the gold rush. He was a miner until he joined the National Museum, Melbourne (1857) as a collector and artist, becoming Assistant Curator (1861) then Curator and Secretary (1864). Bad-tempered, he feuded with the Museum trustees and was dismissed (1874). He refused to accept the dismissal and barricaded himself in his office, having to be carried out of the building, still in his chair, and deposited in the street, the door locked behind him. Feeling he had been unfairly treated, Krefft set up a rival 'Office of the Curator of the Australian Museum' and successfully sued the trustees for a substantial sum of money. That proved the end of his career and he never worked seriously again, but he did write natural history articles for the Sydney press. His most famous discovery was of the Australian Lungfish *Neoceratodus forsteri*. Among over 150 papers and books he wrote *The Snakes of Australia* (1869). Two birds, a mammal and three reptiles are named after him as well as a number of fishes.

Krefft, P

Neotropical Ground Frog sp. *Leptodactylus kreffti* **Werner**, 1904 NCR
[JrSYN *Eupsophus roseus*]
Krefft's River Frog *Phrynobatrachus krefftii* **Boulenger**, 1909
[Alt. Krefft's Puddle Frog]
Krefft's Secret Frog *Callulina kreffti* **Nieden**, 1911

Dr Paul Krefft (1872–1945) was a German physician and herpetologist at the Stuttgart Museum who made three expeditions to Africa, including Tanganyika (Tanzania). He wrote *Reptilien- und Amphibienpflege* (1907). He was the nephew of **Krefft, JLG.**

Krieg

Achala Four-eyed Frog *Pleurodema kriegi* **Müller**, 1926

Professor Dr Hans Krieg (1888–1970) was a physician as well as an ethnographer, anthropologist and zoologist. He was the first President of the DNR (Deutscher Naturschutzring – German Conservation Ring). He was on the German Gran Chaco expedition (1925), made a second expedition to the same region (1932) and a third (1936) after the War of the Chaco. He led an expedition to Patagonia (1937–1938). He wrote *Zwischen Anden und Atlantik. Reisen einen Biologen in Südamerika*. (1948). A reptile and two birds are named after him.

Krishna

Krishna Caecilian *Gegeneophis krishni* **Pillai** & Ravichandran, 1999

This amphibian is named after the type locality - Krishna Farms in Karnataka, India – rather than directly after the Hindu deity.

Krøyer

Kroyer's Dwarf Frog *Physalaemus kroyeri* **Reinhardt** & **Lütken**, 1862

Henrik Nikolai Krøyer (1799–1870) was a Danish marine biologist. In his youth he was idealistic and, like Lord Byron, went south to help the Greeks fight the Turks. He was quickly disillusioned when he found that nineteenth-century Greeks had nothing in common with Pericles, but the journey fired an interest in Mediterranean fauna. He taught school (1827–1830) at Stavanger, Norway. With his wife, he surveyed the Danish coast (1835–1835) in an open boat. He made many collecting expeditions to Norway, Spitzbergen, Madeira and North and South America. He founded the journal *Naturhistorisk Tidsskrift*. Many other taxa are named after him, including some described by Lütken.

Kruger

Kruger Bullfrog *Tomopterna krugerensis* Passmore & Carruthers, 1975
[Alt. Knocking Sand Frog]

This frog is named after the type locality, Kruger National Park, South Africa. The park is named after Paul Kruger (1825–1904) a past President of the Transvaal.

Kubor

Sandy Big-eyed Treefrog *Litoria kubori* **Zweifel**, 1958

This frog is named after the type locality, the Kubor Mountains in Papua New Guinea.

Kuduk

Iagifu Treefrog *Litoria kuduki* **Richards**, 2007
[Syn. *Nyctimystes kuduki*]

Max Kuduk of the Biology Department, University of Papua New Guinea, was a project manager for the World Wide Fund for Nature in PNG. He gave long term support to the author's research in the Kikori Integrated Conservation and Development Project Area.

Kuhl

Kuhl's Frog *Limnonectes kuhlii* **Tschudi**, 1838
[Alt. Big-headed Frog, Kuhl's Creek Frog]

Dr Heinrich Kuhl (1797–1821) was a German naturalist who became an assistant to the famous Dutch zoologist Coenraad Jacob Temminck at the Leiden Museum. He travelled to Java with Johan Coenraad van Hasselt (q.v.) (1820), and sent many specimens back to Leiden, but after a short time in Java he died of a liver infection. Many other taxa, including six mammals, three reptiles and seven birds, are named after him.

Kuliga

Camp Kivu Reed Frog *Hyperolius kuligae* **Mertens**, 1940

Paul Kuliga (1878–1948) was a physician who became Director, Luis Hospital, Düsseldorf, Germany. He was on the expedition to Cameroon (1936) under the leadership of Horst Sieloff, head, Löbbecke Museum, Düsseldorf. The Museum acquired Kuliga's collections after his death.

Kuma

Tree Frog sp. *Litoria kumae* **Menzies** & **Tyler**, 2004

Kuma was the beautiful daughter of Pam, a legendary New Guinea chieftain.

L

Labang

David Labang's Bush Frog *Philautus davidlabangi* **Matsui**, 2009

Encik David Labang collected in Borneo for the University of Malaya (1960s), later becoming an officer of the National Parks and Wildlife Office of Sarawak Forest Department. He now owns and runs a rest house. He supplied support for the describer's field surveys.

Lakim

Indonesian Caecilian sp. *Ichthyophis lakimi* Nishikawa, **Matsui** & Yambun, 2012

Dr Maklarin B. Lakim is in charge of the research division of Sabah Parks, Kota Kinabalu, Malaysia.

La Marca

La Marca's Robber Frog *Pristimantis vanadise* La Marca, 1984

Poison Frog sp. *Mannophryne lamarcai* Mijares-Urrutia & Ahrends-R, 1999

Dr Enrique La Marca is a herpetologist and population biologist at the Department of Biology, University of the Andes, Mérida, Venezuela. He wrote *Crisis de biodiversidad en anfibios de Venezuela: Estudio de casos* (1995).

Lamas

Pasco Poison Frog *Ranitomeya lamasi* Morales, 1992

Dr Gerardo Lamas Müller (b.1948) is a biologist, entomologist and systematist who was Director, Natural History Museum, Universidad Mayor de San Marcos, Lima (1988–1992, re-appointed 2007–2011). His bachelor's degree was awarded in Peru in biological sciences (1969) and he studied for his master's and doctorate (1970–1973) at the University of São Paulo, Brazil. The Smithsonian awarded a second doctorate (1977). He is a commissioner on the International Commission of Zoological Nomenclature.

Lameere

Lameere's Squeaker *Arthroleptis lameerei* **De Witte**, 1921 [Alt. Katanga Screeching Frog]

Professor Auguste Alfred Lucien Lameere (1864–1942) was a Belgian entomologist who was Professor (and Dean 1906–1907) at the Free University, Brussels. He translated a number of Darwin's books into French. He wrote the three-volume *Manuel de la Faune de Belgique* (1895–1900). The Auguste Lameere Zoology Museum, Brussels is named after him.

Lamotte

Ivory Coast Reed Frog *Hyperolius lamottei* **Laurent**, 1958

Rainforest Running Frog *Kassina lamottei* **Schiøtz**, 1967

Mount Oku Caecilian *Crotaphatrema lamottei* **Nussbaum**, 1981

Dr Maxime Lamotte (1920–2007) was a French biologist and geneticist. He collected in West Africa (1942 & 1946), and also trained others to collect and conserve specimens. He was Professor of Zoology at the Faculty of Sciences, Lille (1952–1955), and later at the Faculty of Sciences, Paris. He was President of the Zoological Society of France (1978–1983). Four mammals are named after him.

Lancaster

Lancaster's Treefrog *Isthmohyla lancasteri* **Barbour**, 1928

There is no proper etymology in the original description – just a note that the collector of the holotype was 'C. R. Lancaster' and we have not been able to identify him.

Lancini

Lancini's Robber Frog *Pristimantis lancinii* **Donoso-Barros**, 1965

Dr Abdem Ramon Lancini (1934–2007) was a herpetologist and Director, Museo de Ciencias Naturales de Caracas, Venezuela (1962–1991). He wrote *Serpientes de Venezuela* (1979). Two reptiles are named after him.

Lang

Lang's Reed Frog *Hyperolius langi* **Noble**, 1924

James Paul Lang (1889–1964) was an assistant at the AMNH, New York. He collected the frog holotype during an expedition to the Belgian Congo (1909–1915).

Lange

Marumbi Treefrog *Bokermannohyla langei* **Bokermann**, 1965

Professor Rudolf Bruno Lange (b.1922) is a Brazilian zoologist and botanist who was one of the first curators of the zoological collection at the Paraná Museum, where his collection of arachnids is held. Other taxa, including fishes, are named after him. He collected the frog holotype (1946).

Langer

Screeching Frog sp. *Arthroleptis langeri* Rödel et al., 2009

Detley Langer works as a technical assistant in the herpetological section, Museum für Naturkunde, Berlin.

Langone

Redbelly Toad sp. *Melanophryniscus langonei* Maneyro, Naya & Baldo, 2008

Dr José Antonio Langone Fernández (b.1961) is a Uruguayan herpetologist at the Museo Nacional de Historia Natural, Montevideo. His doctorate (2006) was awarded by the Universidad de la República, Uruguay. He wrote *Anfibios de Aguas Corrientes, Departamento de Canelones, Uruguay (Amphibia)* (1999).

Langsdorff

Ocellated Treefrog *Itapotihyla langsdorffii* **Duméril & Bibron**, 1841

Baron Georg Heinrich (Grigoriy Ivanovich) von Langsdorff (1774–1852) was a German physician, botanist, zoologist, traveller, ethnographer and diplomat. He graduated as a physician at Göttingen (1797). He was elected as a 'corresponding member' of the Academy of Science, St Petersburg (1803), and was on Krusenstern's round-the-world expedition (1803–1806) on board the 'Nadezda'. He continued to travel widely: in Japan (1804–1805), northwestern America (1805–1806) and Kamchatka, Siberia and European Russia (1806–1808). He became Associate Professor in Botany at the Academy (1808), moving to Zoology (1809). He became Russian Consul General in Brazil (1813) and Chargé d'Affaires for Russia to Portugal in Rio de Janeiro (the Portuguese government was in exile during the Napoleonic Wars). He returned to Russia (1821) to organize an expedition in Brazil, which he led (1822–1828), travelling in Nova Briburg, Minas Gerais, São Paulo and Mato Grosso, where he caught a tropical fever leading to a psychological breakdown. He retired to Germany (1831). He wrote *Remarks and Observations on a Voyage Around the World from 1803–1807* (1812). Two birds and a reptile are named after him.

Lantz

Caucasian Smooth Newt *Lissotriton lantzi* **Wolterstorff**, 1914

[Syn. *Triturus vulgaris lantzi*]

Amédée Louis Lantz (1886–1953) was a French herpetologist who provided Wolterstorff with the type specimen of this newt. He studied zoology at Montpellier but became a chemical engineer (1907). He lived in Moscow (from where he sent the type) but moved to Manchester, England, when the Revolution started. He is particularly known for his herpetological study in the Caucasus and French Pyrenees. Among other articles he wrote the paper *Quelques observations nouvelles sur l'herpétologie des Pyrénées centrales* (1927). A bird is also named after him.

Lanza

African Frog genus *Lanzarana* Clarke, 1982

Lanza's Frog *Lanzarana largeni* Lanza, 1978

Lanza's Alpine Salamander *Salamandra lanzai* Nascetti et al., 1988

Dr Benedetto Lanza (b.1924) is an Italian herpetologist and physician, who was Professor of Biology and Director, Museo di Storia Naturale, Università degli Studi di Firenze. He made several expeditions to Somalia (1959–1970). Six reptiles are named after him.

Lapham

Lapham's Warty Frog *Callulina laphami* Loader et al., 2010

Lewis H. Lapham (b.1935) is an American writer who was Editor of *Harper's Magazine* (1976–1981 & 1983–2006) and remains Editor Emeritus. His grandfather was a founder of Texaco. He was educated at Yale and Magdalene College, Cambridge. He started working life as a newspaper reporter, first for the *San Francisco Examiner* and then for the *New York Herald Tribune*. Among his numerous books are *Money and Class in America* (1988), *Theater of War* (2003), *Gag Rule* (2004) and *Pretensions to Empire* (2006). He has made a number of generous donations in support of Tanzanian forest conservation.

LaPlaca

LaPlaca's Andes Frog *Phrynopus laplacai* **Cei**, 1968 NCR

[JrSYN *Psychrophrynella wettsteini*]

Alfredo LaPlaca of La Paz, Bolivia, aided the describer during his expedition to Bolivia.

Largen

Lanza's Frog *Lanzarana largeni* **Lanza**, 1978

Grassland Frog sp. *Ptychadena largeni* **Perret** 1994 NCR

[JrSYN *Ptychadena neumanni*]

Largen's Clawed Frog *Xenopus largeni* Tinsley, 1995

Dr Malcolm John Largen is a herpetologist and a photographer of wildlife. He works at Liverpool Museum and was a member of staff at the Biology

Department, Haile Selassie University, Addis Ababa, Ethiopia (1966–1977). He wrote *Catalogue of the amphibians of Ethiopia, including a key for their identification* (2001). Two reptiles are named after him.

Larsell

Larch Mountain Salamander *Plethodon larselli* **Burns**, 1954

Dr Olof Larsell (1886–1964) was a Swedish-born American biologist and anatomist. He was an instructor at Linfield College, Oregon (1910–1915). He taught zoology at Northwestern University, Illinois (1915–1918), and was at the University of Wisconsin (1918–1921), leaving as Associate Professor to become Professor of Anatomy, University of Oregon Medical School, where he stayed (1921–1952), before becoming Professor of Neuroanatomy, University of Minnesota (1952–1954).

La Salle

Robber Frog sp. *Pristimantis lasalleorum* **Lynch**, 1995

Instituto de La Salle has a number of branches, and the La Salle Christian Brothers - founded over 300 years ago by St John Baptiste de La Salle (1651–1719) - are well established in 80 countries including many in South America. A number of famous naturalists, such as Brother Nicéforo María (q.v.), Daniel González, and Marco Antonio Serna (q.v.) have been members of this order. Lynch's etymology reads: 'The specific epithet honors a group of Colombian Hermanos de las Escuelas Cristianas (Nicéforo María, Daniel González and Marco Antonio Serna) who actively advanced Colombian herpetology during this century through their collecting activities. Colombian herpetology is poorer now that all three have died.'

Lassoalcala

Robber Frog sp. *Pristimantis lassoalcalai* **Barrio-Amorós**, Rojas-Runjaic & Barros, 2010

Oscar Miguel Lasso-Alcalá is an ichthyologist at the Museo de Historia Natural La Salle, Fundación La Salle de Ciencias, Venezuela, where he was Assistant Curator (2001–2005).

Lataste

Lataste's Frog *Rana latastei* **Boulenger**, 1879
[Alt. Italian Agile Frog]
Lataste's Toad *Pseudepidalea latastii* Boulenger, 1882
[Alt. Ladakh Toad]
Mudpuppy sp. *Menobranchus latastei* Garnier, 1888 NCR
[JrSYN *Necturus maculosus*]

Professor Fernand Lataste (1847–1934) was a French zoologist. He made a collection of the reptiles and amphibians of Barbary (Morocco, Algeria and Tunisia) (1880–1884). A few years later he turned his attention to South America, writing on the birds of Chile. He wrote *Etude de la Faune des Vertébrés de Barbarie* (1885). Two reptiles are named after him.

Latiff

Latiff's Slender Toad *Ansonia latiffi* Wood et al., 2008

Professor Dr Abdul Latiff Mohamad of Universiti Kebangsaan Malaysia, Selangor, was honoured in the toad's binomial '...for his extensive contributions to the natural history of Malaysia and to the biologists studying Malaysian natural history.'

La Touche

La Touche's Frog *Hylarana latouchii* **Boulenger**, 1899
[Alt. Brown Wood Frog, Kuatun Frog]

John David Digues La Touche (1861–1935) was a French-born, English-educated ornithologist. He was Inspector of Customs in China (1882–1921), making extensive collections of Chinese fauna. He retired to Ireland, but after a visit to Majorca he died at sea on the way home. He wrote *A Handbook of the Birds of Eastern China* (1925). Nine birds, a reptile and a mammal are named after him.

Laura

Glass Frog sp. *Nymphargus laurae* Cisneros-Heredia & **McDiarmid**, 2007

Laura Heredia is the senior author's grandmother. Cisneros-Heredia wrote of his '...grandmother and friend, who has always fostered Diego's interest for animals and science with infinite love, patience and wisdom.' (See **Maria Elena**, another relative)

Laura Miriam

Vilhena White-lipped Frog *Leptodactylus lauramiriamae* **Heyer** & **Crombie**, 2005

Laura Miriam Heyer is the senior author's daughter.

Laurent

Tree Toad genus *Laurentophryne* Tihen, 1960
Andringitra Bright-eyed Frog *Boophis laurenti* **Guibé**, 1947
Laurent's Reed Frog *Hyperolius rhodesianus* Laurent, 1948
Laurent's Screeching Frog *Arthroleptis discodactylus* Laurent, 1954 NCR
[JrSYN *Arthroleptis schubotzi*]
Schiotz's Reed Frog *Hyperolius laurenti* **Schiøtz**, 1967
Laurent's Night Frog *Astylosternus laurenti* **Amiet**, 1978

Dr Raymond Ferdinand Louis-Philippe Laurent (1917–2005) was a Belgian herpetologist who worked for much of his life in Argentina and started the Herpetology Department, Fundación Miguel Lillo (1975). The University of Brussels awarded his doctorate (1940). He particularly studied African herpetology. He wrote *Diagnoses preliminaires des quelques Serpents venimeux* (1955). Seven reptiles are named after him.

Laurenti

Laurenti's Alpine Newt *Ichthyosaura alpestris* Laurenti, 1768
[Syn. *Mesotriton alpestris*, *Triturus alpestris*]

Josephus Nicolaus Laurenti (1735–1805) was an Austrian zoologist and anatomist who has been described as the father of herpetology. He wrote *Specimen medicum, exhibens synopsin reptilium emendatam cum experimentis circa venena et antidota reptilium austracorum, quod authoritate et consensus* (1768), about the poison function of reptiles and amphibians. In it he defined 30 new genera of reptile and amphibian (including *Bufo*, *Salamandra*, *Gekko*, *Chamaeleo* and *Iguana*). He also wrote *Il Dragone* (1768), a manuscript in Italian describing the Olm (*Proteus anguinus*). A reptile is named after him.

Laurin

Laurin's Rainforest Frog *Albericus laurini* **Günther**, 2000
[Alt. Wandammen Albericus Frog]

Laurin was a king of the dwarves in the *Nibelungenlied*, the main source material for Wagner's Ring Cycle. He was chosen for the binomial to go with Alberich, the dwarf who guarded the Rheingold. (See **Alberich**)

Lauzurica

La Siberia Marsupial Frog *Gastrotheca lauzuricae* **De La Riva**, 1992

Paloma Lauzurica (b.1960) is the describer's wife. She took part in her husband's fieldwork and collected with him during some of his expeditions to the Ecuadorian and Bolivian Andes.

Lava

Pygmy Splayfoot Salamander *Chiropterotriton lavae* **Taylor**, 1942

This salamander is not named after a person but after volcanic lava. Taylor wrote that 'All the specimens were taken from bromelias growing in the stunted forest of the lava fields west of La Joya, Veracruz (Mexico)'.

Lawton

Lawton's Wrinkled Ground Frog *Platymantis lawtoni* Brown & **Alcala,** 1974

E. Lawton Alcala (d.2007) was a biologist with Negros Oriental Environment and Natural Resources Division, Philippines, and a brother of the junior author. Silliman University awarded his bachelor's degree (1966). A reptile is named after him.

Lea

Lea's Frog *Geocrinia leai* **Fletcher**, 1898
[Alt. Fletcher's Ground Froglet]

Arthur Mills Lea (1868–1932) was an Australian entomologist who provided Fletcher with specimens.

Leal

Leal's Treefrog *Dendropsophus leali* **Bokermann**, 1964

Colonel Paulo Nunes Leal was a Brazilian soldier who fought in the Second World War and became Governor of the state of Rondonia (1954–1955 & 1958–1961). He later oversaw the construction of a major highway connecting Brasilia with Acre via Mato Grosso and Rondonia. He settled in Marmelo (Rondonia) where Bokermann met him (1962) and expressed his thanks for all the help Leal had given him.

Lebeau

Lebeau's Frog *Leptopelis lebeaui* **De Witte**, 1933
[Alt. Nyonga Forest Treefrog]

The etymology merely tells us that Lebeau was Chef du Service Provincial de l'Agriculture et des Forêts du Katanga. He may be the same person as H. Lebeau, an entomologist at the Musée Royal de l'Afrique Centrale, Tervuren, Belgium, who was in the Belgian Congo (1920s) and who wrote *La petite colonisation au Congo* (1933).

Leber

Leber's Robber Frog *Eleutherodactylus leberi* **Schwartz**, 1965

Dr David C. Leber was a herpetologist who worked very closely with the describer; they were together on an expedition to Haiti (1962). Leber collected in Hispaniola for some years (1960s), and also illustrated Schwartz and Henderson's *A guide to the identification of the amphibians and reptiles of the West Indies, exclusive of Hispaniola* (1985). He co-wrote *A forest-dwelling species of Eleutherodactylus* (1961). A reptile is named after him.

Leech

Manchurian Salamander *Hynobius leechii* **Boulenger**, 1887

John Henry Leech (1862–1900) was an English explorer, collector and entomologist, interested particularly in coleoptera and lepidoptera. He was educated at Cambridge. Leech collected in Morocco, the Canary Islands, Madeira, Japan, Korea, China and the northwestern Himalayas, but (1887) stayed in Britain to work on his collections, commissioning others to collect for him. He purchased (1889) the entomological periodical 'The Entomologist', partly as a means of publishing his own papers on his collections, but withdrew from it after objections were made about these inclusions. He wrote the three-volume *Butterflies from China, Japan and Corea* (1892–1894). A reptile is named after him.

Legler

Legler's Stream Frog *Ptychohyla legleri* **Taylor**, 1958

Dr John M. Legler was a graduate student curator of the herpetological collections, University of Kansas (1954–1959). He collected in El Salvador (1963), and was at the Department of Biology, University of Utah, Salt Lake City (1975). He co-wrote *A new turtle (genus* Kinosternon*) from northwestern Mexico* (1980).

Lehmann, FC

Robber Frog sp. *Hylodes lehmanni* **Boettger**, 1892 NCR

[JrSYN *Pristimantis unistrigatus*]

Friedrich Carl Lehmann (1850–1903) was a botanist who was German Consul, Popayan, Colombia. He travelled extensively through Central and South America. He was in Costa Rica, Guatemala and Panama in the early 1880s. A reptile is named after him.

Lehmann(-Valencia)

Robber Frog sp. *Eleutherodactylus lehmanvalenciae* **Thornton**, 1965 NCR

[JrSYN *Pristimantis moro*]

Lehmann's Rocket Frog *Hyloxalus lehmanni* **Silverstone**, 1971

Lehmann's Poison Frog *Oophaga lehmanni* **Myers** & **Daly**, 1976

[Alt. Red-banded Poison Frog; Syn. *Dendrobates lehmanni*]

Professor Federico Carlos Lehmann-Valencia (1914–1974) was a Colombian biologist and conservationist. He studied at the University of Cauca (1929–1934), later taking a job with the Institute of Botany, National University of Colombia, Bogotá. He founded the departmental Museum of Natural Sciences (1963) in Cali which is now named after him. Other taxa, including a reptile and two birds, are also named after him.

Lehr

Strabomantid frog sp. *Oreobates lehri* **Padial**, Chaparro & **De la Riva**, 2007

Dr Edgar Lehr (b.1969) is a German herpetologist who joined the Illinois Wesleyan University (2009) as Assistant Professor of Biology. Johannes Gutenberg-Universität Mainz awarded his master's degree (1996) and Johann Wolfgang Goethe University Frankfurt-am-Main his doctorate (2001). He was Feodor Lynen Fellow of the Alexander von Humboldt Foundation, Natural History Museum, University of Kansas (2005–2006). He was manager of the DNA laboratory and Curator of the tissue collection (2001–2004), and herpetologist (2007–2009) at the Natural History State Collections, Dresden. He is affiliate faculty member of Illinois State University, affiliate researcher of the University of Kansas, the Field Museum Chicago, and the Museo de Historia Natural, Universidad Nacional Mayor de San Marcos, Lima, Peru.

Leith

Leith's (Leaping) Frog *Indirana leithii* **Boulenger**, 1888

[Alt. Boulenger's Brown Frog]

Dr Andrew H. Leith was a physician employed in Bombay as Sanitary Commissioner. He started a system of registering deaths recording causes, ages, gender etc. in an effort to identify the prevalence of smallpox (1848). He clearly had an interest in herpetology, as he sent specimens of Indian reptiles and amphibians to the British Museum. Four reptiles are named after him.

Lekagul

Lekagul's Horned Toad *Xenophrys lekaguli* Stuart et al., 2006

[Syn. *Megophrys lekaguli*]

Dr Boonsong Lekagul (1907–1992) was a Thai physician, biologist, conservationist and herpetologist. He qualified as a physician (1933) graduating from Chulalongkorn University, Bangkok. He established the Bangkok Bird Club (1962), and helped to launch nature conservation in Thailand. His work in lobbying for legislation resulted in a National Parks Act (1961). He wrote *Monitors (*Varanus*) of Thailand* (1969). A mammal and a reptile are also named after him.

Leleup

Katanga Metal Frog *Cacosternum leleupi* **Laurent**, 1950

Luvubu Reed Frog *Hyperolius leleupi* Laurent, 1951

Dr Narcisse Leleup (1912–2001) was an entomologist who collected in the Congo for the Tervuren Museum, Belgium (1940s & 1950s), and led the Belgian Zoological Expedition to the Galápagos Islands and Ecuador (1964–1965). He co-wrote *La géographie et l'ecologie des grottes du Bas-Congo. Les habitats de* Caecobarbus geertsi (1954).

Lema

Lema Treefrog *Hypsiboas lemai* **Rivero**, 1972

This frog is named after the type locality, Serranía de Lema, Estado Bolívar, Venezuela.

Lemaire

Lemaire's Toad *Amietophrynus lemairii* **Boulenger**, 1901

Lemaire's White-lipped Frog *Hylarana lemairei* **De Witte**, 1921

Charles François Alexandre Lemaire (1863–1925) was a Belgian army officer, colonial administrator and explorer who led several expeditions in the Belgian Congo (DRC). He is recorded making a collection of fishes from Lake Tanganyika (1898). Lemaire was a keen supporter of Esperanto, writing at least one of his books in it, and published with Esperanto and French texts on opposite pages, thus *Tra Mez-Afriko. A travers l'Afrique centrale. Parolado kun projekcioj donita al la dua Universala kongreso d'Esperanto, en Genevo, je la sabato la de septembro 1906a, conférence avec projections donnée au 2me Congrès universel d'Esperanto, à Genève, le samedi 1er septembre 1906*. His diaries were published later (1986), revealing that he was the leader in murder and maiming of Congolese natives by men under his command. Joseph Conrad's novel *Heart of Darkness*, which is based on his own experiences in the Belgian Congo, has as one of its main characters a terrible man called Kurtz, and it is thought that Lemaire may have been the inspiration for him.

Lemos-Espinal

Lemos-Espinal's Leopard Frog *Lithobates lemosespinali* **Smith** & Chiszar, 2003

[Syn. *Rana lemosespinali*]

Julio Alberto Lemos-Espinal (b.1959) is a Mexican herpetologist who is attached to Universidad Nacional Autónoma de México. He is co-author of *Amphibians and Reptiles of the State of Coahuila, Mexico* (2007). A reptile is named after him.

Leon

Leon's Robber Frog *Pristimantis leoni* **Lynch**, 1976

Sr Francisco (Pancho) León Rodríguez (d.1974) was a field biologist and ornithologist who was a friend of Lynch. He was a member and resident guide for the Field Museum's 'Winter Tour' (1971) to the Andes and Galápagos Islands. He had worked in Galápagos at the Darwin Research Station, before joining the Zoology Department of the Pontificia Universidad Católica del Ecuador in Quito (1965). He was killed in a vehicle accident in Ecuador on 23 July 1974.

Leonard

Olive Striped Frog *Phlyctimantis leonardi* **Boulenger**, 1906

Leonardo Fea. (See **Fea**)

Leonardo

Poison Frog sp. *Mannophryne leonardoi* Manzanilla et al., 2007

Professor Leonardo de Sousa is a medical investigator at the Centro de Investigaciones en Ciencias de la Salud, Universidad de Oriente, Puerto la Cruz, Venezuela. He co-wrote *Natural history, taxonomic notes and conservation status of* Mabuya croizati *Horton, 1973 (Reptilia: Scincidae)* (2007).

Leonce

Hispaniola Robber Frog *Eleutherodactylus leoncei* **Shreve** & **Williams**, 1963

Dr Leonce Bonnefil (fils) (d.1998) was a zoologist and entomologist. He received a Latin American scholarship to the University of Minnesota (1944–1945), where he did a master's degree in entomology. He was at the Department of Entomology, Port-au-Prince, Haiti, being one of Haiti's designated consultants to UNESCO (1956). At the time Shreve and Williams were collecting in Haiti, Leonce was Zoologist at the Department of Agriculture and Natural Resources, Port-au-Prince, and not only was very helpful to them but became their friend.

Leonia

Slender-legged Treefrog sp. *Osteocephalus leoniae* Jungfer & **Lehr**, 2001

Eleonore Stauber was honoured in recognition of supporting taxonomic research and nature conservation in Peru. She is a German patron of BIOPAT.

Leora

Leora's Stream Salamander *Ambystoma leorae* **Taylor**, 1943

Mrs Leora T. Hughes was the wife of Dyfrig McHattie Forbes, who collected the holotype and was an amateur herpetologist. Mr and Mrs Forbes lived near Veracruz, Mexico. Visiting herpetologists were always welcome at their house: Hobart Smith and his wife Rozella stayed with the Forbes (who were planters) while on honeymoon.

Leprieur

Cayenne Slender-legged Treefrog *Osteocephalus leprieurii* **Duméril** & **Bibron**, 1841

Charles-Eugène Leprieur (1814–1892) was a French naval pharmacist and entomologist. He collected in Cayenne (French Guiana) for the Muséum National d'Histoire Naturelle, Paris (1839). He joined (1837) and was twice President of the French Entomological Society.

Lescure

Lescure's Robber Frog *Pristimantis lythrodes* **Lynch** & Lescure, 1980

Lescure's Toad *Rhinella lescurei* Fouquet et al., 2007

Jean Lescure is a herpetologist at the Muséum National d'Histoire Naturelle, Paris. He worked for many years in French Guiana and co-wrote *Atlas des amphibiens de Guyane* (2000).

Lessona

Pool Frog *Pelophylax lessonae* Camerano, 1882

[Syn. *Rana lessonae*]

Professor Dr Michele Lessona (1823–1894) was an Italian zoologist. He graduated as a physician, practicing in Turin (1846) before going to Egypt as Hospital Director at Karnak. While in Egypt he collected reptiles, presenting them to the Turin Museum upon his return. He taught in secondary schools until becoming Professor of Mineralogy and Zoology, University of Genoa (1854), then Professor of Zoology, Bologna (1864). He was an early proponent of Darwinism in Italy and the first to translate *The Descent of Man* into Italian (1871). He entered politics and was a Senator of the Kingdom (1877–1894). Two reptiles are named after him.

Lesueur

Lesueur's Frog *Litoria lesueurii* **Duméril** & **Bibron**, 1841

Charles Alexandre Lesueur (1778–1846) was a French naturalist, artist and explorer. He set sail for Australia and Tasmania aboard 'Le Géographe' as an assistant gunner (1801). When the original artists jumped ship in Mauritius, Lesueur was appointed as an official expedition artist. During the next four years he and Péron (q.v.) collected over 100,000 specimens representing 2,500 new species, and Lesueur made 1,500 drawings, from which he produced a series of watercolours on vellum. These were published (1807–1816) in the expedition's official report. He lived in southern USA (1815–1837). He met Audubon (1824) and so admired his work that he suggested Audubon should try again to get them published in France. He was appointed Curator of the Muséum d'Histoire Naturelle du Havre, which was created to house his drawings and paintings (1845). A mammal, three reptiles and two birds are named after him.

Lewis, CB

Robber frog sp. *Eleutherodactylus lewisi* **Lynn** & Dent, 1942 NCR

[JrSYN. *Eleutherodactylus luteolus*]

Charles Bernard Lewis Jr (b.1913) was Director, Institute of Jamaica, and President and Director, Jamaica Historical Society. His collections of plants (mostly 1940s–1950s) are held there. He was part of a collecting trip to Grand Cayman (1938) and collected the holotypes of a new butterfly and the Grand Cayman Blue Iguana. He co-wrote *The Herpetology of the Cayman Islands, with an Appendix on the Cayman Islands and Marine Turtle* (1940). A reptile is named after him.

Lewis, FB

Neuse River Waterdog *Necturus lewisi* **Brimley**, 1924

Frank Bartto Lewis supplied most of Brimley's specimens of this amphibian. They came from the Neuse River, near Raleigh, North Carolina, and Lewis was probably a local of this area.

Lewis, G

Lewis' Stubfoot Toad *Atelopus chiriquiensis* **Shreve**, 1936

The type specimen is described as being 'from Mr. Lewis' place... Rio Chiriqui, Viejo and branches, Panama República'. We think this is Glenn Lewis (d.1970) who came to Panama (1917) as a locomotive driver for the Panama Canal Company and settled (1920s) in Chiriqui Province. He and his wife retired to the USA (1938), but returned to Panama (1939) and built a house on the Chiriqui Viejo River. He was drowned when the river flooded and washed him and his house away.

Liam

Liam's Bright-eyed Frog *Boophis liami* **Vallan**, Vences & **Glaw**, 2003

Liam Nicolas Vlasimsky's father, Stan M. Vlasimsky, has provided considerable financial support for BIOPAT.

Librarian

Robber Frog sp. *Pristimantis librarius* **Flores** & **Vigle**, 1994

John B. Vigle was the father of the second author, Gregory O. Vigle. In the *Journal of Herpetology* (v. 28, no. 4 (1994), pp. 416–424), authors Glenn Flores and Gregory Vigle write that they have named this species in honor of the late John B. Vigle, lifetime academic librarian. It also pays tribute to all librarians for their facilitation of the study and cataloging of the natural world.

Liebig

Liebig's Frog *Nanorana liebigii* **Günther**, 1860
[Alt. Spiny-armed Frog; Syn. *Chaparana liebigii*]

The etymology merely mentions 'Dr von Liebig Jr.' and we have not been able to identify him with certainty. We think it likely Günther had in mind Baron Justus von Liebig (1803–1873), a German chemist and botanist, whose early experiments were on frogs. Von Liebig was the first to formulate an ecological 'law of the minimum', which hypothesizes that, for plants, growth is limited by a single resource at any one time. Some people may be more impressed by the fact that he created the Oxo Cube!

Liem

Liem's Frog *Taudactylus liemi* **Ingram**, 1980
[Alt. Eungella Tinker Frog, Liem's Tinker Frog]

Dr David S. Liem (b.1937) is a Communication and Information Resource Specialist and Toxicologist/ Pesticide Risk Assessor with the Office of Pesticide Programs at the US Environmental Protection Agency in Washington DC, where he has been since 1990. He has worked at the Department of Lands and Environment in Port Moresby, Papua New Guinea, and with the University of Queensland in Brisbane, Queensland, Australia as an environmental scientist. The University of Illinois awarded his doctorate in zoology (1969). Among other works he co-wrote *Analysis of the intergeneric relationships of the Australian frog family Myobatrachidae* (1976). He famously described the (now extinct) Gastric Brooding Frog, *Rheobatrachus silus* (1973).

Light

Mindanao Flathead Toad *Pelophryne lighti* **Taylor**, 1920

Dr Sol F. Light was a biologist and zoologist who taught at the University of California, Berkeley (1916–1919). He was a professor at the University of the Philippines when Taylor named this toad after him as he '...has manifested much interest and given assistance in this work.' He wrote *Isoptera of Guam* (1946).

Lightfoot

Lightfoot's Moss Frog *Arthroleptella lightfooti* **Boulenger**, 1910
[Alt. Cape Peninsula Moss Frog]

Robert Mark Lightfoot (1864–1921) was a malacologist, entomologist and general factotum at South African Museum, Cape Town, as well as a prolific collector; his first donation to the Museum was an insect (1882). Unfortunately he suffered from kyphosis (commonly known as 'hunchback').

Lili

Lili Treefrog *Hypsiboas liliae* Kok, 2006

Lili Kok is the describer's daughter.

Lilian

Lilian's Bright-eyed Frog *Boophis lilianae* **Köhler, Glaw** & Vences, 2008

Liliane Raharivololoniaina is a Malagasy herpetologist and biologist at the Département de Biologie Animale, Université d'Antananarivo, Madagascar. She co-wrote *Descriptions of the tadpoles of seven species of Malagasy treefrogs, genus Boophis* (2009). She collected two paratypes of this frog.

Lima

São Vicente Treefrog *Dendropsophus limai* **Bokermann**, 1962

João Leonardo Lima (1874–1936) was a Brazilian zoologist. He started working life as an assistant in the Ministry of the Interior, but transferred to the Museum of Pará (1895). He became preparator (1905) and a collector, and then collector-naturalist (1925). A bird is named after him.

Limborg

Limborg's Frog *Limnonectes limborgi* Sclater, 1892

Gustaf Arthur Ossian Limborg (1848–1908) was a Swedish novelist, journalist, ship's captain and poet who collected in Burma (Myanmar) (1877). A bird is named after him.

Lin

Lin's Frog *Babina lini* Chou, 1999

Dr Edgar Jun-yi Lin is a Taiwanese herpetologist and environmentalist. He was Director, Taiwan Agriculture Research Institute (2002). He co-wrote *Tadpoles of Taiwan* (1997).

Lincoln

Lincoln's Mushroomtongue Salamander *Bolitoglossa lincolni* **Stuart**, 1943

Dr Jackson Stewart Lincoln was an ethnographer and archaeologist. Stuart states he named this amphibian after his late friend who had suggested Stuart visit the area where he found this species. Lincoln wrote *An Ethnological study of the Ixil Indians of the Guatemala Highlands*, published posthumously (1945).

Linda

Linda's Treefrog *Hyloscirtus lindae* **Duellman** & Altig, 1978

Santa Isabel Robber Frog *Pristimantis lindae* Duellman, 1978

Murri Beaked Toad *Rhinella lindae* **Rivero** & Castaño, 1990

Dr Linda Trueb (see **Trueb**).

Lindholm

Lindholm's Banana Frog *Afrixalus lindholmi* **Andersson**, 1907

Wasily (Wilhelm) Adolfovitch Lindholm (1874–1935) was a Russian zoologist, herpetologist and malacologist who collected widely. In 1907 Andersson received a total of 135 amphibian specimens from E. Lampe, curator at the Museum Wiesbaden, and interestingly most of the taxa described by him or by Lindholm are housed there still. We conclude that the frog holotype was among Lampe's consignment. The original text refers to him as 'Herrn W. A. Lindholm in Moskau'

Lindner

Lindner's Toad *Mertensophryne lindneri* **Mertens**, 1955

Professor Dr Erwin Lindner (1888–1988) was a German entomologist and head of department at the State Museum of Natural History, Stuttgart (1913–1953). His major interest was diptera and he edited *Die Fliegen der paläarktischen Region*. He collected the toad holotype (1951).

Lindsay

Lindsay's Snouted Treefrog *Scinax lindsayi* **Pyburn**, 1992

Dr Hague L. Lindsay Jr is an American zoologist who has worked on experimental hybridization of anurans. He was awarded both his master's (1951) and doctorate (1958) by the University of Tulsa, Oklahoma, where he taught (1956–1987) with a break (1958–1960) for service in the US Army. He was Associate Professor of Zoology and is now retired.

Linnaeus

Linnaeus' Caecilian *Caecilia tentaculata* Linnaeus, 1758

Rainforest Frog sp. *Cophixalus linnaeus* Kraus & **Allison**, 2009

Carl Linné (1707–1778) is much better known by the Latin form of his name, Carolus Linnaeus, or simply as Linnaeus. Later in life (1761) he was ennobled and so could call himself Carl von Linné. In the natural sciences he was undoubtedly one of the great heavyweights of all time, ranking with Darwin and Wallace. He is thought of primarily as a botanist but he invented the system he called *Systema Naturae* that is still in use today - albeit with modifications - for naming, ranking and classifying living organisms. He entered Lunds Universitet (1727) to study medicine, and (1728) transferred to Uppsala Universitet. At that time the study of botany was part of medical training. His first expedition was to Lapland (1732). He went to the Netherlands (1735) and finished his studies as a physician there before enrolling at Universiteit Leiden. He returned to Sweden (1738), lecturing and practising medicine in Stockholm. He became Professor at Uppsala (1742) and restored the University's botanical garden. He bought the manor estate of Hammarby, outside Uppsala, where he built a small museum for his extensive personal collections (1758). This house and garden still exist and are now run by Uppsala University. His son, also named Carl, succeeded to his professorship at Uppsala, but was never as noteworthy a botanist. When Carl the Younger died (1783) with no heirs, his mother and sisters sold the elder Linnaeus's library, manuscripts and natural history collections to the English natural historian Sir James Edward Smith. He founded the Linnean Society of London to take care of them. Surprisingly few taxa have been named after Linnaeus: those few include five birds, three mammals and four reptiles.

Lips

Toad sp. *Incilius karenlipsae* **Mendelson** & Mulcahy, 2010

Dr Karen R. Lips is a biologist who studies the decline of amphibians and its causes. Her bachelor's degree in zoology was awarded by University of South Florida, Tampa, and her doctorate by the University of Miami (1995). She was Associate Professor of Zoology, Southern Illinois University (1998–2008), and is now Associate Professor Biology at the University of Maryland. She co-wrote *Changes in quality and quantity of stream algae and seston resulting from neotropical amphibian declines* (2008).

Lise

Brauns' Dwarf Frog *Physalaemus lisei* **Braun** & Braun, 1977

Dr Arno Antonio Lise is an arachnologist at the Pontifical Catholic University, Rio Grande do Sul, Brazil, where he is Professor of Zoology. His bachelor's degree (1961) and his doctorate (1962) were both awarded by the Catholic University of Rio Grande do Sul.

Littlejohn

Littlejohn's Toadlet *Uperoleia littlejohni* **Davies, McDonald** & **Corben**, 1986

Littlejohn's Tree Frog *Litoria littlejohni* **White,** Whitford & **Mahony**, 1994

Professor Dr Murray J. Littlejohn (b.1932) is a zoologist and herpetologist at the Zoology Department, University of Melbourne. In the 1950s he concentrated on research into frogs in eastern Australia. He was the first President of the Australian Society for Herpetologists (1964). He wrote *The Frogs of Tasmania Handbook* (1982).

Liu

Hynobiid salamander genus *Liua* **Zhao** & **Hu**, 1983

Rhacophorid frog genus *Liuixalus* Li et al., 2008

Pope's Spiny Toad *Leptobrachium liui* **Pope**, 1947

Liu's Papillae-tongued Frog *Ingerana liui* **Yang**, 1983

Vocal Sacless Spiny Frog *Nanorana liui* **Dubois**, 1987 NCR

[JrSYN *Nanorana yunnanensis*]

Fujian Metacarpal-tubercled Toad *Leptolalax liui* **Fei** & **Ye**, 1990

Professor Ch'eng-chao Liu (1900–1976) was a Chinese herpetologist who graduated at Peking University (1929) and then taught at the University of the North-East in Shenyang. Its libraries were destroyed during the Sino-Japanese war (1931). He studied at Cornell University (1932–1934) and taught at the University of Suzhou (1935–1939), at Chengdu in Sichuan (1939–1950), becoming Head of the Biology Department, University of Yenching (1950–1951). He finally returned to Chengdu (1951) as Director of the Medical School. He wrote *Amphibians of Western China* (1960). A reptile is named after him.

Löding

Löding's Mudpuppy *Necturus lodingi* **Viosca**, 1937 NCR

[Alt. Mobile Waterdog; JrSYN *Necturus alabamensis*]

Peder Henry Löding (1869–1942) was born in Denmark and emigrated to the USA. He lived in Mobile, Alabama, and was described by Blanchard as being a '...pioneer student of Alabama reptiles and amphibians, through whose efforts have come to light the types upon which this species is based.' He wrote the first comprehensive report of the herpetofauna of Alabama. A reptile is named after him.

Loki

Volcan San Martin Robber Frog *Craugastor loki* Shannon & **Werler**, 1955

Loki was the Teutonic god of fire and strife. The holotype was collected on the slopes of a Mexican volcano, which probably inspired the describers.

Lönnberg

Lönnberg's Treefrog *Litoria eucnemis* Lönnberg, 1900

[Alt. Fringed Tree Frog]

Lönnberg's Toad *Mertensophryne lonnbergi* **Andersson**, 1911

Lönnberg's Squeaker *Arthroleptis lonnbergi* **Nieden**, 1915

[Sometimes regarded as synonymous with *Arthroleptis stenodactylus*]

Professor Axel Johan Einar Lönnberg (1865–1942) was a Swedish zoologist who mainly worked at the Vertebrate Department, Swedish Museum of Natural History, Stockholm. He travelled in British East Africa (1910–1911). He was the last prefect of the Kristineberg Marine Zoological Station (1925–1942). He founded the biological journal *Fauna och Flora*. Eight birds are named after him.

Loria

Port Moresby Cross Frog *Oreophryne loriae* **Boulenger**, 1898

Dr Lamberto Loria (1855–1913) was an Italian ethnologist and explorer who collected in New Guinea (1889–1890). He founded the first Italian Museum of Ethnography, Florence (1906). The museum was subsequently transferred to Rome, after he organised the first ethnography exhibition there (1911). Two mammals, three reptiles and three birds are named after him.

Lötters

Harlequin Frog sp. *Atelopus loettersi* **De La Riva** et al., 2011

Dr Stefan Lötters of Trier University, Germany, where he has been a senior lecturer (2007), has made contributions to the taxonomy of *Atelopus* frogs. He graduated from Rheinische Friedrich-Wilhelms-Universität, Bonn (1996) and was awarded his doctorate there (2000). He co-wrote *Assessing the molecular phylogeny of a near extinct group of vertebrates: The Neotropical harlequin frogs* (Bufonidae; Atelopus) (2011).

Lou Jost

Robber Frog sp. *Pristimantis loujosti* Yánez-Muñoz, Cisneros-Heredia & Reyes, 2010

Lou Jost is an ex-physicist who is now a mathematical ecologist and conservation biologist. He is also a painter of nature, an illustrator and a taxonomist specializing in *pleurothallid* orchids (having discovered c.70 species of orchid and other plants, about half of which are published, notably including the world's smallest orchid). He is cofounder and treasurer of the EcoMinga Foundation, which identifies important Andean forests and buys them for preservation, so far protecting 8,000 acres.

Loveridge

Loveridge's Snouted Toad *Mertensophryne micranotis* Loveridge, 1925

Loveridge's Forest Frog *Probreviceps loveridgei* **Parker**, 1931

Loveridge's Screeching Frog *Arthroleptis loveridgei* **De Witte**, 1933

Loveridge's (Mountain) Frog *Philoria loveridgei* Parker, 1940

Frog sp. *Nyctimystes loveridgei* Neill, 1954 [JrSYN *Litoria eucnemis*]

Loveridge's Treefrog *Myersiohyla loveridgei* **Rivero**, 1961

Loveridge's Forest Toad *Mertensophryne loveridgei* **Poynton**, 1991

Arthur Loveridge (1891–1980) was a herpetologist and zoologist who is regarded as the father of East African herpetology. He worked in the Department of Reptiles, Harvard Museum of Comparative Zoology, Massachusetts. He retired to St Helena. He wrote *East African reptiles and amphibians in the United States National Museum* (1929), and the enticing title, *I Drank the Zambesi* (1954). Thirteen reptiles and four birds are named after him.

Lozano

Lozano's Stubfoot Toad *Atelopus lozanoi* Osorno-Muñoz, Ardila-Robayo & Ruiz-Carranza, 2001

Lozano's Salamander *Bolitoglossa lozanoi* Acosta-Galvis & **Restrepo**, 2001

Gustavo Lozano Contreras (1938–2000) was Professor of Botany at the Instituto de Ciencias Naturales de la Universidad Nacional, Bogotá, Colombia.

Lucan

Landana Reed Frog *Hyperolius lucani* **Rochebrune**, 1885

Dr A. Lucan (d.1884) was a physician in the French Congo (1876–1884). He collected the frog holotype.

Lucas

Robber Frog sp. *Pristimantis lucasi* **Duellman** & Chaparro, 2008

Lucas Morciniec of Poland has 'supported field studies in Peru'.

Lucia

Santa Lucia Slender Salamander *Batrachoseps luciae* Jockusch, Yanev & **Wake**, 2001

This salamander is named after the Santa Lucia Mountains in California.

Lucia (De la Riva)

Bright-eyed Frog sp. *Boophis luciae* **Glaw** et al., 2010

Lucia de la Riva (b. 1998) is the daughter of one of Glaw's co-authors; Spanish herpetologist Ignacio de la Riva.

Luciana

Treefrog sp. *Bokermannohyla lucianae* **Napoli** & Pimenta, 2003

Luciana Barreto Nascimento is a Brazilian herpetologist. She graduated in biological sciences at the Pontifícia Universidade Católica de Minas Gerais (1983), and was awarded her master's in biological sciences (zoology) by the National Museum, Federal University of Rio De Janeiro (1991). She is Associate Professor of Biological Sciences at Pontifícia Universidade Católica de Minas Gerais, as well as a curator for the herpetological collection there. She was Associate Professor in the Department of Vertebrates of the National Museum, UFRJ (2004–2006) and first secretary of the Brazilian Herpetological Society (2003–2007). Her research interests are the taxonomy and conservation of amphibians and reptiles.

Lucio

St Nicholas Robber Frog *Eleutherodactylus lucioi* **Schwartz**, 1980

John C. Lucio collected with Schwartz in Haiti (1979).

Lüddecke

Treefrog sp. *Dendropsophus luddeckei* Guarnizo et al., 2012

Dr Horst Lüddecke is a German biologist who has lived and worked in Colombia for many years.

He is at the University of Los Andes, Santa Fe de Bogotá, where he became a professor (2003). Johannes Gutenberg-Universität Mainz awarded both his master's degree (1969) and his doctorate (1974). He wrote *Variation and trade-off in reproductive output of the Andean frog* (2002).

Lüderwaldt

Lüderwaldt's Highland Frog *Holoaden luederwaldti* **Miranda-Ribeiro**, 1920

Hermann Lüderwaldt (1865–1934) was a zoologist at the Museu Paulista, São Paulo, Brazil. He wrote *Os chelonios brasilieros com a lista das espécies do Museu Paulista* (1926). He collected the frog holotype.

Luiz Otávio

Santa Barbara Snouted Treefrog *Scinax luizotavioi* **Caramaschi** & Kisteumacher, 1989

Luiz Otávio Possas Gonçalves of Belo Horizonte, Minas Gerais, Brazil helped the describers in the field. He owns the Green Valley Ecological Park.

Lumad

Litter Frog sp. *Leptobrachium lumadorum* Brown et al., 2010

The Lumad are a group of indigenous peoples who live on Mindanao Island, Philippines.

Lund

Lund's Treefrog *Hypsiboas lundii* **Burmeister**, 1856

Dr Peter Wilhelm Lund (1801–1880) was a Danish physician, botanist, zoologist and palaeontologist who lived and worked in Lagoa Santa, Minas Gerais, Brazil. He first travelled to Brazil (1833) and settled there for health reasons. His interest in fossils led him to explore many of the caves of the area. He assembled one of the most important mammal collections from a single locality in the Neotropics, and made outstanding contributions towards describing the Pleistocene and recent mammal faunas of Brazil. He regularly corresponded with Charles Darwin. Two mammals and a reptile are named after him.

Lundberg

Robber Frog sp. *Hypodactylus lundbergi* **Lehr**, 2005 [Syn. *Eleutherodactylus lundbergi*]

Mikael Lundberg is a Swedish Civil Engineer who is an associate member of Staatliche Naturhistorische Sammlungen Dresden, Germany, and who has collected with Lehr. The two co-wrote *A New Species of* Atelopus *(Anura: Bufonidae) from the Cordillera Oriental of Central Peru* (2008).

Luschan

Luschan's Salamander *Lyciasalamandra luschani* **Steindachner**, 1891

Dr Felix Ritter von Luschan (1854–1924) was an Austrian physician, explorer, anthropologist, archaeologist and ethnographer ('Ritter' is not a name but a title, equivalent to the English baronet). He qualified in medicine in Vienna and then studied anthropology in Paris. He was an Austrian Army surgeon (1878–1879) in Bosnia and visited Dalmatia, Montenegro and Albania. He lectured at the University of Vienna (1882). He was Assistant Director of the Royal Anthropological Museum, Berlin (now the Ethnological Museum) (1885), and Director, Africa and Oceania Department (1904–1911). He became Reader (1904) at Humboldt University, Berlin, then their first Professor of Anthropology. Luschan discovered a late Hittite-period city at Zincirle in Lycia (now in Turkey) and excavated it (1888–1902). He visited South Africa (1905) and Louisiana (1919). He is noted for creating 'von Luschan's chromatic scale': 36 opaque glass tiles for comparison to a subject's skin colour. He wrote *Anthropologie, Ethnographie und Urgeschichte*. (1905).

Luscombe

Robber Frog sp. *Pristimantis luscombei* **Duellman** & **Mendelson**, 1995

B. Anthony Luscombe is an American naturalist who has lived in Peru since 1972 and is President, Asociación de Ecología y Conservación, Lima.

Lütken

Caecilian genus *Luetkenotyphlus* **Taylor**, 1968

Yellow Toad *Incilius luetkenii* **Boulenger**, 1891

Professor Christian Frederik Lütken (1827–1901) was a Danish naturalist. He was a professional soldier in the Danish army, resigning his commission (1852) to concentrate on natural history. He lectured (1856–1862) at the Zoology Department, University of Copenhagen. He then taught at the Polytechnic School (1877–1881) before returning to the University, becoming Professor of Zoology (1885). He suffered a stroke that left him paralysed (1899) and forced him to retire. He co-wrote *Bidrag til Kundskab om Brasiliens Padder og Krybdyr* (1861) and some very successful books popularising natural history.

Lutz, A

Lutz's Poison Frog *Ameerega flavopicta* Lutz, 1925

Lutz's River Frog *Thoropa lutzi* **Cochran**, 1938

Lutz's Rapids Frog *Paratelmatobius lutzii* **Lutz, BMJ** & **Carvalho**, 1958

Lutz's Thin-toed Frog *Leptodactylus lutzi* **Heyer,** 1975

Adolfo Lutz (1855–1940) was a Brazilian medical entomologist and parasitologist whose Swiss parents settled in Brazil (1849). He was taken to Switzerland (1857) to be educated, qualifying at the University of Berne as a physician (1879). He undertook post-graduate work in Leipzig, Prague, Vienna and London, studying under such great names as Lister and Pasteur. He returned to Brazil (1881) and set up in practice in the interior of the state of São Paulo, but returned to Europe (1888) to study dermatology in Hamburg. He was in Hawaii (1889–1991) directing work on leprosy, thence to Brazil (1892), travelling throughout the country before becoming Head, Bacteriological Institute, São Paulo (1893–1908). He was persuaded to move to Rio de Janeiro as Director, Medical Zoology, Instituto de Manguinhos (later renamed the Instituto Oswaldo Cruz) (1908–1938, though officially retiring earlier). A reptile is named after him.

Lutz, BMJ

Lutz's Snouted Treefrog *Scinax longilineus* Lutz, 1968
Lutz's Big-tooth Frog *Megaelosia lutzae* **Izecksohn** & **Gouvêa**, 1987

Bertha Maria Júlia Lutz (1894–1976) (Daughter of **A Lutz** above). She was a zoologist and pioneering feminist, who founded the Brazilian Federation for Feminine Progress (1922). Her father was Swiss and her mother English. She studied zoology at the Sorbonne, Paris, and was a member of the Brazilian Parliament for a short period (1936–1937) until the coup d'état by Getúlio Vargas. Her main interest was in amphibians. Two reptiles are named after her (see **Bertha**).

Lutz, A & BMJ

Lutzs' Lime Treefrog *Sphaenorhynchus orophilus* Lutz & Lutz, 1938
Lutzs' Button Frog *Cycloramphus lutzorum* **Heyer**, 1983
Lutzs' Spinythumb Frog *Crossodactylus lutzorum* Carcerelli & **Caramaschi**, 1993
Snouted Treefrog sp. *Scinax lutzorum* **Cardoso** & **Pombal**, 2010

These frogs are named after both Adolfo and Bertha Lutz.

Lyman

Lyman's Robber Frog *Pristimantis lymani* **Barbour** & **Noble**, 1920

Dr Theodore Lyman (1874–1954) was an American physicist. He graduated from Harvard (1897), which also awarded his doctorate (1900). He joined the faculty at Harvard as Assistant and became Director, Jefferson Physical Laboratory (1908–1917). In 1917 he became full Professor – and the USA declared war on Germany. He served in the US Army in France, returning to Harvard after the First World War. He was wealthy and generous with his money as evidenced by the etymology: 'Named in honor of Prof. Theodore Lyman whose generosity enabled the Museum of Comparative Zoology to send a zoologist with the Harvard Peruvian expedition of 1916.' The crater Lyman on the far side of the moon is also named after him as are two birds.

Lynch

Andes Frog genus *Lynchius* **Hedges, Duellman** & Heinicke, 2008
Lynch's Pilalo Robber Frog *Pristimantis pyrrhomerus* Lynch, 1976
Lynch's Robber Frog *Pristimantis lynchi* Duellman & **Simmons**, 1977
Lynch's Giant Glass Frog *Centrolene lynchi* Duellman, 1980
Lynch's Stubfoot Toad *Atelopus lynchi* **Cannatella**, 1981
Lynch's Swamp Frog *Pseudopaludicola llanera* Lynch, 1989
Lynch's Cochran Frog *Nymphargus ignotus* Lynch, 1990
Lynch's Colombian Treefrog *Hyloscirtus lynchi* **Ruiz-Carranza** & **Ardila-Robayo**, 1991
Lynch's Leaf Frog *Noblella lynchi* Duellman, 1991
Lynch's Rocket Frog *Colostethus lynchi* **Grant**, 1998
Veracruz Green Salamander *Pseudoeurycea lynchi* Parra-Olea, **Papenfuss** & **Wake**, 2001
Lynch's Toad *Rhaebo lynchi* Mueses-Cisneros, 2007

Dr John Douglas Lynch (b.1942) is an American herpetologist who moved to Colombia where he is an associate professor, Instituto de Ciencias Naturales, Universidad Nacional de Colombia, Bogotá. He specializes in amphibians and is Curator of Amphibians at the museum. He first visited Mexico (1964) while he was studying for his master's degree at the University of Illinois. He subsequently went to the University of Kansas for his doctorate and made a trip to Ecuador (1967), with the result that he spent the next 11 years researching Ecuadorian frogs. He visited Colombia (1979) and returned annually (1980–1996). When he decided to move to Colombia, he resigned his position at the University of Nebraska where he had been Professor for 28 years. In 2000 the Colombian rebel National Liberation Army took prisoner a group of environmentalists, including Lynch; he was fortunate to be released after only two days. A reptile is named after him.

Lynn

Lynn's Robber Frog *Eleutherodactylus alticola* Lynn, 1937

Professor Dr William Gardner Lynn (1905–1990) was a naturalist who graduated from Johns Hopkins University, Baltimore, which also awarded his doctorate in biology. He worked at the University as a member of the faculty (1930–1942). He became Professor of Biology, Catholic University, Washington DC (1942–1971) and in retirement taught at a number of institutions including Montgomery College. He co-wrote *The herpetology of Jamaica* (1940). A mammal is named after him.

M

Mabee

Mabee's Salamander *Ambystoma mabeei* **Bishop**, 1928

William Bruce Mabee (b.1897) was an entomologist. He was at North Carolina State College (1920–1926), then Assistant Professor, Entomology, Montana State College (1926–1930) and Entomologist, Pennsylvania State College (1930–1931). After this he ranched cattle in Montana (1932–1938), finally becoming Field Supervisor, Bureau of Entomology and Plant Quarantine, US Department of Agriculture. He collected the salamander holotype (1923) in North Carolina.

Mable

Mable's Robber Frog *Eleutherodactylus jugans* **Cochran**, 1937

Dr Doris Mable Cochran (see **Cochran**).

MacConnell

MacConnell's Bush Toad *Oreophrynella macconnelli* **Boulenger**, 1900

Frederick Vavasour McConnell (1868–1914) was an English traveller who collected in British Guiana (Guyana) (1894–1898). He sometimes collected with John Joseph Quelch (q.v.). The ornithologist Charles Chubb was inspired to write *The Birds of British Guiana: based on the collection of Frederick Vavasour McConnell* (1916–1921), to which McConnell's widow wrote the foreword. Four mammals are named after him.

MacDougall

MacDougall's Pygmy Salamander *Thorius macdougalli* **Taylor**, 1949

Thomas Baillie MacDougall (1896–1973) was a naturalist and collector, born in Scotland. He moved to the USA after the First World War and for many years spent each summer in Mexico. He wrote, or co-wrote, a number of papers, such as *New or unusual Mexican amphibians* (1949). Five reptiles are named after him.

Machado

Machado's Snouted Treefrog *Scinax machadoi* **Bokermann** & **Sazima**, 1973

Ricardo Bomfim Machado is Director, Institut Conservation International do Brasil, Brasilia. His bachelor's degree (1984) in biology is from the Pontifical Catholic University of Minas Gerais, his master's in ecology and wildlife management (1995) from the Federal University of Minas Gerais. The University of Brasília awarded his doctorate in ecology.

Mack

Tree Frog sp. *Litoria macki* **Richards**, 2001

Dr Andrew L. Mack is a conservation biologist with a particular focus on Papua New Guinea. His bachelor's degree is from the University of Delaware (1985) and the University of Miami awarded his doctorate (1995). He was Senior Ecologist, Conservation International, Washington, DC (1995–1999) and then conservation zoologist and co-director, Papua New Guinea Country Program, Wildlife Conservation Society (1999–2007). He was a co-founder of the Crater Mountain Biological Research Station in PNG.

Mackay

Mackay's Forest Treefrog *Leptopelis mackayi* **Köhler** et al., 2006

Alex Duff-MacKay (1939–2003) was a biologist and herpetologist. He lived his entire life in Africa. He went to school in Nairobi and took a bachelor's degree in zoology and entomology at Rhodes University, Grahamstown, South Africa. He was a forest entomologist with the East African Agricultural and Forestry Research Institute until 1964, when he joined the National Museum of Kenya. Here he became (1980) Head of the Herpetology Department until retirement (1995). He wrote *Amphibia: Conservation status report No.1* (1980).

Macrinius

Oaxacan Mushroomtongue Salamander *Bolitoglossa macrinii* Lafrentz, 1930

Emil Macrinius of Cafetal Concordia, Oaxaca, Mexico, assisted the describer, Dr K. Lafrentz, when he was collecting in Mexico (1926–1928). A reptile is named after him.

Madhava

Mudur Caecilian *Gegeneophis madhavai* Bhatta & Srinivasa, 2004

Madhava Bhat, Madhavarao Bhide, Dr Tonse Madhava Anantha Pai and Professor Madhava Gadgil (b.1942) are an apparently disparate group of people with the name Madhava in common and, as the description says, they are honoured '...for supporting the first author's research into caecilians.' Dr Tonse Madhava Anantha Pai (1898–1979) was an educationalist, banker and philanthropist.

Professor Madhava Gadgil (b.1942) is a distinguished biologist and ecologist whose first degrees were from the Universities of Poona and Bombay and whose doctorate was awarded by Harvard. We have not been able to trace any details of Madhava Bhat and Madhavarao Bhide. It would have been more correct to use the plural form *madhavaorum* for the binomial.

Madidi

Madidi Robber Frog *Oreobates madidi* **Padial**, Gonzáles & **De la Riva**, 2005

This frog is named after the Madidi National Park, Bolivia.

Magalhães

São Paulo Tree Toad *Hylodes magalhaesi* **Bokermann**, 1964

Reis de Magalhães (1921–2002) was a Brazilian farmer, coffee grower and ornithologist. He collected many thousands of bird skins that are now in the Museum of Zoology, University of São Paulo. He was among those who started the Association of Environmental Protection (1954). He ensured that his own estate was preserved; at one time it held some of the most endangered primates in Brazil. He was also a pioneer in recording bird song. He led expeditions, including taking Bokermann to see birds in their natural environment, as well as helping him interpret his estate, about which Bokermann wrote his master's thesis. He was an amateur naturalist who added greatly to scientific knowledge.

Magnusson

Magnusson's Toad *Rhinella magnussoni* **Lima**, Menin & Araújo, 2007

Dr William E. Magnusson is a herpetologist in Manaus, Brazil, where he is Co-ordinator of Ecological Research with the National Institute of Research of Amazonia. He was previously (1980s) in the Department of Zoology, University of Sydney, New South Wales.

Mahnert

Mahnert's Ridged Frog *Ptychadena mahnerti* **Perret,** 1996
[Alt. Highland Ridged Frog]

Dr Volker Mahnert (b.1943) is an Austrian ichthyologist, arachnologist and parasitologist whose doctorate was awarded by the University of Innsbruck (1971). He was Curator, Department of Herpetology. Geneva Museum of Natural Sciences, where he became Director (1989). He became (1990) Associate Professor of Zoology and Animal Biology, University of Geneva, and (1991) a member of the International Commission on Zoological Nomenclature. He and Perret collected ants together in East Africa.

Main

Main's Ground Froglet *Geocrinia lutea* Main, 1963
[Alt. Walpole's Frog, Nornalup Frog]
Main's Frog *Litoria maini* **Tyler** & **Martin**, 1977
[Alt. Sheep Frog; Syn. *Cyclorana maini*]

Dr Albert (Bert) Russell Main (1919–2009) was a well-known Australian ecologist. He served in the RAAF (1939–1942). After the Second World War he studied at the University of Chicago. He was a researcher at the Department of Zoology, University of Western Australia (1952–1967), completing his doctorate there (1956) and becoming Professor of Zoology (1967). He served on a number of public bodies, including being President of the National Parks Commission. A mammal and two reptiles are named after him.

Majikthise

Treefrog sp. *Litoria majikthise* Johnston & **Richards**, 1994

Majikthise (pronounced 'magic thighs') was a member of the Amalgamated Union of Philosophers in the humorous science fiction novel *The Hitchhiker's Guide to the Galaxy*. The name apparently reflects the frog's 'vividly coloured thighs and groin'.

Major

Ambohimitombo Bright-eyed Frog *Boophis majori* **Boulenger** 1896

Charles Immanuel Forsyth Major (1844–1923) was born in Glasgow and educated in Switzerland, graduating in medicine at Basel (1868). However, his real interest was in fossils and he is remembered as a vertebrate palaeontologist rather than as a physician. He did field work in Italy and Corsica, but is most noted for his expedition to Madagascar (1894–1896). This was the first systematic survey of Madagascar's mammalian fauna. Four mammals are named after him.

Malcolm

Malcolm's Ethiopia Toad *Altiphrynoides malcolmi* **Grandison**, 1978

Dr James R. Malcolm is a British biologist who is Professor of Biology, Redlands University, California. His bachelor's degree was from Oxford and his doctorate from Harvard. His interests are animal behaviour and evolutionary biology. In the early 1970s he studied African wild dogs in Tanzania, and since 1975 he has made at least five visits to Ethiopia.

Malcolm Smith

Malcolm Smith's Shrub Frog *Pseudophilautus malcolmsmithi* **Ahl**, 1927

(See under **Smith, MA**)

Malkin

Rio Ampiyacu Robber Frog *Pristimantis malkini* **Lynch**, 1980

Borys Malkin (1917–2009) was a Polish entomologist, ethnographer, herpetologist, traveller and filmmaker. He emigrated to the USA (1938) and served in US Air Force in the Second World War in the Pacific. He studied in London after being demobilized. He lectured on anthropology and biology (1956–1957) at the University of Washington, Seattle. He had begun field studies among the Seri Indians in Mexico (1953), and then spent nearly all his time (1957–1994) in South America, especially Ecuadorian Amazonia. He died in Warsaw, where much of his collection is housed.

Mamusi

Wrinkled Ground Frog sp. *Platymantis mamusiorum* Foufopoulos & Brown, 2004

The Mamusi are a highland tribe who inhabit the area of New Britain, Papua New Guinea, where the holotype was discovered.

Mandel

Mount Turumiquire Rocket Frog *Allobates mandelorum* **Schmidt**, 1932

Leon Mandel II and Fred L. Mandel Jr sponsored the Field Museum's Mandel Venezuela Expedition (1931–1932). They were members of the family that owned Chicago's department store Mandel Brothers, which traded for over a century (1855–1960). Leon also sponsored expeditions to Guatemala (1936) and the Galápagos Islands (1941), on which he and his wife Carola were members of the expedition, using his yacht 'Carola' for transport and as a base. A bird is named after Leon.

Mangyan

Litter Frog sp. *Leptobrachium mangyanorum* Brown et al., 2010

The Mangyan are a group of eight indigenous peoples who live on Mindoro Island, Philippines.

Manoharan

Manohar's Bush Frog *Raorchestes manohari* Zachariah et al., 2011

Shri T. M. Manoharan was Principal Chief Conservator of Forests, Kerala, India. He co-wrote *Silent valley: Whispers of reason* (1999) with others including S. D. Biju (q.v.). The etymology reads: 'This elegant species is named after T. M. Manoharan, Principal Chief Conservator of Forests, Kerala, India for his outstanding efforts in conserving the forests of Kerala and for his untiring support for batrachological studies. The name 'manohari' in Malayalam language means 'beautiful'.'

Manrique

Snouted Treefrog sp. *Scinax manriquei* Barrio-Amorós, Orellana & Chacón, 2004

Róger Manrique is a Venezuelan entomologist, lepidopterist, artist, birding guide and expert in Andean fauna. He has spent many years studying Venezuelan butterflies, having one of the largest and most important collections in the country. He has directed major entomological expeditions to the Amazon and Andean regions. He has discovered and described at least one new butterfly and is commemorated in the names of two others.

Manthey

Manthey's Chorus Frog *Microhyla mantheyi* **Das**, Yaakob & **Sukumaran**, 2007

Ulrich Manthey (b.1946) trained as an engineer, but since the early 1980s has been a freelance herpetologist. Since 1990 he has been a collaborator of the Museum für Naturkunde, Berlin. He co-wrote *Amphibians and Reptiles of Mount Kinabalu (North Borneo)* (2002). Two reptiles are named after him.

Mantze

Sichuan Torrent Frog *Amolops mantzorum* **David**, 1872

[Alt. Mouping Sucker Frog]

The Mantze is how the ethnic Chinese in the nineteenth century described the area and people of Sichuan, which was inhabited by 'independent barbarian' tribes of Tibetan origin. Marco Polo never visited northwest Sichuan but he too referred to the Mantzi. Père David, who travelled extensively among the Mantze in Moupin, where he collected the frog holotype (1869), wrote that he was leaving '…le pays des Mantze, après y avoir souffert plus de fatigues, de peines, de privations et de maladies, qu'il n'est opportun de le dire ici.' He appears to have had a rough time there!

Marani

Microhylid frog sp. *Metamagnusia marani* **Günther,** 2009

Salom Marani is a Papuan friend of the describer. He lives at Yeretuar near the coast of Umar Bay, West Papua (Irian Jaya), Indonesia. He and the describer have collected together for many years.

Marcus

Villa Tunari Caecilian *Caecilia marcusi* **Wake**, 1985

Dr Harry Marcus (1880–1976) was a German, born in Egypt, who trained as a physician in Hamburg and served as a doctor in the German army in the First World War. He became Professor of Anatomy at Munich University (1915). He, his English wife and seven children were emigrating to South America when the Second World War broke out. They were interned in Marseilles but later released. They travelled to Cochabamba, Bolivia, where he became a professor at Cochabamba University. He returned to Munich (1954).

Mareku

Mareku's Treefrog *Litoria mareku* **Günther**, 2008

Genus Mareku lives in Yeretuar village, West Papua, and is a friend of the describer. He collected specimens of this treefrog and was invaluable with his support and help in the Wondiwoi Mountains.

Margareta

Margareta's Frog *Odorrana margaretae* **Liu,** 1950

Mrs Margaret W. Schmidt was the wife of Karl Patterson Schmidt (q.v.). Liu wrote that that he held her 'in affectionate regard'.

Maria

Maria's Giant Glass Frog *Nymphargus mariae* **Duellman** & **Toft**, 1979

[Syn. *Centrolene mariae*]

Maria Koepcke (see **Koepcke**) obtained the type specimen of this frog (1969).

Maria Elena

Glass Frog sp. *Chimerella mariaelenae* Cisneros-Heredia & **McDiarmid**, 2006

[Syn. *Centrolene mariaelenae*]

María Elena Heredia is the senior author's mother, of whom he said that he '...will always be grateful for her permanent support of his work in herpetology, her friendship, field companionship, and infinite love.' In another context, he stated that she 'provided love, support and financial assistance.' Cisneros-Heredia is noted for his rather flowery etymologies (see **Laura**).

Marian

Spaldings Treefrog *Osteopilus marianae* **Dunn**, 1926

[Alt. Yellow Bromeliad Frog]

Dunn collected the holotype (1925) in Jamaica. He always categorically refused to say who Marian was, and his description is notable for the absence of any clue!

Maridad

Beautiful Forest Toad *Churamiti maridadi* **Channing** & **Stanley**, 2002

Though it looks like an eponym, the binomial comes from the Swahili word for 'beautiful'.

Marinkelle

Stubfoot Toad sp. *Atelopus marinkellei* **Cochran** & **Goin**, 1970

Dr Cornelis Johanes Marinkelle (b.1925) is Emeritus Professor of the Laboratory of Microbiology and Parasitology at the University of the Andes in Bogotá. He was born in Vienna, is a Dutch citizen and lives in Colombia. The University of Utrecht awarded his doctorate in biology (1963). He worked and taught at many institutions in many places in England, Belgium, Indonesia, El Salvador, Iraq and Sudan before going to Colombia. The University of Michigan Museum of Zoology holds 454 specimens he collected. It is clear that he undertook a number of collecting trips (1950s–1960s), as other specimens from that period appear in other collections. He also amassed a collection of 25,000 birds' eggs, which he donated to the Humboldt Institute in 2001. He did not collect most of these personally, but accumulated the collections of others, originating as early as 1871. Others were purchased, either by him, his father or grandfather. He wrote a number of scientific papers, mostly on diseases and epidemiology of both humans and other mammals, (1970s–1990s). He has a mammal named after him.

Mark

Mark's Bush Frog *Raorchestes marki* **Biju** & Bossuyt, 2009

Dr Mark Wilkinson has worked as a researcher (since 2003) in the Herpetology Research Group at the BMNH, London, which he joined (1998). Oxford awarded his bachelor's degree in zoology (1985), London University his master's (1988), and the University of Bristol his doctorate (1992). He lectured in taxonomy at the University of Glasgow

(1993–1994) and in biological sciences at Bristol (1995–1998). He co-wrote *The conservation biology of caecilians* (2005).

Marnock

Marnock's Frog *Eleutherodactylus marnockii* **Cope**, 1878
[Alt. Cliff Chirping Frog]

Gabriel Wilson Marnock (1838–1920) emigrated from Scotland and settled in Texas c.1867. He collected natural history specimens for Cope (1877). He was Postmaster, Bexar, Texas (1904–1919). His widow sold his collection to John Strecker and it is now at Baylor University, Waco, Texas.

Mars

Western Tiger Salamander *Ambystoma mavortium* **Baird**, 1850
[Alt. Barred Tiger Salamander]
Mars Robber Frog *Pristimantis mars* **Lynch** & **Ruiz-Carranza**, 1996

Mars was the Roman god of war. 'Mavors' is another version of his name, and *mavortium* means 'of Mars'. Baird did not explain his choice of this binomial. It may have been inspired by the fact that the type specimen was taken in New Mexico during the Mexican-American War (1846–1848).

Marsha

Robber Frog sp. *Eleutherodactylus marshae* **Lynch**, 1964
[JrSYN *Pristimantis cruentus*]

Marsha Chalene Lynch née Fritts is the describer's wife.

Martens

Martens' Bush Frog *Pseudophilautus leucorhinus* Lichtenstein & Martens, 1856
[Alt. White-nosed Bush Frog]
Martens' (Puddle) Frog *Occidozyga martensii* **Peters**, 1867
[Alt. White-lined Flood Frog]

Eduard Carl von Martens (1831–1904) was a German zoologist and malacologist. He studied at the Universities of Stuttgart, Munich and Tübingen, graduating in 1855. He was zoologist on an expedition to the Far East (1860) and stayed on in the Malay Archipelago for a further 15 months after the expedition was over. He was Director, Department of Malacology and other invertebrates, Natural History Museum, Berlin (1862–1904), acting as the Museum's Director (1883–1887). Martens and Wilhelm Karl Peters (q.v.) worked together at the Museum, and Martens became Director when Peters died.

Martia

Martia's Robber Frog *Pristimantis martiae* **Lynch**, 1974

Professor Dr Martha ('Marty') L. Crump is currently (2011) Adjunct Professor at Northern Arizona University and a Conservation Fellow with the Wildlife Conservation Society. She received her doctorate from the University of Kansas (1974) and conducted post-doctoral work at Brooklyn College (1974–1976). She was a professor at the University of Florida, Department of Zoology (1976–1992). She has worked in Argentina, Brazil, Chile, Costa Rica and Ecuador and has written some 50 publications, including being one of six co-authors of the textbook *Herpetology* by Pough et al. She is also author of a children's book, *Amphibians, Reptiles, and their Conservation*. Perhaps her best-known work is *In Search of the Golden Frog* (2000). The etymology reads: 'Named for Martha ('Marty') L. Crump whose quest for ecologic knowledge of Amazonian frogs has resulted in an outstanding collection of Amazonian eleutherodactylines.'

Martienssen

Martienssen's Torrent Frog *Petropedetes martiensseni* **Nieden**, 1911
[Alt. Usambara Torrent Frog]

Herr Martienssen was a German colonial district administrator ('Stationsleiter' – station manager) at Tanga, Tanganyika (Tanzania). He collected the frog holotype. He also supplied the Museum für Naturkunde, Berlin, with the type specimen of a mammal (a bat) that is named after him.

Martin

Martin's Toadlet *Uperoleia martini* **Davies** & **Littlejohn**, 1986
[Alt. Coastal Toadlet]

Angus A. Martin (b.1940) is (2009) Adjunct Professor, Macquarie University, Sydney, Australia. He was formerly at the Department of Zoology, University of Melbourne. He was a founder member and first treasurer (1964) of the Australian Society of Herpetologists. Among his many papers and longer works is the book *Tasmanian Amphibians* (1982), written with Littlejohn.

Martínez

Martínez's Tropical Bullfrog *Leptodactylus martinezi* **Bokermann**, 1956

Antonio Martínez (1922–1993) was an Argentine entomologist and an expert on scarab beetles. He was Director, Catedra de Microbiología y Parasitología, Universidad de Buenos Aires, and a Principal Investigator with the Consejo Nacional de Investigaciones Científicas. In retirement he worked with the Instituto de Investigaciones Entomológicas Salta. He and his wife were killed in a motor accident in Bolivia, whilst on a scientific excursion.

Martins

Martins' Treefrog *Bokermannohyla martinsi* **Bokermann**, 1964

Dr Ubirajara Ribeiro Martins de Souza is a Brazilian entomologist at the Museu de Zoologia, Universidade de São Paulo, where he was a professor and a colleague of Bokermann. The Universidade Federal de Viçosa awarded his bachelor's degree in agronomy (1954), and the Universidade de São Paulo his doctorate (1975). He co-wrote *Notas sobre Calliini (Coleoptera, Cerambycidae, Lamiinae)* (1990).

Marty

Guianan Leaf-litter Toad *Rhinella martyi* Fouquet et al., 2007

Dr Christian Marty is a dermatologist, naturalist and herpetologist who works in French Guiana. He co-wrote *Atlas des Amphibiens de Guyane* (2000).

Mary (Inger)

Mary's Frog *Ingerana mariae* **Inger**, 1954
[Alt. Palawan Eastern Frog]

Mary Lee Inger née Ballew (1918–1985) was the first wife of the describer and accompanied him on many expeditions. They met in the refectory of Chicago University when he returned there after the Second World War.

Mary (Loveridge)

Mary's Reed Frog *Hyperolius mariae* **Barbour** & **Loveridge**, 1928
[Alt Usambara Reed Frog]

Mrs Mary V. Loveridge collected the holotype (1926). She was a botanist and collected (1920s–1930s) in Tanganyika (Tanzania), Uganda, Rhodesia (Zimbabwe), Burundi and Congo (DRC). She was the wife of Arthur Loveridge, the junior describer.

Maskey

Maskey's Burrowing Frog *Sphaerotheca maskeyi* Schleich & Anders, 1998
[Perhaps a junior synonym of *Sphaerotheca breviceps*]

Dr Tirtha M. Maskey (d. 2006) was Director General, Department of National Parks and Wildlife Conservation, Nepal. He was killed in a helicopter crash. He wrote *Movement and survival of captive reared gharial (*Gavialis gangeticus*) in the Narayani River, Nepal* (1999).

Maslin

Maslin's Wood Frog *Rana maslini* **Porter**, 1969
[Generally considered synonymous with *Lithobates sylvaticus*]

Dr Thomas Paul Maslin (1909–1984) was born to American missionary parents in China, where he lived until entering high school in Los Angeles (1927). He was at the University of California, Berkeley (1928–1933) and acquired a life-long interest in herpetology. He taught at an American school in China (1934–1936) and researched the local herpetofauna. He returned to the USA (1936) to do graduate study at Berkeley, taking his master's degree (1941). He was at Stanford (1941–1945), first teaching and then becoming Curator at the Natural History Museum (1943). Stanford awarded his doctorate in zoology (1945). He taught in Colorado at the State University and at the University of Colorado, Boulder, combining the roles of Assistant Professor and Curator of Zoology (1945–1974). He became Emeritus Professor (1975). He revisited China (1981). His papers and his collection of over 60,000 specimens are at the University of Colorado. A reptile is named after him.

Mason

Javan Torrent Frog *Huia masonii* **Boulenger**, 1884

George E. Mason sent a number of specimens to the BMNH, London from various locations in Japan, Burma (Myanmar) and the Dutch East Indies (Indonesia). This frog was collected near Batavia (Jakarta). He wrote *Description of a new earth-snake of the genus* Silybura *from the Bombay Presidency with remarks on little known Uropeltidae* (1888).

Mass

Madagascar Frog sp. *Spinomantis massi* **Glaw** & Vences, 1994
[Syn. *Mantidactylus massi*]

Robert Mass and Andrea Mass née Glaw are a married couple. Her father, the senior describer, named this frog after them to congratulate them on their marriage. (It has been suggested that the binomial should be amended to the plural form, *massorum*.)

Massart

Brazilian Big-tooth Frog *Megaelosia massarti* **De Witte**, 1930

Dr Jean Massart (1865–1925) was a Belgian physician and botanist at the University of Brussels, where he was Professor. He was also Curator (1902–1906), Jardin Botanique de l'État. The *Experimental Garden Jean Massart* at the Université Libre de Bruxelles commemorates him.

Mathiasson

Mathiasson's Treefrog *Dendropsophus mathiassoni* **Cochran** & **Goin**, 1970

Dr Sven Mathiasson (1925–2009) was a Swedish zoologist at Naturhistoriska Museet, Göteborg, where he was Head of the Vertrebrate Department (1993). He was formerly a lecturer at the University of Khartoum (Sudan) and led three zoological expeditions in Africa. The authors acknowledge their gratitude for being allowed to examine specimens in his care. He wrote *Naturhistoriska museet, Oljan och Fåglarna* (1981).

Matsui

Matsui's Borneo Frog *Meristogenys macrophthalmus* Matsui, 1986

Professor Dr Masafumi Matsui of Kyoto University (b.1950) is a Japanese herpetologist. He is President of the Herpetological Society of Japan and co-wrote *Frogs and Toads of Japan* (1999). Two reptiles are named after him.

Matuda

Matuda's Robber Frog *Craugastor matudai* **Taylor**, 1941

Matuda's Spikethumb Frog *Plectrohyla matudai* **Hartweg**, 1941

Dr Eizi Matuda (1894–1978) was a Japanese botanist who moved to Mexico (1922) and became a Mexican citizen (1928). All his university education was in Formosa (Taiwan), then under Japanese control. The University of Tokyo awarded his doctorate (1962). He taught in Japan and travelled to study flora in mainland Asia (1914–1921). In Mexico he eventually became Head of the Department of Botany, National Institute of Forestry. Three reptiles and three birds are named after him.

Mayorga

Mayorga Rocket Frog *Aromobates mayorgai* **Rivero**, 1980

Horacio Mayorga is a herpetologist at the Biology Department, University of Puerto Rico, where he is a colleague of the describer. He wrote *A rapid method for clearing and staining amphibian skeletons* (1965).

McCoy

McCoy's Toad *Incilius mccoyi* Santos-Barrera & Flores Villela, 2011

Dr Clarence John 'Jack 'McCoy (1935–1993) took his bachelor's (1957) and master's degrees (1960) at Oklahoma State University and was awarded his doctorate by the University of Colorado (1965) while working as a research assistant at the University's museum. Known for his phenomenal memory he could not only recall where he collected a specimen but, to quote his obituary in *Herpetologica*: '...he could tell you the rock it was collected under, the temperature that day, and what he had for lunch!' He joined the staff of the Carnegie Museum (1964), becoming Associate Curator (1970) and then Curator (1972). During his time there he more than doubled the size of the herpetological collection. He made seven collecting trips to Mexico and then more to Algeria, Uruguay, Paraguay, Argentina, Brazil and Belize. He was also Governor of the American Society of Ichthyologists and Herpetologists (1967–1972) and Vice-President (1984–1985) then President (1986–1987) of the Herpetologists' League. His publications include *Amphibians and Reptiles in Pennsylvania* (1982). A reptile is named after him.

McCranie

McCranie's Robber Frog *Craugastor chrysozetetes* McCranie**, Savage** & **Wilson**, 1989

James Randy McCranie is a postal worker and amateur herpetologist who lives in Miami and is an acknowledged expert on Honduran herpetology. His first fieldwork as a field biologist was (1976) in Honduras. He co-wrote *The Amphibians and Reptiles of the Honduran Mosquitia* (2006).

McDiarmid

McDiarmid's Rocket Frog *Allobates mcdiarmidi* Reynolds & Foster, 1992

McDiarmid's Thorius *Thorius munificus* **Hanken** & **Wake**, 1998

[Alt. McDiarmid Minute Salamander]

Glass Frog sp. *Rulyrana mcdiarmidi* Cisneros-Heredia et al., 2008

[Syn. *Cochranella mcdiarmidi*]

Dr Roy Wallace McDiarmid (b.1940) is a research zoologist and Curator, Amphibians and Reptiles, US Geological Survey, Patuxent Wildlife Research Center, and AMNH, New York. He trained to become a physician but swiftly changed course and decided on zoology. He received his master's degree (1966) and doctorate (1968) from the University of Southern California. He has worked and researched in Central and South America, particularly Mexico, Costa Rica, Peru and Ecuador. He taught at the University of South Florida, Tampa (1968–1978). He co-edited *Tadpoles: The Biology of Anuran Larvae* (1999). A reptile is also named after him.

McDonald

McDonald's Frog *Cophixalus mcdonaldi* **Zweifel**, 1985

Dr Keith R. McDonald (b.1950) is Vice-President (Projects), Wildlife Preservation Society, Queensland, and is Principal Senior Technical Officer, Threatened Species Unit, Queensland Department of Environment and Resource Management. He is an Honorary Research Associate of the Queensland Museum and has nearly 40 years' experience in conservation for Queensland National Parks and Wildlife Service. His main interests are the documentation and conservation of Queensland's flora and fauna, with special emphasis on rare and threatened frogs and reptiles, and distribution patterns of earthworms. He received the Queensland Museum Medal and was co-recipient of CSIRO Medal (2000) for studies on the chytrid fungus causing amphibian declines. A reptile is named after him.

McGregor

McGregor's Stream Toad *Ansonia mcgregori* **Taylor**, 1922

Richard Crittenden McGregor (1871–1936) was an Australian who first went to the USA as a child with his American mother, after the death of his father. He studied zoology at Stanford, but his academic career was interrupted by a fish-collecting expedition to Panama. He finally received his bachelor's degree (1898) in philosophy. He was appointed Ornithologist to the Manila Bureau of Science, Philippines. He wrote *A Manual of Philippine Birds* (1909) but never published anything on herpetology. He collected the toad holotype (1907). Three birds and a reptile are named after him.

Medem

Medem's Giant Glass Frog *Centrolene medemi* **Cochran** & **Goin**, 1970

Finca Chibigui Salamander *Bolitoglossa medemi* **Brame** & **Wake**, 1972

Robber Frog sp. *Pristimantis medemi* **Lynch**, 1994

Professor Dr Federico Medem (1912–1984) was born in Riga as Friedrich Johann Comte von Medem. He was of German origin but considered himself a Latvian. His family left Latvia after the Russian Revolution (1917) and moved to Germany. Medem studied at the Humboldt University, Berlin and at Tübingen. He worked for his doctorate at the marine biology station in Naples, run by Gustav Kramer. He served in the Wehrmacht in the Second World War and fought on the Russian front. After the war he worked in Germany and Switzerland, from where he moved to Colombia (1950), changed his name and became a Colombian herpetologist and ardent conservationist at the research station at Villavicencio and at the National University, Bogotá. There is a herbarium named after him at the Instituto Alexander von Humboldt. He wrote numerous scientific papers (1950s–1980s), mostly on Colombian reptiles. He published two volumes that make up his *Los Crocodylia de Sur America* (1981–1983). The crocodilians he collected (1955–1966) are in the Field Museum, Chicago. A mammal and four reptiles are named after him.

Medina

Rancho Grande Leaf Frog *Agalychnis medinae* Funkhouser, 1962

[Syn. *Phyllomedusa medinae*]

Dr Gonzalo Medina Padilla (d.2009) was a Venezuelan biologist who was Chief of the Rancho Grande Biological Station and Curator of the museum there (1957–1981). He collected the holotype and lent specimens to Funkhouser.

Megson

Stump-toed Frog sp. *Stumpffia megsoni* **Köhler** et al., 2010

Steven Megson is a research assistant at the environmental exploration company Frontier Madagascar. He collected the holotype.

Méhely

Méhely's Fanged Frog *Xenorhina mehelyi* **Boulenger,** 1898

Méhely's Land Frog *Austrochaperina mehelyi* **Parker**, 1934

Méhely's Humming Frog *Chiasmocleis mehelyi* **Caramaschi** & **Cruz**, 1997

Lajos Ludwig von Méhely (1862–1946 / 1953) was a Hungarian biologist, herpetologist and anthropologist. (He was a political prisoner after the Second World War and the date of his death seems unclear, although one authority gives it as 4 February 1953) He became (1913) Head of the Zoological Department at the Hungarian National Museum, and also Professor of Zoology and Anatomy at the University Pázmány Péter, Budapest. He was extremely racist and embraced fascist ideology, leading to his imprisonment after WW2. He wrote a seminal, unpublished monograph called *Herpetologia Hungarica*. The Museum's excellent herpetology collection was almost entirely destroyed in the Budapest uprising (1956). One mammal and two reptiles are named after him.

Mello

Southeastern Brazil Snouted Treefrog *Scinax melloi* **Peixoto**, 1989

Dr Rubens Pinto de Mello is an entomologist at the Instituto Oswaldo Cruz, Rio de Janeiro, Brazil, where he is a researcher. The Universidade Federal Rural do Rio de Janeiro awarded all his degrees: bachelor's (1961), master's (1967) and doctorate in veterinary parasitology (1989).

Mendelson

Water Frog sp. *Telmatobius mendelsoni* **De la Riva**, **Trueb** & **Duellman**, 2012

Dr Joseph 'Joe' Ralph Mendelson III (b.1964) is Curator of Herpetology, Zoo Atlanta, Adjunct Associate Professor (Biology) Utah State University and President of the Society for the Study of Amphibians and Reptiles. His research is mostly focused on the systematics of Central American anurans.

Menon

Menon's Caecilian *Uraeotyphlus menoni* **Annandale**, 1913

Professor K. Ramunni Menon of Madras collected the holotype and presented it to the Indian Museum. He became Vice-Chancellor of the University of Madras (1929).

Menzies

Iarowari School Frog *Pherohapsis menziesi* **Zweifel**, 1972

Dr James Ian Menzies (b.1928) was a professor in the Biology Department of the University of Papua New Guinea, which awarded him an honorary doctorate in science (2003). He is now Emeritus and Editor of *Science in New Guinea*. He taught biology at Bo School in Sierra Leone (1956–1961), and was the first Education Officer at Paignton Zoo, Devon (1961–1962). He then worked in Nigeria, before moving to Papua New Guinea (1967–1977), returning there (1980) after working in Swaziland and Lesotho. He was Curator of Natural History, National Museum, Port Moresby (1983–1987). He was a visiting research fellow, University of Adelaide, South Australia, where he has lived since 2000. He co-wrote *The Flora of Motupure Island–Papua New Guinea* (1995). Two mammals are named after him.

Mercedes

Mercedes' Robber Frog *Pristimantis mercedesae* **Lynch** & **McDiarmid**, 1987

[Syn. *Yunganastes mercedesae*]

Dr Mercedes S. Foster McDiarmid (b.1942) is a research zoologist with the Smithsonian at the Patuxent Research Centre. She works closely with Roy McDiarmid and (1987) they co-wrote *Additions to the Reptile Fauna of Paraguay with Notes on a Small Herpetological Collection from Amambay*. We believe that socially she is known as Mrs McDiarmid. A bird is named after her.

Merian

Toad sp. *Rhinella merianae* **Gallardo**, 1965

Anna Maria Sibylla Graff née Merian (1647–1717) was born in Frankfurt into a Swiss family and lived most of her life in Germany, then latterly in the Netherlands. She was a pioneering female naturalist, scientific illustrator and trader in insect specimens. She was recognized as talented when she was 13, when she began studying insects and plants, painting what she saw. She married (1665) Johann Andreas Graff, but continued to paint and teach students. She was the first person to make clear the life cycle of a butterfly and used her illustrations for a book that she published, *Der Raupen wunderbare Verwandlung und sonderbare Blumennahrung (The Caterpillar, Marvellous Transformation and Strange Floral Food)* (1679). She left her husband (1685), he divorced her (1692), and she took her mother and two daughters to a Labadist religious community in Friesland, Netherlands, moving (1691) to Amsterdam. When her elder daughter married and moved to Surinam, Anna went too (1699). She stayed for two years, studying South American fauna and flora. She was extremely critical of the Dutch planters' treatment of their black slaves and the local Amerindians. She contracted malaria and had to return to Amsterdam (1701). She suffered a stroke (1715), became partially paralyzed and died a pauper. Recently she has become something of a cult figure in Germany where her portrait appeared on a bank note, schools are named after her, and a research ship 'Maria S Merian' was named in her honour, as was a reptile.

Merlin

Merlin's Clawed Frog *Pseudhymenochirus merlini* **Chabanaud**, 1920

Martial Henri Merlin (1860–1935) was a French colonial administrator. He was Governor-General of Senegal (1907–1908 & 1919–1923), French Congo (1908–1910), French Equatorial Africa (1910–1917), Madagascar (1917–1918) and Indochina (1923–19250. A bird is named after him.

Mertens

Salamander genus *Mertensiella* **Wolterstorff**, 1925
Toad genus *Mertensophryne* Tihen, 1960
Mertens' Leaf Frog *Phyllomedusa rohdei* Mertens, 1926
[Alt. Rohde's Leaf Frog]
Mertens' Yellow Treefrog *Dendropsophus robertmertensi* **Taylor**, 1937
Mertens' Running Frog *Kassina mertensi* **Laurent**, 1952
Mertens' Tree Toad *Hylodes mertensi* **Bokermann**, 1956
Mertens' Egg Frog *Leptodactylodon mertensi* **Perret**, 1959
Mertens' Rocket Frog *Colostethus mertensi* **Cochran** & **Goin,** 1964
Marsupial Frog sp. *Gastrotheca mertensi* Cochran & Goin, 1970 NCR
[JrSYN *Gastrotheca aureomaculata*]
Mertens' Caecilian *Caecilia mertensi* Taylor, 1973
Mertens' Smalltongue Toad *Werneria mertensiana* **Amiet**, 1976
Yellow Cururu Toad ssp. *Rhinella icterica mertensi* Cochran, 1950
Black-striped Sedge Frog *Hyperolius quinquevittatus mertensi* **Poynton**, 1964

Robert Friedrich Wilhelm Mertens (1894–1975) was a German zoologist and herpetologist who was born in St Petersburg. He left Russia (1912) to study medicine and natural history, obtaining his doctorate from the University of Leipzig (1915). After serving in the German army (WW1), he worked at the Senckenberg Museum in Frankfurt as an assistant (1919–1920). He was appointed (1920) to replace the dismissed Sternfeld (q.v.) in charge of herpetology, becoming Curator (1925) and Director (1947). He was a man of prodigious energy; his wife was his sole assistant (1920–1943), and during that period he was additionally in entire charge of mammals (1919–1953), birds (1923–1947) and fishes (1920–1954). He was Chairman of the Zoology Section (1934–1955). He retired as Director Emeritus (1960). He lectured at the University of Frankfurt-am-Main (1932–1939) and became Professor (1939). In spite of these incredible responsibilities and huge workload, he still found time to publish c.800 scientific paper and 13 books. He visited Indonesia in the early 1920s and published *Über einige Reptilien aus Borneo* (1924). During the Second World War he had most of the Museum collection evacuated to small towns, where its contents were set up in such locations as dance halls for use and study. He encouraged German soldiers fighting outside Germany to collect specimens for him, and a regular supply of reptiles and other creatures reached him by courtesy of the German Field Post Office system. He died, aged 81, after a bite from a Forest Vine Snake *Thelotornis kirtlandi,* a South African snake that he had long kept as a pet. No antivenin then existed for this species. He took 18 very painful days to die and kept a diary of each day's events, remarking with true gallows humour, 'für eined Herpetologen einzig angemessene Ende' (a singularly appropriate end for a herpetologist). A mammal, a bird and seven reptiles are named after him.

Merton

Merton's Cross Frog *Oreophryne mertoni* **Roux**, 1910

Dr Hugo Merton (1879–1939) was a German zoologist. He visited the Kai and Aru Islands in eastern Indonesia with Roux (1907–1908). He took a teaching post (1912) at Universität Heidelberg, becoming Professor of Zoology (1920), but was dismissed (1935) when the Nazis 'cleansed' the universities under the provisions of the Nuremberg citizenship laws. He moved to Britain (1937) to work at the Institute of Animal Genetics, University of Edinburgh. He tried to visit Heidelberg (late 1938), but was arrested and sent to Dachau concentration camp. Although he was later released and returned to Edinburgh, his health was broken and he died soon afterwards. A reptile and two birds are named after him.

Metcalf

Metcalf's Salamander *Plethodon metcalfi* **Brimley**, 1912
[Alt. Southern Grey-cheeked Salamander]

Professor Dr Maynard Mayo Metcalf (1868–1940) was an American zoologist. His doctorate was awarded by Johns Hopkins University (1893). He was Associate Professor and then full Professor of Biology, Goucher College, Baltimore (1893–1906). He was Professor of Zoology, Oberlin College, Ohio (1906–1914) and was able to get time off to undertake zoological studies in Germany and Italy (1907–1908). He was Chairman, Division of Agriculture, National Research Council (1924–1925 and 1926–1933) and a trustee of the Marine Biology Laboratory, Woods Hole (1896–1932). He wrote *An Outline of the Theory of Organic Evolution* (1904). He was the only scientist allowed to testify on the stand at the 1925 Scopes anti-evolution trial (*aka* the Scopes Monkey Trial).

Methner

Methner's Toad *Spelaeophryne methneri* **Ahl**, 1924
[Alt. Scarlet-snouted Frog]

Wilhelm Methner (b.1871) was a colonial administrator in German East Africa (Tanzania) (1892–1908). He became administrator of the southern part (1905) and was District Chief of Moschi (1906–1908). During the First World War he was a lieutenant in the German army in German East Africa. His collection of zoological specimens, including the holotype of the toad, is in the Museum für Naturkunde, Berlin. He wrote *Unter Drei Gouverneuren – 16 Jahren Dienst in deutschen Tropen* (1938).

Mette

Mette's Reed Frog *Hyperolius pseudargus* **Schiøtz** & Westergaard, 1999

Mette Mejsen Westergaard, the junior describer, is a Danish biologist and photographer who is interested in amphibians and works very closely with Arne Schiøtz. The University of Copenhagen awarded her master's degree (1998). She collected the holotype.

Meyer, Adolf

Meyer's Wrinkled Ground Frog *Platymantis meyeri* **Günther**, 1873 NCR
[JrSYN *Platymantis dorsalis*]

Dr Adolf Bernhard Meyer (1840–1911) was a German anthropologist and ornithologist who collected in the East Indies, New Guinea and the Philippines. He was Professor at the Ethnographische Museum, Dresden, becoming Director, Staatlisches Museum für Tierkunde, Dresden (1872). He wrote *The Birds of the Celebes and Neighbouring Islands* (1898). He was very interested in the evolution debate and corresponded with Wallace. Seventeen birds, a mammal and two reptiles are named after him.

Meyer, Axel

Tsaratanana Bright-eyed Frog *Boophis axelmeyeri* Vence, **Andreone** & Vieites, 2005

Dr Axel Meyer (b.1960) is a German evolutionary biologist and ichthyologist. Since 1997 he has been Professor of Zoology and Evolutionary Biology, Universität Konstanz, Baden-Württemberg, Germany. He studied biology at Christian-Albrechts-Universität zu Kiel and (1982) moved to the USA and studied at the University of Miami, Florida. He received his master's and doctorate from the University of California, Berkeley, where he also did post-doctoral work (1988–1990). He was at Stony Brook University, New York, as Assistant (1990–1993) and Associate Professor (1993–1997). He wrote *Molecular approaches to the phylogenetic study of vertebrates* (1993).

Michael

Michael Rainforest Frog *Cophixalus nubicola* **Zweifel**, 1962

This frog is named after the type locality – Mount Michael, Eastern Highlands Province, Papua New Guinea.

Midas

Santa Cecilia Cochran Frog *Teratohyla midas* **Lynch** & **Duellman**, 1973
[Syn. *Cochranella midas*]

King Midas of Phrygia is a figure in Greek mythology. The god Dionysus granted Midas's wish that anything he touched be turned to gold. According to the describers' etymology, the name was associated with this Amazonian frog as it is found 'along the Rio Aguarico, meaning 'rich water', in reference to gold found in the river, and in allusion to the gold flecks on the frogs.' A bird and two mammals carry the name.

Miguel (Formas)

Miguel's Ground Frog *Eupsophus migueli* Formas, 1978

Miguel Formas is the describer's son.

Miguel (Rodrigues)

Miguel's Button Frog *Cycloramphus migueli* **Heyer**, 1988

Dr Miguel Trefaut Urbano Rodrigues (b.1953) is a Brazilian herpetologist. He obtained his doctorate (1984) from the University of São Paulo. He became (1996) Professor of Biological Sciences, Universidad de São Paulo and was Director of the University's Zoological Museum (1997–2001). The frog's etymology says that Rodrigues '…has made significant collections of frogs throughout much of Brazil.' He collected the holotype in 1986. A fish and five reptiles are named after him.

Miles

Miles' Robber Frog *Craugastor milesi* **Schmidt**, 1933

M. S. Miles was an American resident in Honduras. He and his wife played host to Schmidt as their house was very convenient for collecting in the local area. Schmidt's etymology says the frog was 'Named for M S Miles, our genial and helpful host at the Hacienda Santa Ana in 1928.'

Miller

Miller's Treefrog *Hyla evittata* Miller, 1899 NCR
[JrSYN *Hyla cinerea*]

Gerrit Smith Miller Jr (1869–1956) was an American zoologist. He grew up in relative isolation on a country estate and through the influence of his great uncle, an ornithologist, he developed an early interest in natural history. After graduating from Harvard (1894) he joined the Biological Survey, Department of Agriculture. He joined the Smithsonian (1898) as Assistant Curator, Mammals, becoming Curator (1909–1940). He remained an associate in biology at the Smithsonian (1940–1956). He was an early critic of the claimed discovery of 'Piltdown Man' in England, concluding that the jaw actually came from an ape (which it had!) His main interest was in mammals and ten are named after him.

Millet

Millet's Frog *Hylarana milleti* **Smith**, 1921

Fernand Millet was Superintendent of Forests, Annam, French colonial Vietnam. A professional hunter since 1902, he appears to have been responsible for considerable slaughter. He acted as a professional guide for important visitors, such as the French journalist Albert Londres, who wrote that by the time he met him in Indochina, Millet had already shot 47 tigers. He wrote (1930) *Les grands animaux sauvages de l'Annam, leurs moeurs, leur chasse, et leur tir*, a title that can be rendered as *The Large Wild Animals of Annam, their Manners, their Hunting, and their Shooting*. A mammal and a bird are named after him.

Millet Horsin

Mali Screeching Frog *Tomopterna milletihorsini* **Angel**, 1922
[Syn. *Arthroleptis milletihorsini*]

Major Dr H. P. Millet-Horsin was a medical officer in the French colonial army, whose various postings in peace and war largely governed where he could collect. He was primarily an ornithologist but made a collection of amphibians (including this frog) and reptiles in French Sudan (1921). He was one of the first people seriously to study the birds of Ivory Coast, where he made two extended visits (1906–1907 & 1913–1914). He collected in Tunisia (1912) and Greece (1915). After the First World War he was in Senegal and Togo. He wrote *Contribution à l'étude de la faune ornithologique du Bas-Togo* (1923). The binomial of this frog's name is unusual in that both parts of the doctor's surname are in the genitive.

Millot

Millot's Froglet *Platypelis milloti* **Guibé**, 1950

Professor Dr Jacques Millot (1897–1980) was a French physician and arachnologist at the Muséum National d'Histoire Naturelle, Paris. As well as publishing on spiders, he wrote *Biology of the Human Races* (1952). He spent many years in Madagascar in charge of the Scientific Institute of Madagascar and (1953) started the periodical *Le Naturaliste malgache*. A reptile is named after him.

Millson

Niger Forest Treefrog *Leptopelis millsoni* **Boulenger,** 1895

Alvan W. Millson (1861–1896) was Assistant Colonial Secretary at Lagos. As Commissioner to the Interior he was sent to Ibadan to resolve the conflict there. Among other papers, he wrote *The Yoruba country, West Africa* in the Proceedings of the Royal Geographical Society (1891). In it he described Ibadan as the 'London of Negroland' with a population of c.120,000 and an economy based on slave trading. He collected a large number of earthworm species in the area around the Niger, which he bought back to England – one of which was named in his honour in both the generic and specific names: *Alvania millsoni*. He also sent specimens of other flora and fauna taxa to Boulenger at the BMNH.

Milne

Reed Frog sp. *Hyperolius milnei* **Loveridge**, 1935 NCR
[JrSYN *Hyperolius pusillus*]

R. D. Milne was a planter at Witu, Kenya. He was Loveridge's host for five days (1934), during which he helped Loveridge hunt for frogs.

Miranda

Miranda's White-lipped Frog *Leptodactylus macrosternum* Miranda-Ribeiro, 1926
Toad sp. *Rhinella mirandaribeiroi* **Gallardo**, 1965
Miranda's Disc Frog *Synapturanus mirandaribeiroi* **Nelson** & **Lescure**, 1975
Miranda's Button Frog *Cycloramphus mirandaribeiroi* **Heyer**, 1983

Dr Alipio de Miranda-Ribeiro (see **Alipio**). Two birds are named after him.

Miré

Mount Okou Wolterstorff Toad *Wolterstorffina mirei* **Perret**, 1971

Philippe Bruneau de Miré (b.1921) was a French entomologist who was in French West Africa and the Sahara. After the collapse of France in the Second World War, he was in Algeria (1940), but came back to France (1941) for studies in botany, zoology and geology at the Sorbonne, Paris. He appears to have annoyed the Occupation authorities, was forbidden entrance to the Sorbonne and was about to be arrested. He acquired a new identity card in the name of Vauquelin and (1942) was at the Entomology Department, Muséum National d'Histoire Naturelle, Paris. He eked out a living as a freelance journalist in Paris until the Liberation. During this time he became friendly with Baron Breunig, an officer in the Wehrmacht and a fellow entomologist. Together they prevailed upon General von Chölitz, German Military Governor, Paris, to forbid the destruction of the best trees in the forest of Fontainbleu. Miré became an official at the National Institute for Agriculture (1945), investigating desert locusts. He worked in Africa (1946–1956), conducting expeditions and surveys in Morocco, Mauretania, Chad and Niger. He resigned (1957) and took a year's sabbatical at the Muséum National d'Histoire Naturelle, Paris. He became a correspondent for the Museum, which involved further travels in Africa, visiting (1958–1961) Chad and Cameroon, followed by time (1962–1963) at the Museum. He was entomologist for the French Institute of Coffee and Cocoa Research, Centre Agricultural, Nkolbisson, Yaounde, Cameroon (1964–1974). He started a laboratory at Montpellier (1974) devoted to identifying pests of tropical crops. He retired (1984) but continued his involvement in Africa and his concerns for French forests. A bird is named after him.

Miriam

Miriam's Frog *Phyzelaphryne miriamae* **Heyer**, 1977

Mrs Miriam Heyer is the wife of the describer, William Heyer. A reptile is named after them both.

Miroslawa

Miroslawa's Andes Frog *Phrynopus miroslawae* Chaparro, **Padial** & **De la Riva**, 2008

Miroslawa Jagielko of Poland is a BIOPAT supporter and recognized for '…her friendship and her support of taxonomic research and nature conservation in Peru.'

Mitchell, BL

Mitchell's Reed Frog *Hyperolius mitchelli* **Loveridge**, 1953

B. L. Mitchell was a biologist and naturalist working for the Department of Game, Fish and Tsetse Control in Nyasaland (Malawi) in the 1940s. He worked for the Rhodes-Livingstone Museum (1960). He wrote *Some reptiles and amphibians of Nyasaland* (1950). A reptile is named after him.

Mitchell, JD

Mitchell's Narrow-mouthed Toad *Gastrophryne areolata* **Strecker**, 1909 NCR

[Type specimen may be a hybrid: *Gastrophryne olivacea* x *G. carolinensis*]

Joseph Daniel Mitchell (1848–1922) was a pioneer entomologist and conchologist. He worked as a cowboy as a teenager, then became a rancher (1867–1887). His interest in entomology sprang from his research to eradicate cattle ticks. He retired to study natural history and began collecting Indian relics, minerals and various animal taxa. For four decades he sent specimens to the Smithsonian. He began collaborating with the US Bureau of Entomology on cotton pests and cactus-feeding insects. He was known as the leading Texan herpetologist and the type specimen of the toad came from his collection.

Mittermeier

Mittermeier's Rocket Frog *Hyloxalus mittermeieri* **Rivero**, 1991

Mittermeier's Shrub Frog *Pseudophilautus mittermeieri* Meegaskumbura & Manamendra-Arachchi, 2005

Mittermeier's Stubfoot Toad *Atelopus mittermeieri* Acosta-Galvis et al., 2006

Dr Russell Alan Mittermeier (b.1949) is an American primatologist, herpetologist and conservationist. He has been President of Conservation International since 1989. Previously he worked for 11 years for the World Wildlife Fund–USA. He received his bachelor's degree (1971) at Dartmouth, New Hampshire and his doctorate from Harvard (1977). He has been Adjunct Professor, State University of New York at Stony Brook, since 1978. His fieldwork in Brazil, Madagascar and other tropical locations has led to the discovery of several new primate species. He co-wrote *Lemurs of Madagascar* (1994). Two mammals are named after him.

Miyata

Miyata's Robber Frog *Pristimantis miyatai* **Lynch**, 1984

Hosteria La Selva Treefrog *Dendropsophus miyatai* **Vigle** & Goberdhan-Vigle, 1990

Dr Kenneth Ichiro Miyata (1951–1983) was an American herpetologist of Japanese descent who worked at the Museum of Comparative Zoology at Harvard. He was also a noted fly-fisherman, but this passion led to disaster: he drowned whilst fishing alone on the Big Horn River, Montana. Dr Kenneth Miyata Scholarships are awarded to students who are continuing to promote and preserve cultural heritage unique to Japanese ancestry. Two reptiles are named after him.

Mjöberg

Mjöberg's Toadlet *Uperoleia mjobergii* **Andersson**, 1913
Mjöberg's Dwarf Litter Frog *Leptobrachella mjobergi* **Smith**, 1925
Murud Bubble-nest Frog *Philautus mjobergi* Smith, 1925

Dr Eric Georg Mjöberg (1882–1938) was a Swedish naturalist, ethnographer and explorer. He took his initial degree at Stockholm University (1908) and his master's at Lund (1912). He had various jobs in Sweden, including working at the National Museum and teaching in high schools (1903–1909). He led Swedish scientific expeditions in northwest Australia (1910–1911) and Queensland (1912–1913). He worked in Sumatra at an experimental station (1919–1922), which he combined with being Swedish Consul. He was Curator of the Sarawak Museum (1922–1924) and led a scientific expedition to Borneo (1925–1926). He died in poverty in Stockholm after suffering a long, undiagnosed illness. Two reptiles and two birds are named after him.

Mo

Chinese Newt sp. *Pachytriton moi* Nishikawa, Jiang & **Matsui**, 2011

Yun-Ming Mo is an associate professor at the Guangxi Natural History Museum. The etymology relates that he '...actively elucidated faunal diversity in Guangxi and helped the describers with their field surveys.' He has written papers on new amphibians, including co-writing *A new species of* Rhacophorus (Anura: Ranidae) *from China* (2008).

Mocquard

Mocquard's Rain Frog *Scaphiophryne calcarata* Mocquard, 1895
Sulawesi frog sp. *Hylarana mocquardii* **Werner**, 1901
Mocquard's Toad *Mertensophryne mocquardi* **Angel**, 1924
Mocquard's Madagascar Frog *Mantidactylus mocquardi* Angel, 1929

François Mocquard (1834–1917) was a French herpetologist. He gained degrees in mathematical sciences (1865) and medicine (1873), but then changed his career to study natural history. He wrote *Synopsis des familles, genres et espèces des reptiles écailleux et des batraciens de Madagascar* (1909). After obtaining his doctorate (1884) he worked in the ichthyology and herpetology department of the Muséum National d'Histoire Naturelle, Paris. Eighteen reptiles are named after him.

Modigliani

Modigliani's Huia Frog *Huia modiglianii* Doria, Salvido & Tavano, 1999

Emilio Modigliani (1860–1932) was an Italian zoologist and anthropologist who collected in Sumatra (1886–1894). This Sumatran frog was described on the basis of museum specimens known to pre-date 1920. Four reptiles and three birds are named after him.

Moehring

Moehring's Smooth Horned Frog *Proceratophrys moehringi* **Weygoldt** & **Peixoto**, 1985

Karl Heinz Möhring (d.2003) and his family were very hospitable to the authors during their time in Vitoria, Brasil and this they acknowledged. Möhring returned home to Germany at the end of the Second World War and moved (1947) from the Russian zone to Hamburg, where he started a wholesale veneer business. Over the years the company became the Moehring group of companies and bought timbers from all over the world. He moved to North Carolina and started Atlantic Veneer there (1964). They had an office in Vitoria when (1968) the Brazilian government imposed an embargo on the export of logs and only allowed wood products to be exported. Möhring therefore decided to build a veneer factory in Vitoria and moved his family to Brazil, living and working there until they moved back to the USA (1989). The business is now run by his grandsons, who have the surname Weygoldt; as the senior author was Research Associate and Visiting Professor at Duke Marine Laboratory, North Carolina (1965–1967), we suspect a family connection!

Moellendorff

Culion Frog *Hylarana moellendorffi* **Böttger**, 189

Otto Franz von Möllendorff (1848–1903) was a German expert on living and fossil molluscs. He started his career in China, where he had gone (1873) to learn to be an interpreter. His elder brother, Paul Georg von Möllendorff, had been in China since 1869 and the brothers appear not to have left China until 1882. He wrote a number of articles including *On the supposed New Zealand species of Leptopoma* (1893). Two mammals and a reptile are named after him.

Moffett

Sarisariñama Rocket Frog *Anomaloglossus moffetti* Barrio-Amorós & **Brewer**-Carias, 2008

Mark W. Moffett (b.1958) is a biologist, entomologist and photographer famous for his ant photography. Beloit Colege, Wisconsin awarded his bachelor's degree (1979) and Harvard his doctorate (1989). He became Curator of Ants at the Museum of Comparative Zoology, Harvard (1989), later becoming Research Associate at the Department of Anthropology (1997–2000) and Visiting Scholar at the Museum of Vertebrate Zoology, University of California, Berkeley (1998–2005). He is currently Research Associate at the Department of Entomology at the Smithsonian. He has travelled and explored in many parts of the world, including Myanmar where he was with Joseph Bruno Slowinski (2001) when Slowinski was bitten by a krait and died from the bite.

Molinari

Las Playitas Rocket Frog *Aromobates molinarii* **La Marca**, 1985

Dr Jesús Molinari is at the Department of Biology, Faculty of Sciences, University of the Andes, Mérida, Venezuela and Research Associate at the AMNH, New York. La Marca named the frog after Molinari '...in acknowledgement for his invaluable help and unmatched companionship in the field'.

Moller

Iberian Treefrog *Hyla molleri* **Bedriaga**, 1890

Moller's Gulf Frog *Hyperolius molleri* Bedriaga, 1892

Adolphe F. Moller (1842–1920) was a Portuguese botanist who collected on São Tomé. He published *Notes sur les Amphibiens et Reptiles recueillis par M. Adolphe F. Moller aux Îles de la Guinée* (1892). A bird is also named after him.

Molleturo

Molleturo Robber Frog *Pristimantis ruidus* **Lynch**, 1979

This frog is named after the type locality, Molleturo, Ecuador.

Moloch

Black-spotted Frog *Theloderma moloch* **Annandale**, 1912

[Alt. Eerie Tree Frog]

Moloch was the sun god of the ancient Canaanites to whom children were sacrificed. Later Christians viewed him as a hideous 'fallen angel'. Milton's *Paradise Lost* calls Moloch 'horrid King besmear'd with blood of human sacrifice, and parents tears'. Two mammals and reptile are also named after the deity.

Moltrech

Moltrecht's Green Treefrog *Rhacophorus moltrechti* **Boulenger**, 1908

Dr Arnold Christian-Alexander Moltrecht (1873–1952) was a Latvian entomologist and lepidopterist who qualified at Tartu University, Estonia as a physician and ophthalmologist (1899). He worked in St Petersburg before being put in charge (1906) of mobile eye clinics in Russia's Far East. Based in Vladivostock, he travelled in the Amur region, parts of China, and visited Formosa (Taiwan) to collect birds, butterflies and fishes (1907).

Mondolfi

Mondolfi's Robber Frog *Pristimantis mondolfii* **Rivero**, 1984

Glass Frog sp. *Hyalinobatrachium mondolfii* **Señaris** & **Ayarzagüena**, 2001

Professor Dr Edgardo Mondolfi (1918–1999) was a Venezuelan mammalogist, ecologist and biologist. He was Professor, Metropolitan University, Caracas. He had a lifelong interest in manatees and made studies of their distribution. He was on the board of several conservation bodies and, in later life, was Venezuela's Ambassador to Kenya. He co-wrote *Jaguar* (1993). He died from dengue fever. A mammal is named after him.

Monjerai

Gunung Jerai Odorous Frog *Odorrana monjerai* **Matsui** & Jaafar, 2006

The binomial is derived from the name of the type locality, Gunung Jerai in Malaysia. The describers state: 'It is also a dedication to the Kedah State Government, the Kedah Forestry Department, and the Malaysian Forestry Department, which are the governing bodies of the Gunung Jerai Forest Reserve.'

Monniche

Dunn's Robber Frog *Craugastor monnichorum* **Dunn**, 1940

Tollef Bache Monniche (1874–1958) of Boquete was a Canal Zone engineer (he designed the auxiliary locks). Together with his wife, Julia Huger Monniche (1884–1975), he later owned a coffee plantation Finca Lerida (known as Monniche's Plantation) close to Volcan Chiriquí, Panama's highest mountain. There, in a region partly under cultivation but with much of the original forest intact, Monniche studied birds and amassed a collection of some 2,000 meticulously labelled specimens representing more than 250 distinct forms – later acquired by the Field Museum,

Chicago. Julia seems to have collected all sorts of taxa, including two species of tick new to science, which she removed from her dog!

Mono Hernández

Hernández's Stubfoot Toad *Atelopus monohernandezii* Ardile-Robayo, Osorno-Muñoz & **Ruiz-Carranza**, 2002

Jorge Ignacio Hernández-Camacho (see **Hernandez** - he was apparently known by the nickname 'El Mono Hernández').

Montalenti

Corsican Painted Frog *Discoglossus montalentii* **Lanza** et al., 1984

Professor Giuseppe Montalenti (1904–1991) was an Italian geneticist. He graduated in Rome (1926) and studied genetics in Montpellier (1929), becoming an assistant professor at the University of Bologna (1933–1937). After a short time in the USA he was appointed the first Professor of Genetics at Naples University (1940–1960). He was appointed Secretary General (1953–1958) and later (1958–1961) President of the International Union of Biological Sciences.

Montandon

Montandon's Newt *Lissotriton montandoni* **Boulenger**, 1880

[Alt. Carpathian Newt]

Dr Arnold Lucien Montandon (1852–1922) was a French naturalist, primarily an entomologist, at the Natural History Museum 'Grigore Antipa', Bucharest (1896–1907). He worked in Romania and the Carpathian region from the late nineteenth century until his death. He wrote *Espèces d'Hémiptères-Hétéroptères d'Algérie et de Tunisie* (1897).

Montezuma

Montezuma Leopard Frog *Lithobates montezumae* **Baird**, 1854

Montezuma II (1480–1520) was the last Aztec emperor. During his reign the Aztec Empire reached its zenith. He was killed during the initial stages of the Spanish conquest of Mexico. Two birds and a reptile are named after him.

Monzón

Monzón's Moss Salamander *Cryptotriton monzoni* **Campbell** & Smith, 1998

José Monzón Sierra, an entomologist at the Universidad del Valle de Guatemala, collected with the junior author (after whom he has named a beetle). The etymology states the salamander is named '...in appreciation of his tireless efforts during the course of our field investigations and other assistance to us in Guatemala.' He wrote *Three new species of* Chrysina *Kirby (Coleoptera: Scarabaeidae; Rutelinae) from Guatemala and Mexico* (2010).

Moodie

Luzon Wart Frog *Fejervarya moodiei* **Taylor**, 1920

Dr Roy Lee (né Louis Roy) Moodie (1880–1934) was an American palaeopathologist to the Wellcome Historical Medical Museum, and was a professor of palaeontology at the University of California, Los Angeles. His bachelor's degree was awarded by the University of Kansas, Lawrence. In 1908 he was Assistant Professor of Zoology at the University of Chicago. He wrote *New or little known forms of Carboniferous amphibia in the American Museum of Natural History* (1909).

Moore, JA

Moore's Frog *Litoria moorei* **Copland** 1957

[Alt. Motorbike Frog, Western (Green & Golden) Bell Frog]

Moore's Frog *Lithobates johni* **Blair**, 1965

[Replacement name for *Rana moorei* Blair, 1947]

Dr John Alexander Moore (1915–2002) was an American herpetologist and biologist. His bachelor's, master's and doctorate were all from Columbia University, New York. He taught at Brooklyn and Queens College and became Assistant Professor, Barnard College (1943–1949) and Professor of Zoology, Columbia University (1949–1952). He went to Australia (1952) as a Fulbright Scholar, collecting specimens and eventually writing *Frogs of Eastern New South Wales* (1961). He was Professor of Biology, University College Riverside (1969–1982), retiring as Emeritus Professor. He was Research Associate, Department of Herpetology, AMNH (1942–2002). He wrote *Principles of Zoology* (1957).

Moore, JP

Moore's Salamander *Desmognathus marmoratus* Moore, 1899

[Alt. Shovel-nosed Salamander]

John Percy Moore (1869–1965) was an American zoologist. The University of Pennsylvania awarded his bachelor's degree (1892) and doctorate (1896) and he worked there as an instructor in zoology (1892–1907), Assistant Professor (1907–1909), and thereafter Professor. He was also Assistant Curator (1902) at the Academy of Natural Sciences, Philadelphia, and was employed off and on by the US Fish Commission (1890).

Mooreorum

Moores' Shrub Frog *Pseudophilautus mooreorum* Meegaskumbura & Manamendra-Arachchi, 2005

Dr Gordon E. Moore and his wife Betty are Americans who have made significant contributions to conservation. They made a generous donation to the University of Cambridge for the construction of a new £7.5m library for physical sciences, technology and mathematics. Dr Moore was Director of Development, Fairchild Semiconductor (1960s), and made very accurate predictions concerning growth of computing power. His predictions have become known as Moore's Law. This, combined with Intel Corporation's microprocessor, is the basis for today's microcomputer revolution. Dr Moore is Chairman Emeritus, Intel Corporation, which he co-founded (1968). Dr and Mrs Moore are deeply involved with Conservation International and set up the Gordon E and Betty I Moore Foundation (2000) to fund scientific, educational and environmental ventures. A bird and a mammal are named after them and four other members of their family. (The binomial *mooreorum* means 'of the Moores'.)

Morafka

Morafka's Frog *Odorrana morafkai* Bain et al., 2003

Dr David Joseph Morafka (1947–2004) was an American herpetologist with a great interest in chelonians. He worked at California State University (1972–2001), and on retirement was made Emeritus Professor of Biology. The University of California, Berkeley awarded his bachelor's degree (1967) and his doctorate (1974) was from the University of Southern California. At the time of the frog's description he was Research Associate of the Royal Ontario Museum and California Academy of Sciences; the etymology says the naming was in recognition of his unfailing friendship, his unselfish development and perpetuation of multiple, independent research programs, and for his catapulting the careers of many conservation biologists. He died of pancreatic cancer. A reptile is named after him.

Morato

Botucatu Escuerzo *Proceratophrys moratoi* **Jim & Caramaschi**, 1980

Dr Celso Morato de Carvalho is a Brazilian biologist and herpetologist. The Universidade Estadual Paulista in Botucatu (the place where the holotype was collected) awarded his bachelor's degree (1978) and his doctorate (1992) was from the Instituto Nacional de Pesquisas da Amazônia. He was a post-doctoral fellow at the Museu de Zoologia da Universidade de São Paulo (2002). He was Head of Research at the Instituto Nacional de Pesquisas da Amazônia, Roraima (1985–1995), researcher at the Universidade Federal de Sergipe (1995–2005), and is co-ordinator of aquatic research at the Instituto Nacional de Pesquisas da Amazônia, Manaus. A reptile is named after him.

Moreira

Maldonado Redbelly Toad *Melanophryniscus moreirae* **Miranda**-Ribeiro, 1920

Sorocaba Dwarf Frog *Physalaemus moreirae* Miranda-Ribeiro, 1937

Professor Carlos Moreira (1869–1946) was a Brazilian zoologist who was a specialist in crustaceans. He became Director, Museu Nacional, Rio de Janeiro (1916). A bird is named after him.

Morelet

Morelet's Leaf Frog *Agalychnis moreletii* **Duméril**, 1853

[Alt. Black-eyed Leaf Frog]

Pierre Marie Arthur Morelet (1809–1892) was a French naturalist and malacologist who collected in Cuba, Guatemala and Mexico. He was sent (1846) to Central America by the French Academy of Sciences in order to collect specimens for the Muséum National d'Histoire Naturelle, Paris. Two birds and two reptiles are named after him.

Morère

Morère's Spiny Reed Frog *Afrixalus morerei* **Dubois**, 1986

Jean-Jacques Morère is a herpetologist at the Ecology and Biodiversity Department, Muséum National d'Histoire Naturelle, Paris. He has worked with Dubois, with whom he wrote *Quelques problèmes rencontrés à l'occasion d'une enquête sur la répartition des Amphibiens en France* (1983).

Mortensen

Mortensen's Frog *Hylarana mortenseni* **Boulenger**, 1903

Dr Ole Theodor Jensen Mortensen (1868–1952) was a Danish Professor of Zoology. He graduated in theology (1890) and took a doctorate (1898). He made zoological expeditions to the Faeroe Islands and to Siam (Thailand) (1899–1900); during the latter he collected the frog holotype. He was in New Zealand waters (1914–1915) on government vessels.

Mosauer

Cave Splayfoot Salamander *Chiropterotriton mosaueri* Woodall, 1941

Dr Walter Mosauer (1905–1937) was an Austrian physician who qualified at the University of Vienna (1929). While still a student at Vienna he made a reptile-collecting trip to Tunisia for the Museum, University of Vienna. He went to the USA (1929) undertaking a doctorate at the University of Michigan (1931). He joined the faculty of the University of California, Los Angeles as Instructor in Zoology (1932) and (1932–1937) made a number of collecting trips to Mexico, Arizona and Colorado. A keen skier, he introduced the sport to students in Los Angeles, organizing and training student ski teams. He wrote *The amphibians and reptiles of the Guadalupe Mountains of New Mexico and Texas* (1932). He died of blood poisoning whilst on a field trip collecting reptiles in Mexico.

Moser

Moser's Madagascar Frog *Gephyromantis moseri* **Glaw** & Vences, 2002

Felix Moser of Hamburg provides financial support to biodiversity research and nature conservation through the BIOPAT programme.

Mouhot

Litter Frog sp. *Leptobrachium mouhoti* **Stuart**, Sok, & Neang, 2006

Alexandre Henri Mouhot (1826–1861) was a French traveller, naturalist, and the earliest scientific collector of amphibian and reptile specimens in eastern Cambodia. He was a philologer and taught languages in Russia (1844–1854). He started studying natural science (1856) and married a descendant, probably the granddaughter, of the Scottish explorer Mungo Park – a connection that helped him when he decided to go to Indochina to collect botanical specimens (1857). The French authorities rejected his proposals, but the Royal Geographical Society and the Zoological Society of London, for both of which he collected specimens, supported the trip. He wrote *Voyage dans les royaumes de Siam, de Cambodge, de Laos et autres parties centrales de l'Indochine* (posthumously published, 1864). He died in Laos. Two reptiles and two birds are named after him.

Mozart

Mozart's Frog *Eleutherodactylus amadeus* **Hedges**, **Thomas** & Franz, 1987

[Alt. Haitian Robber Frog]

Wolfgang Amadeus Mozart (see **Amadeus**).

Müller

Müller's Mushroomtongue Salamander *Bolitoglossa mulleri* **Brocchi**, 1883

The original description simply refers to 'M. Muller', and the M probably only stands for Monsieur. He thus cannot be identified.

Müller, A

Müller's Termite Frog *Dermatonotus muelleri* **Böttger**, 1885

Dr August Müller gave a collection of reptiles and amphibians from Paraguay, including the holotype of this frog, to Böttger.

Müller, F

Müller's Smooth Horned Frog *Proceratophrys cristiceps* Müller, 1883

Müller's Shrub Frog *Pseudophilautus sarasinorum* Müller, 1887

Müller's Stream Toad *Ansonia muelleri* **Boulenger,** 1887

Dr Friedrich 'Fritz' Müller (1834–1895) was a Swiss physician and zoologist. He originally studied at the University of Basel and then at Würzburg and Prague where he qualified as a physician (1857). He returned to Basel to practise medicine after further studies in Vienna, Paris and Berlin. He gave public lectures in zoology at the University of Basel from 1868 onwards. His main interests were arachnids, crustaceans and herpetology. Five reptiles are named after him.

Müller, JP

Müller's Clawed Toad *Xenopus muelleri* **Peters**, 1844

[Alt. Savanna Clawed Frog, Sago-bellied Clawed Frog]

Professor Johannes Peter Müller (1801–1858) was a German physiologist, ichthyologist and comparative anatomist. He entered the University of Bonn (1819), graduated and started teaching (1824), becoming Professor (1830). He was Professor of Anatomy and Physiology at Humboldt University, Berlin (1833–1858) and a member of the Royal Prussian Academy of Science, Berlin. He wrote *Handbuch der Physiologie des Menschen* (1833–1840). Among his discoveries was (1835) that caecilians are amphibians and not snakes. A reptile is named after him.

Müller, K

Greek Marsh Frog *Pelophylax kurtmuelleri* Gayda, 1940

Kurt Müller collected the holotype in Albania (1938). There is no more detail in the description to help identify him further.

Müller, L

Müller's Spinythumb Frog *Crossodactylus aeneus* Müller, 1924

Lorenz Müller (1868–1953) was a German herpetologist who trained as an artist in Paris and the Low Countries and worked as a scientific illustrator at the State Zoological Collection, Munich. As he was particularly interested in herpetology, and a vacancy existed, he became de facto Curator (1903). He was a member of the Museum's expedition to the Lower Amazon (1909–1910). He served in the German Army in the First World War, but was able to get posted to the Balkans and spent most of his time collecting specimens! Afterwards he returned to Munich and became Chief Curator of Zoology (1928). During the Second World War both his private collection and the Museum's were largely destroyed in air raids, but he set to rebuilding the collections. During his life he published over 100 articles, monographs and papers on herpetology. Two birds and a reptile are named after him.

Müller, P

Müller's Giant Glass Frog *Centrolene muelleri* **Duellman** & **Schulte**, 1993

Professor Paul S. Müller is Head, Institute of Biogeography, Saarland University, Germany. He was honoured in the frog's binomial as he supported Schulte's field studies in Peru.

Muñoz

Upper Amazon Glass Frog *Hyalinobatrachium munozorum* **Lynch** & **Duellman**, 1973

Idelfonso Muñoz and Senora Blanca Muñoz were the describers' hosts at Santa Cecilia, Ecuador, where the frog holotype was collected.

Murthy

Murthy's Frog *Fejervarya murthii* **Pillai**, 1979

Dr T. S. N. Murthy (b.1936) is a herpetologist who was formerly Deputy Director, Zoological Survey of India. He is the author of over 130 publications, his best-known work being *Illustrated Encylopaedia of the Reptiles of India* (1995). He co-wrote several papers with Pillai, including *Herpetofauna of Eastern Ghats* (1982).

Musters

Paghman Mountain Salamander *Paradactylodon mustersi* **Smith**, 1940

James Lawrence Chaworth-Musters (1901–1948) was an Englishman who spent some of his early life in Norway where his family had an estate. His main interests were ornithology, botany, exploring and travelling. As an undergraduate at Cambridge (1921) he assembled an expedition to Jan Mayen Island in the Arctic and (1934) led a Norwegian expedition to Utsira (an island off the Rogaland coast, Norway). His other expeditions included Greece, Morocco, North Africa, Western Asia and Afghanistan, which (1939) was his last expedition and during which he collected the salamander holotype. During the Second World War he served in the Royal Navy and was a member of Kompani Linge (Norwegian resistance), being seconded from SOE (Special Operations Executive) and operating in Norway as a secret agent (1941–1943). The Norwegians set up a memorial to him (2004) at Vaulen, where his family home and estate had been.

Myers, CW

Treefrog genus *Myersiohyla* **Faivovich** et al., 2005
Myers' Poison Frog *Allobates myersi* **Pyburn**, 1981
Myers' Surinam Toad *Pipa myersi* **Trueb**, 1984
Myers' Thin-toed Frog *Leptodactylus myersi* **Heyer**, 1995

Dr Charles William Myers (b.1936) is an American herpetologist. He was a research assistant at the University of Florida (1958–1960) when he took his master's degree. He has published many herpetological papers and articles, including *An enigmatic new snake from the Peruvian Andes, with notes on the Xenodontini (Colubridae: Xenodontinae)* (1986). He replaced Bogert (q.v.) at the AMNH, New York (1968). He retired (1999) from active involvement, having been Chairman for two terms (1980–1987 & 1993–1998), but still continues to research there as Curator Emeritus. Six reptiles are named after him. (See **Adelphos**)

Myers, GS

Myers' Wrinkled Ground Frog *Platymantis myersi* **Brown**, 1949
Myers' Robber Frog *Pristimantis myersi* **Goin** & **Cochran**, 1963

Dr George Sprague Myers (1905–1985), a biogeographer, herpetologist and ichthyologist, was Professor Emeritus of Biological Sciences at Stanford, California. He was a keen natural historian with a life-long interest in fish and amphibians. He published his first paper on ichthyology aged 15, eventually writing over six hundred scientific papers and articles. He was a volunteer assistant at the AMNH, New York (1922–1924). He enrolled (1924) at Indiana University part-time, but when his sponsor fell ill he transferred to Stanford and graduated from there (1930), eventually completing his master's and doctorate (1933). He was Assistant Curator at the Smithsonian but was invited (1936) to return to Stanford as Assistant Professor in Biological Sciences and Curator of Zoological Collections. He developed courses in systematics for ichthyology and vertebrate palaeontology and was appointed full Professor (1938). During the Second World War he spent over two years in Brazil on US State Department funds to aid the Museo Nacional and Divisão de Caça e Pesca – a programme to maintain good relations with Latin America. He amassed an extensive library on ichthyology, herpetology, biogeography, the history of biology and exploration, and, as a sidelight, the American Civil War. After retirement (1970) he became Visiting Professor of Ichthyology at the Museum of Comparative Zoology, Harvard. A reptile is named after him.

N

Nadkarni

Nadkarni's Caecilian *Gegeneophis nadkarnii* Bhatta & Prashanth, 2004

[Alt. Bondla Caecilian]

Vinayaka Baburao Nadkarni is a biologist at the Department of Zoology, Karnatak University, Dharwad, India. His contributions to the field of comparative vertebrate endocrinology and biology of reproduction were acknowledged in the etymology.

Naeyaka

Sri Lanka Tribal Rock-frog *Nannophrys naeyakai* Fernando et al., 2007

[Alt. Kokagala Streamlined Frog]

This is not an eponym, though it looks like one. Naeyaka (or Nae Yaka) among the Vedda tribal people of Sri Lanka represents a belief in life after death.

Nagel

Nagel's Long-fingered Frog *Cardioglossa gratiosa peternageli* **Böhme** & **Schneider**, 1987

Dr Peter Nagel is an entomologist who is based at the University of Basel, Switzerland, where he is Dean of Studies, Faculty of Science. He studied biology and geography (1968–1975) at the University of the Saarland, Saarbrücken, Germany, which awarded his doctorate in biogeography, zoology and physical geography (1975). He was Professor for Biogeography at the University of the Saarland and a lecturer at Chiang Mai University, Thailand (1991–1996). In 1996 he joined the faculty at the University of Basel (1996) as Head of the Institute and Professor, Nature and Landscape Conservation and Environmental Protection and Biogeography.

Nagao

Nagao's Pug-snout Frog *Ramanella nagaoi* Manamendra-Arachchi & Pethiyagoda, 2001

Eijiro Nagao is President of Marusan Securities Co. of Japan. The describers expressed their thanks to him 'for the Nagao Environmental Foundation's generous support of our research into the Amphibia of Sri Lanka.'

Nahderer

Estrada Saraiva Treefrog *Dendropsophus nahdereri* **Lutz** & **Bokermann**, 1963

Karl (or Carlos) Nahderer collected a specimen of this frog on the Saraiva Road, Santa Catarina, Brazil (1949). Lutz described him as 'a charming naturalist'.

Nahuma

Stubfoot Toad sp. *Atelopus nahumae* **Ruiz-Carranza**, Ardila-Robayo & Hernandez-Camacho, 1994

The etymology describes Nahuma as a 'personaje épico' (legendary character) of the indigenous peoples in the area of Colombia where the holotype was collected.

Nakanai

Nakanai Wrinkled Ground Frog *Platymantis nakanaiorum* Brown, Foufopoulos & **Richards**, 2006

The Nakanai is the name of the indigenous people in whose territory in New Britain, Papua New Guinea, the frog holotype was collected.

Namdapha

Namdapha's Tree Frog *Rhacophorus namdaphaensis* Sarkar & Sanyal, 1985 NCR

[JrSYN *Rhacophorus rhodopus*]

This frog was named after the type locality, Namdapha camp, Arunachal Pradesh, India.

Namiye

Namiye's Frog *Limnonectes namiyei* **Stejneger**, 1901

[Alt. Okinawa Wart Frog]

Motoyoshi Namiye (1854–1918) was a Japanese naturalist and herpetologist. He was a member of the faculty of zoology at the Tokyo Educational Museum. Stejneger wrote a report on a collection of birds Namiye had made in the Riu Kiu Islands, Japan. He wrote *Oviposition of a blind snake from Okinawa* (1912). He is remembered in the names of other taxa, including a fish and six birds.

Nanuza

Jaboticatubas Treefrog *Bokermannohyla nanuzae* **Bokermann** & **Sazima**, 1973

Dr Nanuza Luiza de Menezes (b.1934) is a botanist at the University of São Paulo. She took all her degrees at São Paulo: bachelor's (1960), master's (1969) and doctorate (1971). She became a professor (1984), is now officially retired but still teaches as an associate. She spent time at the Royal Botanical Gardens, Kew (1979). She has spent much time investigating the trees of the Amazon basin. A reptile is also named after her.

Naomi

Naomi's Forest Frog *Platymantis naomii* **Alcala, Brown** & **Diesmos**, 1998

Mrs Naomi Alcala is the wife of the senior author, who ought to have known better than to give his wife a masculine binomial. Some attempts have been made (but rejected) to amend the binomial to *naomiae*.

Napoli

Treefrog sp. *Bokermannohyla napolii* **Carvalho, Giaretta** & Magrini, 2012

Dr Marcelo Felgueiras Napoli is a Brazilian herpetologist who has been Professor of Vertebrate Zoology at the Universidade Federal da Bahia since 2002. He is Curator of Amphibians at the University's zoological museum. He specialises in neotropical anurans. His doctoral thesis was a taxonomic revision of *Bokermannohyla* species.

Narain

Narain's Bush Frog *Micrixalus narainensis* **Rao**, 1937
[Alt. Narain's Torrent Frog]

Dr Prahlad Narain Mathur was an Indian anatomist and herpetologist. He was (1913–1947) Chairman of the Zoology Department and Principal of the College at Government College, Ajmer (founded 1836 by the Honourable East India Company and regarding itself as 'the Eton of the East'). He wrote *The anatomy of the reptilian heart. Part II. Serpentes, Testudinata and Loricata* (1944).

Narayan

Narayan's Caecilian *Uraeotyphlus narayani* **Seshachar**, 1939
[Alt. Kannan Caecilian]

Professor C. R. Narayan Rao. (See **Rao**)

Natterer

Cuyaba Dwarf Frog *Eupemphix nattereri* **Steindachner**, 1863
Toad sp. *Rhinella nattereri* **Bokermann**, 1967

Dr Johann Natterer (1787–1843) was an Austrian naturalist, collector and explorer. He studied botany, zoology, mineralogy, chemistry and anatomy and was appointed as a taxidermist at the Natural History Museum, Vienna. Along with Spix and others, he went on the expedition to Brazil (1817) which started on the occasion of the Archduchess Leopoldina's wedding to the Brazilian Crown Prince. He explored a potential river route to Paraguay (1818–1819) and went on five expeditions, exploring the Mato Grosso and the Amazon Basin (1821–1835). He returned to Vienna with a huge collection of specimens (37 crates), despite losing most of his possessions in the Civil War in Brazil, and deposited 12,293 birds and c.24,000 insects with the Vienna Museum, where they can still be seen. He died of a lung ailment. He did not publish any account and unfortunately his notebooks and diary were destroyed by fire (1848). He never received the credit he was due in Austria, but was held in the highest esteem abroad. Among many taxa named after him are two mammals, twenty-two birds, three reptiles and a piranha.

Neblina

Treefrog sp. *Stefania neblinae* **Carvalho** et al., 2010

This species is named after the type locality, Pico da Neblina National Park in northern Brazil.

Necker

Necker's Wrinkled Ground Frog *Platymantis neckeri* **Brown** & **Myers**, 1949

Walter Ludwig Necker (1913–1979) was a librarian and herpetologist. He was born in Germany and arrived in the USA (1923). He worked for the Chicago Academy of Sciences (1930s) and joined the Zoology Department, Field Museum, Chicago (1943) but was then enlisted in the US Navy. He served in the Pacific (1943–1945), after which he and his family founded Aardvark Books, dealing in antique books relating to natural sciences. The University of Chicago awarded his bachelor's degree (1940). He studied in England (1952–1953) for a doctorate, which he never completed. He was elected (1939) to be the first historian of the American Society of Ichthyologists and Herpetologists. He wrote *Contribution to the Herpetology of the Smoky Mountains of Tennessee* (1934).

Neild

Leaf Frog sp. *Phyllomedusa neildi* Barrio-Amorós, 2006

Andrew F. Neild (b.1966) is an entomologist and lepidopterist at the Natural History Museum, London. He lived in Venezuela (1978–1981), where he was captivated by the number and beauty of its butterflies. He is writing *The Butterflies of Venezuela* (2 volumes, 1996 & 2008, plus 2 more volumes to come).

Nelson, CE

Microhylid frog genus *Nelsonophryne* **Frost**, 1987

Professor Dr Craig E. Nelson was honoured in this name for his contributions to microhylid systematics. He is Professor Emeritus of Biology at Indiana

University where he has been on the faculty since 1966 (retiring 2004). The University of Kansas awarded his bachelor's degree (1962) and the University of Texas his master's (1964) and PhD (1966). His biological research was on evolution and ecology, initially on frogs, most recently on sex-determination in turtles. He has published more than 60 papers, including some designed to make you smile as well as think, such as *How can students who are reasonably bright and who are trying hard to do the work still flunk?* (2000).

Nelson, EW

Amargosa Toad *Anaxyrus nelsoni* **Stejneger**, 1893

Dr Edward William Nelson (1855–1934) was an American naturalist and ethnologist. He joined the US Army Signal Corps (1877) and was posted to Alaska. While taking part in the search for the missing arctic exploration vessel 'USS Jeanette', his expedition was the first to reach and explore Wrangel Island (1881). Nelson accepted an appointment as a special field agent with the Death Valley Expedition (1890), following which he collected in Mexico (1892–1902). He was chief field naturalist (1907–1912), in charge of biological investigations (1912–1913), Assistant Chief (1914–1915), then Chief (1916–1927) of the US Biological Survey. One of his greatest contributions was the establishment of the Migratory Bird Treaty, which is still in force today. Twenty-four birds, five reptiles and fifteen mammals are named after him.

Netting

Netting's Salamander *Plethodon nettingi* **Green**, 1938
[Alt. Cheat Mountain Salamander]
Western Lesser Siren *Siren intermedia nettingi* **Goin**, 1942

Morris Graham Netting (1904–1996) was a herpetologist and museum director. He took his bachelor's degree (1926) at the University of Pittsburgh, when he was an assistant in the Herpetology Department, Carnegie Museum, Pittsburgh. The University of Michigan awarded his master's degree (1929) but work on his doctorate was interrupted by the Wall Street Crash (1929). He worked at the Carnegie Museum (1928–1975), rising from Assistant Curator to Director (1954). He became Emeritus Director, Carnegie Museum (1975). He did not just work in an office: he led the Carnegie Museum Expedition to Venezuela (1929–1930). He taught zoology, herpetology and geography (part-time) at the University of Pittsburgh (1944–1963). He wrote *Hand list of the Amphibians and Reptiles of Pennsylvania* (1939).

Neumann

Neumann's Grassland Frog *Ptychadena neumanni* **Ahl**, 1924

Professor Oskar Rudolph Neumann (1867–1946) was a German ornithologist who collected in East Africa (1892–1894). He was in Somaliland and Ethiopia with Carlo von Erlanger (1899–1901). In the early 1900s he studied the birds and mammals of Rothschild's collection. Later in life he moved to Chicago to escape Nazi persecution and worked at the Field Museum. Thirty-four birds, two reptiles and three mammals are named after him.

Newton

Newton's Grassland Frog *Ptychadena newtoni* **Bocage**, 1886
Newton's Water Frog *Petropedetes newtoni* Bocage, 1895 NCR
[JrSYN *Petropedetes johnstoni*]

Colonel Francisco Xavier Oakley de Aguiar Newton (1864–1909) was a Portuguese botanist who collected in West Africa (1880s), São Tomé (1888) and Timor (1896). He wrote accounts of his travels and findings, being meticulous in recording detailed information on localities and ecology, to a standard exceptional for his time. Three birds and two reptiles are named after him.

Nibaldo

Nibaldo's Wood Frog *Batrachyla nibaldoi* Formas, 1997

Professor Jorge Nibaldo Bahamonde Navarro (b.1924) is a Chilean marine biologist. He graduated (1943) and then took a teaching qualification. By 1946 he was Professor, Biological & Chemical Sciences, University of Chile, Santiago. He carried out research on marine fauna in Europe including at the University of Bergen, Norway (1950s) and continued to teach, with his interests turning to ecology and conservation (1960s). He went to the Juan Fernández Islands (1965) to investigate a rumour, which fortunately turned out to be true, that the Juan Fernández Fur Seal *Arctocephalus philippii*, then thought extirpated, was in fact still extant. He became Professor Emeritus, University of Chile (2004). He co-wrote *Peces de Chile, Lista sistemática* (1975).

Nicéforo

Strabomantid frog genus *Niceforonia* **Cochran** & **Goin**, 1963
Honda Caecilian *Parvicaecilia nicefori* **Barbour**, 1924
Leaf Frog sp. *Phyllomedusa nicefori* Barbour, 1926 NCR
[JrSYN *Phyllomedusa tarsius*]

Nicéforo's Marsupial Frog *Gastrotheca nicefori* **Gaige**, 1933

Nicéforo's Stubfoot Toad *Atelopus nicefori* **Rivero**, 1963

San Gil Mushroomtongue Salamander *Bolitoglossa nicefori* **Brame** & **Wake**, 1963

Colombian Backpack Frog *'Cryptobatrachus' nicefori* Cochran & Goin, 1970

[Taxonomic placement uncertain: perhaps belongs in *Hyloscirtus*]

Colombian Beaked Toad *Rhinella nicefori* Cochran & Goin, 1970

Nicéforo's Poison Frog *Ameerega ingeri* Cochran & Goin, 1970

[Alt. Inger's Poison Frog, Brother Nicéforo's Poison-arrow Frog]

Nicéforo's Robber Frog *Pristimantis nicefori* Cochran & Goin, 1970

Brother Nicéforo María (1888–1980), né Antoine Rouhaire Siauzade, was born in France and became a Catholic missionary in Colombia under his monastic name. After religious training in Luxembourg and Belgium, he went to Medellín, Colombia (1908) and was given the task of forming a natural history museum. He was transferred (1922) to the Instituto de La Salle in Bogotá and spent the rest of his life there. He was interested in all vertebrates, but was primarily a herpetologist and an excellent taxidermist. Many taxa are named after him, including seven reptiles, three birds and a mammal.

Nicholls

Nicholls' Toadlet *Metacrinia nichollsi* Harrison, 1927

Desert Spadefoot Toad *Notaden nichollsi* **Parker**, 1940

Dr George Edward Nicholls (1878–1953) was Professor of Biology, University of Western Australia, Perth. He was Scientific Assistant, University of London, Lecturer at King's College and King's College for Women, London, and later Professor of Biology, Agra College, India, before his posting at Perth. He collected the holotype of the toadlet (1925) and sent 28 examples for Harrison to study at Sydney University. His other specialisms were entomology and crustaceans – he and his students made a notable collection of insects in a single location in Western Australia (1933). He made regular collecting expeditions in Western Australia as well as on the Antarctic Expedition (1911–1914). He collected many crustaceans, some of which were named after him, and his collections were housed at the University and used for teaching. He wrote the report (1938) *Crustacea Amphipoda from the Australian Antarctic Expeditions (1911–1914)* and *The structure of the vertebral column in the Anura Phaneroglossa and its importance as a basis of classification* (1916).

Nicole

Nicole's Andes Frog *Phrynopus nicoleae* Chaparro, **Padial** & **De la Riva**, 2008

Nicole Morciniec of Poland is a supporter of BIOPAT and recognized for '…her friendship and her support of taxonomic research and nature conservation in Peru.'

Nieden

Sagalla Caecilian *Boulengerula niedeni* Müller et al., 2005

Fritz Nieden (1883–1942) was a German zoologist at the Museum für Naturkunde, Berlin. He specialized in African herpetofauna. He wrote *Die Reptilien (außer den Schlangen) und Amphibien Kamerum* (1910). A reptile is named after him.

Nike

Nike's Squeaker *Arthroleptis nikeae* **Poynton**, 2003

Nike Doggart is an environmentalist and conservation activist who has a master's degree from the University of London. She was Project Director (1999–2005) for the Frontier Tanzania Forest Research Programme, Dar es Salaam. She is editor of the Tanzania Forest Conservation Group's newsletter *The Arc Journal*. She helped promote the introduction of sustainable butterfly farming for women living adjacent to the Amani Nature Reserve. She co-wrote *Kwamgumi Forest Reserve – a biodiversity survey* (1999).

Nito

Rio Negro Frog *Atelognathus nitoi* **Barrio**, 1973

Nito Barrio is the author's son, who discovered the species and brought specimens for his father to study.

Noble

Strabomantid frog genus *Noblella* **Barbour**, 1930

Noble's Robber Frog *Craugastor noblei* Barbour & **Dunn**, 1921

Dr Gladwyn Kingsley Noble (1894–1940) was an American biologist and herpetologist. His father was a co-founder of the publishers Barnes & Noble. He went to Peru, Newfoundland and Guadeloupe (1914–1916). Harvard awarded his bachelor's (1917) and master's (1919) degrees; Columbia University, New York awarded his doctorate (1922). He served in the US Navy (1918–1919). He joined the AMNH, New York (1919) as an assistant, becoming Assistant Curator (1922), and was Curator of Herpetology (1923–1940). He was also Curator of Experimental Biology (1928) and Visiting Professor, University of Chicago (1931). He led expeditions to Santo Domingo (1922) and Cuba (1937). Among his publications is *The Experimental Animal from the Naturalist's Point of View* (1939). He died from a streptococcal infection. Three reptiles are named after him.

Noellert

Noellert's Toad *Duttaphrynus noellerti* Manamendra-Arachchi & Pethiyagoda, 1998

Andreas Nöllert is a photographer and herpetologist at the State Office for Environmental Protection, Jena, Thuringia, Germany. He edits the magazine *Nature and Landscape in Thuringia*. He co-wrote *Die Amphibien Europas* (1992).

Nora

Oncol Spiny-chest Frog *Alsodes norae* Cuevas, 2008

Nora was the author's grandmother.

Nora Lotta

Madagascar Frog sp. *Mantidactylus noralottae* Mercurio & **Andreone**, 2007

Nora Lotta Mercurio née Fröhder is the senior author's wife.

Norhayati

Norhayati's Gliding Frog *Rhacophorus norhayatii* Onn & Grismer, 2010

Dr Norhayati Ahmad is a zoologist, herpetologist and conservation biologist whose bachelor's degree in biology (1992) was awarded by Bishop's University, Lennoxville, Quebec, Canada. Her master's (1996) in conservation biology and her doctorate in zoology (2001) were both awarded by Universiti Kebangsaan Malaysia, Selangor, where she is a member of the Faculty of Science and Technology and an associate professor.

Norton

Norton's Robber Frog *Eleutherodactylus nortoni* **Schwartz**, 1976

James W. Norton accompanied Schwartz on the expedition (1974) to Haiti and the Dominican Republic, where he collected the holotype.

Novais

Bokermann's Treefrog *Dendropsophus novaisi* **Bokermann**, 1968

Alfredo Novais owned a ranch or farm called Fazenda Santo Onofre, Bahia, Brazil. He was Bokermann's host during his party's collecting trip.

Nussbaum

Madagascar Frog sp. *Spinomantis nussbaumi* Cramer, Rabibisoa & Raxworthy, 2008

Mushroomtongue Salamander sp. *Bolitoglossa nussbaumi* **Campbell** et al., 2010

Dr Ronald Archie Nussbaum (b.1942) is Professor of Biology and Curator of the Museum of Zoology, University of Michigan. The University of Idaho awarded his bachelor's degree (1967), Central Washington University his master's (1968), and Oregon State University his doctorate (1972), after which he stayed on as a research associate until 1974 when he moved to Michigan. He was Director, Edwin S. George Reserve, University of Michigan (1983–2006). He wrote *Montane amphibian and reptile communities in Madagascar* (1996).

Nyika

Nyika Dwarf Toad *Mertensophryne nyikae* **Loveridge**, 1953

This toad is named after the type locality, Nyika Plateau in northern Malawi.

O

Oates

Toungoo Frog *Humerana oatesii* **Boulenger**, 1892

Eugene William Oates (1845–1911) was a civil servant in British colonial India and Burma (Myanmar), and an amateur naturalist. When he returned to England, he was Secretary of the British Ornithologists' Union (1898–1901). He wrote *The Fauna of British India* (1889). Seven birds and a reptile are named after him.

Obst

Obst's Mehely Frog *Copiula obsti* **Günther**, 2002

Professor Dr Fritz Jürgen Obst (b.1939) is a German herpetologist. He studied psychology and biology at Heidelberg University and at Hohenheim University, which awarded his doctorate (1996). He was Curator of Lower Vertebrates and Insects at the Zoological Gardens, Stuttgart (1990–1996) and became Herpetologist at the Staatliche Museum für Naturkunde, Dresden, rising to Deputy Director (1997), then Director (2001). Since 1997 he has also taught at Leipzig University. He wrote *Turtles, Tortoises and Terrapins* (1988). A bird and three reptiles are named after him.

Ochoa, A

Marsupial Frog sp. *Gastrotheca antoniiochoai* **De la Riva** & Chaparro, 2005

Dr José Antonio Ochoa is a friend of the describers. He was at one time devoted to herpetology but is now an expert on scorpions. His bachelor's degree (1996) was awarded by the Universidad Nacional de San Antonio Abad del Cusco, Peru, and his doctorate by the Universidad Nacional de Córdoba, Argentina (2003). He was a post-doctoral fellow (2007–2009) at the AMNH, New York.

Ochoa, O

Chilca Marsupial Frog *Gastrotheca ochoai* **Duellman** & Fritts, 1972

Professor Oscar Ochoa Mendieta of the Departamento de Biología, Universidad Nacional San Antonio Abad del Cusco, Peru, was honoured in the frog's name as he '…provided facilities at his home at Hacienda Chilca and aided in collecting specimens.'

Ockenden

Carabaya Robber Frog *Pristimantis ockendeni* **Boulenger**, 1912

George Richard Ockenden (1868–1906) was a professional collector, especially of lepidoptera, who was regarded very highly by his contemporaries. He undertook an expedition (1900–1906) to the Peruvian Andes to collect zoological specimens of all kinds for W. F. H. Rosenberg (q.v.). He was taken ill with typhoid and, being a long way from any hospital, he died before any medicine could help him. A bird is named after him.

Octavio

Octavio's Robber Frog *Ischnocnema octavioi* **Bokermann**, 1965

Octavio de Oliveira was a Brazilian collector. Bokermann wrote that Octavio had 'made special efforts to obtain the series of specimens' of this frog.

O'Donnell

O'Donnell's Salamander *Bolitoglossa odonnelli* **Stuart**, 1943

Reginald O'Donnell was Stuart's host at Finca Volcán, Guatemala, and drew his attention to this species.

Oedipus

Worm Salamander genus *Oedipina* **Keferstein**, 1868

This genus was at one time known as *Oedipus*, but the name was found to be invalid owing to its prior usage for an orthopteran genus. The meaning of *oedipus* is 'swollen foot', and it would seem the salamanders were named for this feature rather than after the mythical Greek king of Thebes who married his own mother.

Ohaus

Petropolis Treefrog *Flectonotus ohausi* **Wandolleck**, 1907

Wandolleck's Button Frog *Cycloramphus ohausi* Wandolleck, 1907

Dr Friedrich Ohaus (1864–1946) was a German physician and entomologist. He made use of his medical qualification to travel as a ship's doctor to South America three times, visiting Brazil (1898–1899 & 1926) and Ecuador (1904–1906), travelling overland to Iquitos on the Peruvian Amazon, and down the Amazon to Pará, Brazil. After the First World War he worked as a scientific assistant in the Natural History Museum, Mainz (1919–1946). He was President and Editor of Deutsche Entomologische Gesellschaft in Berlin (1911–1914).

Olalla

Tandayapa Andes Toad *Andinophryne olallai* **Hoogmoed**, 1985

Jorge Olalla, who collected the holotype (1970) and after whom this toad is named, was one of the famous Olalla family of Ecuador, who were all professional collectors. Other members of the family included Carlos Olalla and his three sons, Alfonso Manuel Olalla, Ramón Olalla and Rosalino Olalla. Jorge collected from the 1950s.

Olfers

Atlantic Forest Dwarf Frog *Physalaemus olfersii* Lichtenstein & **Martens**, 1856

Ignaz Franz Werner Maria von Olfers (1793–1871) was a German naturalist, historian and diplomat. He was posted to Brazil (1816). He became Director of the Imperial Museum, Berlin (1838). A reptile is named after him.

Oliveira

Xeric Treefrog *Dendropsophus oliveirai* **Bokermann**, 1963

F. M. Oliveira and the describer, Bokermann, collected the frog holotype (1963).

Olmon

Bloody Bay Poison Frog *Mannophryne olmonae* **Hardy**, 1983

Janet E. Olmon is an American biologist who worked closely with Hardy. They co-wrote *Restriction of the range of the frosted salamander,* Ambystoma cingulatum *based on a comparison of the larvae of* Ambystoma cingulatum *and* Ambystoma mabeei (1974). The Virginia Institute of Marine Science awarded her bachelor's (1972) and master's (1972) degrees and she was a Graduate Assistant in the Department of Environmental Physiology there (1972–1974). She was the first to point out to Hardy the different call of this Tobago frog compared to others heard in Trinidad.

Omiltem

Omiltemi Minute Salamander *Thorius omiltemi* **Hanken, Wake** & Freeman, 1999

This salamander is named after the Omiltemi State Ecological Park in Mexico, close to where the holotype was collected.

Onore

El Reventador Caecilian *Chthonerpeton onorei* **Nussbaum**, 1986

Stubfoot Toad sp. *Atelopus onorei* **Coloma** et al., 2007

Father Giovanni Onore (b.1941) is an Italian Marianist missionary priest, entomologist and now retired Professor, Pontifical Catholic University of Ecuador. He graduated in Agricultural Sciences at the University of Turin, entered the priesthood and served as a missionary at Brazzaville, Congo. He first went to Ecuador (1980) and has been living in Quito ever since. He is very active in ecology and founded and is President of the Otonga Foundation, which has preserved an area of primary rainforest. He wrote *Edible insects in Ecuador* (1997).

Oommen

Oommen's Caecilian *Uraeotyphlus oommeni* Gower & Wilkinson, 2007

Dr Oommen V. Oommen is Professor and Head of Zoology, University of Kerala, where he took his bachelor's degree (1969). The University of Varanasi, India awarded his master's (1973) and doctorate (1977). He co-wrote *Ultrastructural observations of previtellogenic ovarian follicles of the Caecilians* Ichthyophis tricolor *and* Gegeneophis ramaswamii (2007).

Orcés

Bolivar Robber Frog *Pristimantis orcesi* **Lynch**, 1972

Stubfoot Toad sp. *Atelopus orcesi* **Coloma** at al., 2010

Professor Gustavo Orcés (1902–1999) was an Ecuadorian zoologist and herpetologist who worked at the Polytechnic, Quito. The Fundación Herpetológica Gustavo Orces at the Museum of Natural History was founded (1989) to hold his collection. A bird, five reptiles and two mammals are named after him.

Orcutt

Orcutt's Coqui *Eleutherodactylus orcutti* **Dunn**, 1928 [Alt. Arntully Robber Frog]

Charles Russell Orcutt (1864–1929) was primarily a botanist and malacologist. He combined collecting with publishing scientific journals. He had no formal schooling, being taught on the farm by his parents. The family moved (1879) from Vermont to San Diego. He started (1884) to publish *The West American Scientist* to get his own work and notes before the public. It continued to appear, sporadically, until 1919. He accumulated a large, if eclectic, collection, which ended up with the San Diego Society of Natural History. He collected for the Smithsonian (1927–1929) in Baja California, Mexico, Central America and the Caribbean, including Jamaica. A reptile is named after him.

Orejas-Miranda

Orejas-Miranda's Redbelly Toad *Melanophryniscus orejasmirandai* Prigioni & **Langone**, 1987

Braulio Rubí Orejas-Miranda (1933–1985) was a Uruguayan biologist and herpetologist. After taking his bachelor's degree (1952) he taught in secondary schools (1953–1967). He was Director of the Caracas

Zoo and taught biology at the Universidad Central de Venezuela, Caracas (1960–1961). He also taught biology at the Universidad del Valle, Cali, Colombia (1964) and the National Institute of Sciences, Bogotá (1970). He went on expeditions for the Museo Nacional de Historia Natural y Antropología, Montevideo, to Mato Grosso (1955) and the Orinoco (1957). He was a research associate at the Smithsonian (1967–1985).

Orestes

Culata Mushroomtongue Salamander *Bolitoglossa orestes* **Brame** & **Wake**, 1962
Urdaneta Robber Frog *Pristimantis orestes* **Lynch**, 1979
Blue Ridge Dusky Salamander *Desmognathus orestes* Tilley & Mahoney, 1996

In Greek mythology, Orestes was the son of King Agamemnon and his wife Clytemnestra. The meaning of the name is 'he who stands on the mountain', which seems to have inspired use of his name for montane amphibians.

Organ

Northern Pygmy Salamander *Desmognathus organi* Crespi, Browne & Rissler, 2010

Dr James A. Organ is a biologist who worked on salamanders (1950s–1990s). He was Professor and Chairman, City College, New York when he retired. He had previously been in the Department of Zoology, University of Michigan, Ann Arbor. He wrote *Studies of the local distribution, life history, and population dynamics of the salamander genus* Desmognathus *in Virginia* (1961).

Orlov

Orlov's Tree Frog *Rhacophorus orlovi* Ziegler & **Köhler**, 2001

Dr Nikolai Lusteranovich Orlov (b.1952) is a Russian zoologist, herpetologist and a Senior Research Scientist, Herpetology Department, Zoological Institute, Russian Academy of Sciences, St Petersburg. He has made over 60 field trips, mainly to countries of the former USSR, but also to the Indian sub-continent and Southeast Asia. He co-wrote *A new species of mountain stream snake, genus* Opistotropis *Gunther, 1872 (Serpentes: Colubridae: Natricinae), from the tropical rain forests of Southern Vietnam* (1998). Two reptiles are named after him.

Ortiz

Ortiz's Robber Frog *Pristimantis ortizi* Guayasamin, Almeida-Reinoso & Nogales-Sornosa, 2004

Professor Dr Fernando Ignacio Ortiz-Crespo (1942–2001) was an Ecuadorian ornithologist at the Catholic University in Quito. He took his bachelor's degree at the University of California, Berkeley. The etymology says: 'The specific name...is a patronym for Fernando Ortiz-Crespo, one of the most noted Ecuadorian ornithologists. Fernando Ortiz-Crespo was well known for his research in the Galápagos Islands and the Andes. He drowned while conducting fieldwork in the Laguna de la Mica in the high Andes of Ecuador on 13 September 2001.' He was surveying migrating ducks when his boat overturned. Among his innumerable publications his best-known was *Aves del Ecuador – continente y archipiélago de Galápagos*.

Ory

Ory's Tree Frog *Leptopelis oryi* **Inger**, 1968
[Alt. Garamba Forest Treefrog]

Albert Ory was the warden in charge of Garamba National Park, Democratic Republic of Congo, at the time of the civil war in that country.

Osa

Airstrip Caecilian *Oscaecilia osae* Lahanas & **Savage**, 1992

This caecilian is named after the type locality, Península de Osa, Costa Rica.

Osgood

Osgood's Ethiopian Toad *Altiphrynoides osgoodi* **Loveridge**, 1932

Wilfred Hudson Osgood (1875–1947) was an American ornithologist and mammalogist who began his career working as a biologist in the US Department of Agriculture (1897–1909), in charge of the US biological investigation in Canada. He worked at the Field Museum, Chicago, as Assistant Curator, Mammalogy and Ornithology (1909–1921) and Curator of Zoology (1921–1940). He conducted biological explorations and surveys of many areas of North and South America, Ethiopia and Indo-China. He spent 1906, 1910 and 1930 studying in European museums. He led the Field Museum Abyssinian Expedition (1926–1927) and Magellanic Expedition (1939–1940). He co-wrote *Artist and Naturalist in Ethiopia* (1936), and wrote *Mammals of Chile* (1943). Seven mammals and six birds are named after him.

Osório

Osório's Spiny Reed Frog *Afrixalus osorioi* Ferreira, 1906
[Alt. Angola Banana Frog]

Balthazar Osório (fl.1917) was a Portuguese ichthyologist working on the fauna of Portuguese

Africa at the end of the nineteenth and beginning of the twentieth centuries as Curator at the Museu e Laboratório Zoológico e Antropológico. He undertook an expedition during the 1890s to São Tomé. He described a number of fishes new to science, as well as other, mainly marine, organisms. His writings include such papers as *Da distribuição geográfica dos peixes e crustáceos colhidos nas possessões portuguesas da África Occidental e existentes no Museu Nacional de Lisboa* (1898), mostly published in the 'Jornal de Sciencias. Mathematicas, Physicas e Naturaes' of Lisbon.

Otávio

Otávio's Tree Toad *Hylodes otavioi* **Sazima** & **Bokermann**, 1983

Dr Otávio Augusto Vuolo Marques is a biologist and zoologist at the Instituto Butantan, São Paulo, Brazil, where he is Director, Special Laboratory of Ecology and Evolution. The University of São Paulo awarded his bachelor's degree (1987), his master's in ecology (1992) and doctorate in zoology. He works with Sazima and they co-wrote *Natural History of Reptiles of the Ecological Station, Juréia-Itatins* (2004). A reptile is named after him.

Ouachita

Rich Mountain Salamander *Plethodon ouachitae* **Dunn** & Heinze, 1933

This salamander is named after the type locality, Ouachita National Forest, on Rich Mountain, Arkansas.

Owston

Owston's Tree Frog *Rhacophorus owstoni* **Stejneger**, 1907

[Alt. Japanese Flying Frog]

Alan Owston (1853–1915) was an English collector of Asian wildlife, as well as a businessman and yachtsman; he founded Yokohama Yacht Club and his business was as a dealer in wildlife specimens. He left for Asia when still quite young. He married two Japanese ladies (1880 & 1893) and had nine children by them. His most active collecting period was the early twentieth century. He died of lung cancer in Yokohama, Japan. Eleven birds, a shark and a mammal are among the taxa named after him.

Oxapampa

Oxapampa Harlequin Frog *Atelopus oxapampae* **Lehr**, **Lötters** & **Lundberg**, 2008

This frog is named after a place in Peru close to where the holotype was collected.

P

Pacchamama

Ayacucho Marsupial Frog *Gastrotheca pacchamama* **Duellman**, 1987

Pacchamama is a mythological goddess, worshipped by the Quecchuan Indians in the Andes.

Padial

Robber Frog sp. *Pristimantis padiali* Moravec et al., 2010

Dr José Manuel Padial-Fregenal (b.1977) is a Spanish biologist and herpetologist. The University of Granada, Spain awarded both his bachelor's degree (2001) and doctorate (2007). He is presently a post-doctoral fellow at the AMNH.

Padre Carlos

Robber Frog sp. *Pristimantis padrecarlosi* Mueses-Cisneros, 2006

Padre Carlos Fernando Duarte Ribeiro is the priest at the church dedicated to San Gil at Santander, Colombia, where the holotype was collected.

Padre Luna

Treefrog sp. *Dendropsophus padreluna* **Kaplan** & **Ruiz-Carranza**, 1997

While this sounds like an eponym, the binomial is actually derived from the local name of the type-locality in Colombia.

Paeng

Paeng's Wrinkled Ground Frog *Platymantis paengi* Siler et al., 2007

'Paeng' is a Tagalog (Philippines) nickname for people called Rafael, and in this case it is applied to Dr Rafe M. Brown, who is Chief Curator of Herpetology and Assistant Professor in Ecology and Evolutionary Biology at the University of Kansas. He co-wrote *New Species of Narrow-Mouthed Frog (Amphibia: Anura: Microhylidae; Genus* Kaloula*) from the Mountains of Southern Luzon and Polillo Islands, Philippines* (2002).

Paez

Colombian Giant Glass Frog *Centrolene paezorum* **Ruiz-Carranza**, **Hernández-Camacho** & **Ardila-Robayo**, 1986

The Paez people, also known as the Nasa, are an indigenous group in Colombia.

Pageot

Tonkin Toad *Bufo pageoti* **Bourret**, 1937

Lieutenant Pageot was the officer in charge at Chapa, Tonkin (Vietnam). He collected the toad holotype.

Pakenham

Pakenham's River Frog *Phrynobatrachus pakenhami* **Loveridge**, 1941

Richard Hercules Wingfield Pakenham (1906–1993) was Senior Commissioner to Zanzibar (1948–1956). He is recorded collecting in Angola (1936). He described a number of birds from Pemba Island. He co-wrote *The Land Vertebrates of Pemba, Zanzibar, and Mafia: A Zoogeographical Study* (1940).

Palmer

Palmer's Treefrog *Hyloscirtus palmeri* **Boulenger**, 1908

Palmer's Robber Frog *Pristimantis palmeri* Boulenger, 1912

Mervyn George Palmer (1882–1954) was an English naturalist, traveller and collector in Central and South America. After graduating he became an analytical chemist before deciding upon a career as a freelance collector and naturalist. He collected for the Natural History Museum, London (1904–1910) in Colombia, Ecuador and Nicaragua. During this time he discovered over 60 species new to science, learnt to speak Spanish and two South American Indian languages, married a South American woman, undertook archaeological digs, and explored and mapped the Río Segovia between Nicaragua and Honduras. He worked for commercial concerns in Ecuador (1910–1918) before moving to London with the same company. After suffering malaria and yellow fever, he was declared unfit for overseas army service in the First World War. He was in Venezuela (1919–1921), later being based in London but with frequent visits to South America. He lived in Ilfracombe, Devon, England (1932–1954), where he founded and was curator of a museum and ran a library and field club, as he 'wanted something to do'. He was also at one time the Editor of the Natural Science Gazette. He wrote *Through Unknown Nicaragua – The Adventures of a Naturalist on a Wild-Goose Chase* (1945). A reptile and a bird are named after him.

Pancho

Guatemala Stream Frog *Ptychohyla panchoi* **Duellman** & **Campbell**, 1982

Dr Laurence Cooper (Don Pancho) Stuart (1907–1983) was a research associate at the Museum of Zoology, Michigan. He published widely from the 1930s to the 1960s, with works such as *A checklist of the herpetofauna of Guatemala* (1963). The describers wrote: 'We are pleased to name this frog in honor of Dr. L. C. (Don Pancho) Stuart, who has helped both of us in our careers and whose contributions to the knowledge of Middle American herpetogeography are legendary.' (See also **Stuart, LC**)

Panha

Prachuap Sucker Frog *Amolops panhai* **Matsui** & Nabhitabhata, 2006

Dr Somsak Panha is a zoologist and Associate Professor, Biology Department, Chulalonkorn University, Thailand. His bachelor's degree was awarded by Srinakarinwirote University (1977), his master's by Chulalonkorn (1981), and his doctorate by the University of Kyoto (1988). He is particularly interested in malacology and mollusc taxonomy and phylogeny. He has written over 70 scientific papers and articles.

Panton

Panton's Robber Frog *Eleutherodactylus pantoni* **Dunn**, 1926

Edward Stuart Panton (1866–1962) was a Jamaican lepidopterist, naturalist and property owner who raised livestock on his property 'Hopeton' near Mandeville. He was educated in England and was a member of staff at the Institute of Jamaica. Dunn spent ten days in Jamaica (1925) and in that time discovered seven new species of the genus *Eleutherodactylus,* all endemic to Jamaica, and named most of them after Jamaican naturalists or worthies.

Papalo

Papalo Minute Salamander *Thorius papaloae* **Hanken** & **Wake**, 2001

This salamander is named after Concepción Pápalo, Oaxaca, Mexico, a location close to where the holotype was collected.

Papenfuss

Muscular Salamander *Pseudoeurycea papenfussi* Parra-Olea et al., 2005

Dr Theodore Johnstone Papenfuss (b.1941) is a research specialist in herpetology at the University of California, Berkeley. He co-wrote *Karyotypes of Chinese species of the genus* Teratoscincus *(Gekkonidae)* (1998). Three reptiles are named after him.

Paraewing

Victorian Frog *Litoria paraewingi* Watson, Loftus-Hills & **Littlejohn**, 1971

This frog is named *paraewingi* because of its similarity to *Litoria ewingi* – see entry for **Ewing**.

Parbocagi

Cryptic Tree Frog *Leptopelis parbocagii* **Poynton** & **Broadley**, 1987

This species is not named directly after a person but is a reference to the species being very similar to another: *Leptopelis bocagii* Günther, 1865 – see entry for **Bocage**.

Paresh

Paresh's Caecilian *Gegeneophis pareshi* Giri et al., 2011

Paresh Porob is a range forest officer in Goa, India.

Parima

Rocket Frog sp. *Anomaloglossus parimae* **La Marca**, 1997

This frog is named after its area of distribution, the Serranía Parima mountain range on the Brazilian-Venezuelan border.

Parker, F

Parker's Wrinkled Ground Frog *Platymantis parkeri* **Brown**, 1965

Imigabip Snouted Frog *Xenorhina parkerorum* **Zweifel,** 1972*

Parker's Land Frog *Austrochaperina parkeri* Zweifel, 2000

Fred Parker (b.1941) of Kirwan, Queensland is an Australian naturalist and explorer. He was on Bougainville, Solomon Islands (1966) and in the 1970s he worked for the Wildlife Section, District Administration, Konebobu, Papua New Guinea. He co-wrote *A new species of* Cyrtodactylus *(Gekkonidae) from New Guinea with a key to species from the island* (1973). A reptile is named after him.

* The snouted frog is named after him and Hampton Wildman Parker (see below).

Parker, HW

Parker's Running Frog *Kassina maculifer* **Ahl**, 1924 [*Kassina parkeri* (**Scortecci**, 1932) is a junior synonym]

Parker's Mantella *Mantella pulchra* Parker, 1925 [Alt. Beautiful Mantella]

Parker's Slow Frog *Nanorana parkeri* **Stejneger**, 1927

Parker's Tree Frog *Leptopelis parkeri* **Barbour** & **Loveridge**, 1928

Snouted Treefrog sp. *Scinax parkeri* **Gaige**, 1929

Parker's Toad *Poyntonophrynus parkeri* Loveridge, 1932 [Alt. Mangasini Toad]

Parker's Reed Frog *Hyperolius parkeri* Loveridge, 1933

Parker's River Frog *Phrynobatrachus parkeri* **De Witte**, 1933

Parker's Brazilian Dumpy Frog *Stereocyclops parkeri* **Wettstein**, 1934

Parker's Dot Frog *Ramanella palmata* Parker, 1934
Parker's Snouted Frog *Xenorhina minima* Parker, 1934
Parker's White-lipped Frog *Hylarana parkeriana* **Mertens**, 1938
Parker's Caecilian *Epicrionops parkeri* **Dunn**, 1942
Parker's Rainforest Frog *Cophixalus parkeri* Loveridge, 1948
Parker's Tree Toad *Laurentophryne parkeri* **Laurent**, 1950
Parker's Cross Frog *Oreophryne parkeri* Loveridge, 1955
Imigabip Snouted Frog *Xenorhina parkerorum* **Zweifel**, 1972*
Parker's Andes Frog *Lynchius parkeri* **Lynch**, 1975
Parker's Water Frog *Petropedetes parkeri* **Amiet**, 1983

Dr Hampton Wildman Parker (1897–1968) was an English zoologist and herpetologist. He took both his bachelor's (1923) and master's (1935) degrees at Cambridge. He joined the Natural History Museum, London (1923). After war service in the Admiralty, he returned (1945) and became Keeper of Zoology (1947–1957). He wrote *A monograph of the frogs of the family Microhylidae* (1934). Sixteen reptiles are named after him.

* The snouted frog is named after him and Fred Parker (above).

Parker, N

Rocket Frog sp. *Anomaloglossus parkerae* Meinhardt & Parmalee, 1996

Dr Nancy Parker acted as undergraduate adviser to the senior author, Daniel Meinhardt.

Passarelli

Passarelli's Frog *Arcovomer passarellii* **Carvalho**, 1954

A. Passarelli Filho was a commercial collector who provided material for the Museu Nacional, Rio de Janeiro, Brazil (1940s). A catfish is also named after him.

Patita

Don Simon Rocket Frog *Hyloxalus patitae* **Lötters**, Morales & Proy, 2003

Patita is the nickname of one Patricia Vargas, honoured 'for her valuable and innocent courage as a herpetologist assistant in the hard slopes of the Cordillera El Sira'.

Patricia

Patricia's Robber Frog *Eleutherodactylus patriciae* **Schwartz**, 1965

Patricia A. Heinlein, along with Ronald Klinikowski, collected the type specimens (1962) when assisting Schwartz on a field trip to the Dominican Republic.

Paucar

Andes Frog sp. *Phrynopus paucari* **Lehr, Lundberg** & **Aguilar**, 2005

Paucar was a Quechuan god who participated in the legend of the origin of Paucartambo. The name is used in reference to the type locality, which is located within the Distrito de Paucartambo, Peru.

Paula

Forest Toad sp. *Nectophrynoides paulae* Menegon et al., 2007

Paola Magagnato (d. 2004) was a Tanzanian teacher of natural sciences.

Paulian

Pauliani's Dagger Frog *Boophis pauliani* **Guibé**, 1953
Madagascar Frog sp. *Mantidactylus pauliani* Guibé, 1974

Dr Renaud Paulian (1913–2003) was the leading European expert on scarab beetles. He became Assistant, Laboratory of Entomology, Muséum National d'Histoire Naturelle, Paris (1937). He was at the Institut de Recherche Scientifique de Madagascar, Tananarive (1947–1966), eventually becoming Director, and founding the publication *Faune de Madagascar* (1956). He was Director, Institut Scientifique de Congo-Brazzaville (1961–1966) and head of the local university, moving to be Head, Université d'Abidjan, Ivory Coast (1966–1969). He returned (1969) to France becoming, successively, Rector of the Academy of Amiens and the Academy of Bordeaux. Among the taxa named after him are a bird, a mammal and two reptiles.

Paulo Alvin

Paulo's Lime Treefrog *Sphaenorhynchus pauloalvini* **Bokermann**, 1973

Dr Paulo de Tarso Alvim Carneiro (1919–2011) was a Brazilian plant biologist and expert on cocoa who graduated in agriculture and veterinary medicine from the Universidade Federal de Viçosa, Brazil (1940). Cornell awarded his doctorate (1948) and the Universidade Federal da Bahia, Brazil, awarded him an honorary doctorate (1985). He was Technical Director, Centro de Pesquisas do Cacau, Ilhéus (1973) and director of the organization covering all agricultural research in Brazil (1991). A number of plants, including the bamboo genus *Alvimia*, are named after him.

Paulo Dutra

Paulo's Robber Frog *Ischnocnema paulodutrai* **Bokermann**, 1975

Dr Paulo Coutinho Dutra was an agricultural engineer at the Centro de Pesquisas do Cacau, Ihéus, Bahia, Brazil – the type locality of this frog.

Paulson

Paulson's Robber Frog *Eleutherodactylus paulsoni* **Schwartz**, 1964

Dr Dennis Roy Paulson (b.1937) is an American zoologist who was Director, Slater Museum of Natural History, University of Puget Sound, Tacoma, Washington State (1990–2004). His bachelor's degree (1958) and doctorate (1966) were both awarded by the University of Miami. He was a member of Schwartz's collecting party in Haiti (1961), and helped obtain the type series of this frog. He wrote *Shorebirds of North America* (2005), *Dragonflies and Damselflies of the West* (2008) and *Dragonflies and Damselflies of the East* (2011).

Pavioti

Pavioti's Smooth Horned Frog *Proceratophrys paviotii* **Cruz**, Prado & **Izecksohn**, 2005

Antônio Pavioti was a Brazilian naturalist, collector and taxidermist who specialized in the fauna of the Santa Teresa region of Espiritu Santo state.

Pearse

Colombian Plump Frog *Relictivomer pearsei* **Ruthven**, 1914

Arthur Sperry Pearse (1877–1956) was an American zoologist. The University of Nebraska awarded his bachelor's degree (1900) and his master's (1904) and doctorate (1908) were from Harvard. He taught (1900–1904) in Omaha High School, as an assistant at Harvard (1904–1907), taught zoology at Lake View High School, Chicago (1907), at Harvard (1908), the University of Michigan (1909–1910), in the Philippines (1911), and the University of Wisconsin (1912–1927), becoming Professor (1919). He was at Duke University (1929–1945), where he was Director, Marine Biology Laboratory (1938–1945). He directed the journal *Ecology Monographs* (1931–1951) and wrote *The Migration of Animals from Sea to Land* (1936).

Pearson, JC

Pearson's Frog *Litoria pearsoniana* **Copland**, 1961

Dr John C. Pearson collected the holotype (1958). He was a professor at the Department of Parasitology, University of Queensland, Brisbane, (1991).

Pearson, NE

Pearson's Slender-legged Treefrog *Osteocephalus pearsoni* **Gaige**, 1929

Dr Nathan Everett Pearson (1895–1982) was an ichthyologist at Indiana University, which had awarded his doctorate. He was on the Mulford Expedition (1921–1922) to Bolivia, during which he collected the holotype on the Upper Beni River (1921). He wrote *Fishes of the Rio Beni Basin* (1924).

Pech

Robber Frog sp. *Craugastor pechorum* **McCranie** & Wilson, 1999

The Pech are an indigenous people of Honduras. Their population has been reduced to less than 1,500 persons.

Peck

Peck's Robber Frog *Pristimantis pecki* **Duellman** & **Lynch**, 1988

Dr Stewart Blaine Peck is an evolutionary biologist, speleologist and entomologist who joined Carleton University, Ottawa, Canada (1971). His doctorate is from Harvard. He has conducted fieldwork throughout North and South America, the West Indies, Australia and parts of Asia and Africa. He wrote *The Beetles of the Galápagos Islands, Ecuador, Evolution, Ecology, and Diversity (Insecta: Coleoptera)* (2006).

Pedro Medina

Henle's Snouted Treefrog *Scinax pedromedinae** **Henle**, 1991

Pedro Medina Pizango was from Puerto Maldonado, Peru.

* As the treefrog is named after a man, there is a school of thought that the binomial should be *pedromedinai*, but this amendment is viewed as unjustified, much to the pleasure of traditionalists!

Péfaur

Arico Water Frog *Telmatobius pefauri* Veloso & **Trueb**, 1976

Péfaur's Rocket Frog *Aromobates serranus* Péfaur, 1985

Dr Jaime Eduardo Péfaur Vega (b.1944) is a Chilean ecologist and herpetologist who became (1995) Professor at the Department of Animal Ecology, Faculty of Sciences, University of the Andes, Mérida, Venezuela. He qualified as a veterinary surgeon at the Universidad de Chile, Santiago de Chile (1966). His master's (1971) and doctorate (1978) were both awarded by the University of Kansas, Lawrence. He wrote *Checklist and Bibliography (1960–1985) of the Venezuelan Herpetofauna* (1992).

Peixoto

Peixoto's Snouted Treefrog *Scinax atratus* Peixoto, 1989

Snouted Treefrog sp. *Scinax peixotoi* Brasileiro et al., 2007

Dr Oswaldo Luiz Peixoto (b.1947) is a Brazilian herpetologist who specializes in amphibian taxonomy and biology. He has worked at the Department of Animal Biology, Institute of Biology, Federal Rural University of Rio de Janeiro since 1972, where he is now Professor and where he took a bachelor's degree in natural history (1970) and a master's in zoology (1979). The University of São Paulo awarded his doctorate in zoology (1987). He co-wrote *Decimations, Extinctions, and Colonizations of Frog Populations in Southwest Brazil and Their Evolutionary Implications* (1988).

Penan

Penan Tree Frog *Rhacophorus penanorum* Dehling, 2008

This frog is named after the Penan people who live in the Gunung Mulu area of Sarawak, Malaysia, where the holotype was collected.

Penther

Penther's Rain Frog *Breviceps adspersus pentheri* **Werner**, 1899

Dr Arnold Penther (1865–1931) was an Italian-born Austrian zoologist, conchologist and botanist who collected plants and molluscs in South Africa and Southern Rhodesia (now Zimbabwe) (1896).

Penton

Penton's Toad *'Bufo' pentoni* **Anderson**, 1893
[Alt. Shaata Gardens Toad; placed in *Bufo* as true genus unassigned]

Colonel Richard Hugh Penton (1863–1934) was a British army surgeon, who was seconded to the Egyptian army (1892–1905), being Principal Medical Officer (1898–1905). He took part in Kitchener's campaigns to conquer Sudan, being with the 1896 Dongala Expeditionary Force, served in both Nile Expeditions (1898–1899), and was present at the Battles of Atbara and Khartoum. He was Assistant Director of Medical Services of 1st Indian Cavalry Division in Flanders (1915) and later fulfilled the same function at Dunkirk. A bird is named after him.

Pepper

Pepper's Poison Frog *Ameerega pepperi* Brown & Twomey, 2009

Mark Pepper is a Canadian explorer and conservationist, who since 2004 has been based in Iquitos, Peru. He co-wrote (with Brown and Twomey) *Revision of the* Ranitomeya fantastica *species complex with description of two new species from Central Peru (Anura: Dendrobatidae)*. He helped design a project for the conservation of poison frogs through habitat purchase and protection. By selling captive-bred Peruvian frogs, Mark generates funds, which are used for the purchase of land critical for frog conservation.

Peracca

Peracca's Madagascar Frog *Spinomantis peraccae* **Boulenger**, 1896
Peracca's Clawed Frog *Xenopus clivii* Peracca, 1898
[Alt. Eritrean Clawed Frog]
Peracca's River Frog *Ptychadena tellinii* Peracca, 1904
[Alt. Tellini's Grass Frog]
Peracca's Plump Toad *Osornophryne bufoniformis* Peracca, 1904
Peracca's Andes Frog *Hypodactylus peraccai* **Lynch**, 1975

Count Dr Mario Giacinto Peracca (1861–1923) was an Italian herpetologist who kept a collection of Galápagos Giant Tortoises and other reptiles and amphibians in his greenhouse in Turin. He abandoned medical studies in favour of zoology. He took his doctorate (1886) and joined the Institute of Zoology, University of Turin, as an assistant. He worked there until his retirement (1920). He visited the Cape Verde Islands (1891), bringing back to Italy 40 living specimens of the now-extinct Cape Verde Giant Skink. Seven reptiles are named after him.

Pérez

Pérez's Snouted Frog *Edalorhina perezi* **Jiménez de la Espada**, 1870
Pérez's Frog *Pelophylax perezi* Lopez-Seoane, 1885
[Alt. Iberian Green Frog]

Professor Laureano Pérez Arcas (1824–1894) was a Spanish entomologist and malacologist who became Professor of Zoology at the University of Madrid. He was co-founder of the Spanish Society of Natural History.

Perinet

Perinet's Bright-eyed Frog *Boophis hillenii* **Blommers-Schlösser**, 1979
[Perhaps a synonym of *B. idae*]

Not an eponym but a toponym: the type locality was 'near Perinet' in eastern Madagascar, which also has three reptiles named after it.

Perkins

Perkins' Treefrog *Exerodonta perkinsi* **Campbell** & **Brodie**, 1992

Bob F. Perkins (1930–1999) was a geologist and palaeontologist. The Southern Methodist University (SMU) in Dallas awarded his bachelor's and master's degrees. He spent two years at the University of Michigan, then taught at SMU and the University of Houston whilst undertaking his doctorate. He joined Shell Development Company (1957), researching carbonates. He went on to Louisiana State University's School of Geosciences (1966) where he eventually became Chairman. He spent a great deal of time editing and publishing books there. He left to become Professor of Geology and Dean of the Graduate School at the University of Texas at Arlington (1975). The describers wrote: 'This species is named in honor of Dean Bob F. Perkins of the Graduate School of the University of Texas at Arlington, who has strongly and consistently supported our studies on Neotropical herpetofauna'.

Peron

Péron's Tree Frog *Litoria peronii* **Tschudi**, 1838
Striped Marsh Frog *Limnodynastes peronii* **Duméril** & **Bibron**, 1841

François Péron (1775–1810) was a French voyager and naturalist. He was a member of Baudin's scientific expedition with the ships 'Geographe' and 'Naturaliste' (1800–1804), which visited New Holland, Maria Island, Van Diemen's Land (now Tasmania) and Timor, Indonesia. As well as collecting zoological specimens, Péron also conducted pioneering experiments on seawater temperatures at depth. He died at the age of 35 of tuberculosis. The Peron Peninsula, Western Australia is named after him as are two mammals, six birds and three reptiles.

Perret

Perret's Grassland Frog *Ptychadena perreti* **Guibé** & **Lamotte**, 1958
Perret's Toad *Amietophrynus perreti* **Schiøtz**, 1963
Perret's Egg Frog *Leptodactylodon perreti* **Amiet**, 1971
Perret's Shovelnose Frog *Hemisus perreti* **Laurent**, 1972
Perret's Water Frog *Petropedetes perreti* Amiet, 1973
Perret's Night Frog *Astylosternus perreti* Amiet, 1978
Perret's Squeaker Frog *Arthroleptis perreti* Blackburn et al., 2009

Dr Jean-Luc Perret (b.1925) is a Swiss herpetologist who specializes in African amphibians. The Université de Neuchâtel awarded his doctorate (1963). He is Honorary Curator, Department of Herpetology and Ichthyology, Natural History Museum, Geneva, where he was a research assistant (1970–1987). He wrote *The Amphibians of Cameroon* (1966). Four reptiles are named after him.

Pesantes

Andes Frog sp. *Phrynopus pesantesi* **Lehr, Lundberg** & **Aguilar**, 2005

Dr Omar Pesantes Segura (d. 2004) was a Peruvian biologist and herpetologist at the Museo de Historia Natural, Universidad Nacional Mayor de San Marcos, Lima. The describers wrote that he was their friend and colleague.

Peters, G

Tsaratanana Digging Frog *Rhombophryne guentherpetersi* **Guibé**, 1974

Dr Günther Peters (b.1932) is a German herpetologist and Director, Institute for Systematic Zoology, Humboldt Museum, Berlin. He studied for his doctorate in Leningrad (St Petersburg) (1957). He wrote *Eine neue Wirtelschwanzagame aus Ostafrika (Agamidae: Agama)* (1987). Three reptiles are named after him.

Peters, JA

Peters' Caecilian *Epicrionops petersi* **Taylor**, 1968
Peruvian Poison Frog *Ameerega petersi* **Silverstone**, 1976
Jacobus Peters' Poison Frog *Colostethus jacobuspetersi* **Rivero**, 1991
[Alt. Quito Rocket Frog]
Peters' Stubfoot Toad *Atelopus petersi* **Coloma** et al., 2007
Peters' Shiny Peeping Frog *Eleutherodactylus nitidus petersi* **Duellman**, 1954

Dr James Arthur Peters (1922–1972) was an American zoologist who specialized in Ecuadorean herpetofauna. He attended the University of Michigan and was awarded a bachelor's degree (1948), a master's (1950) and a doctorate (1952). He taught at Brown University, Rhode Island (1952–1958), being an associate professor. He left to become the Fulbright Lecturer at Universidad Centrale de Ecuador (1958–1959). He was a professor at San Fernando Valley State College (1959–1964) and at the Smithsonian was Assistant Curator, Reptiles and Amphibians (1965–1972), and Curator for the last year of his life. He co-wrote *Catalogue of the Neotropical Squamata* (1970). Nine reptiles are named after him.

Peters, WKH

Peters' Bullfrog *Pyxicephalus edulis* Peters, 1854
[Alt. Edible Bullfrog]
Peters' Mushroomtongue Salamander *Bolitoglossa adspersa* Peters, 1863
Peters' Thin-toed Frog *Leptodactylus petersii*

Steindachner, 1864
Peters' Frog *Litoria inermis* Peters, 1867
[Alt. Bumpy Rocket Frog]
Peters' Four-eyed Frog *Pleurodema diplolister* Peters, 1870
Peters' River Frog *Phrynobatrachus dispar* Peters, 1870
Peters' Frog *Fejervarya brevipalmata* Peters, 1871
Peters' Malaysian Frog *Hylarana raniceps* Peters, 1871
Peters' Dwarf Frog *Engystomops petersi* **Jiménez de la Espada**, 1872
Peters' Smooth Horned Frog *Proceratophrys bigibbosa* Peters, 1872
Peters' Andes Frog *Phrynopus peruanus* Peters, 1873
Peters' Reed Frog *Hyperolius glandicolor* Peters, 1878
Peters' Foam-nest Treefrog *Chiromantis petersii* **Boulenger**, 1882
Peters' Rain Frog *Breviceps adspersus* Peters, 1882
[Alt. Bushveld Rain Frog, Common Rain Frog]
Upper Amazon Caecilian *Nectocaecilia petersii* Boulenger, 1882
Peters' Clawed Frog *Xenopus petersii* **Bocage**, 1895
African Caecilian sp. *Geotrypetes petersii* Boulenger, 1895 NCR
[JrSYN *Geotrypetes seraphini*]
Peters' Bubble-nest Frog *Philautus petersi* Boulenger, 1900
[Alt. Peters' Bushfrog]

Wilhelm Karl Hartwig Peters (1815–1883) was a German zoologist and traveller who made a major expedition to Africa (1842–1847), travelling from Angola to Mozambique. He returned to Berlin with very important zoological collections. For many years he was Curator of the Berlin Zoology Museum. He was elected as a corresponding member of the Russian Academy of Sciences (1876). Thirty-nine reptiles, twenty-three mammals and two birds are named after him.

Peters, WKH & JA

Peters' Robber Frog *Pristimantis petersorum* **Lynch**, 1991

Lynch's etymology states that he is naming the frog after 'James A. Peters and Wilhelm Peters, each of whom worked on frogs from northwestern South America.' (See **Peters, JA** and **Peters, WKH**)

Pewzow

Pewzow's Toad *Pseudepidalea pewzowi* **Bedriaga**, 1898
[Syn. *Bufo pewzowi*]

General Michail Wassiljewitsch Pewzow (1843–1902) was a geographer, cartographer and explorer who led an expedition to Kuen Lun Mountains and Yarkand (1889–1890). A bird is named after him.

Philip

Robber Frog sp. *Pristimantis philipi* **Lynch** & **Duellman,** 1995

Dr Philip Strong Humphrey (1926–2009) was Professor Emeritus, Systematics and Ecology, Curator Emeritus and former Director, Natural History Museum, University of Kansas. After service (1944–1947) in the US Air Force, he earned a bachelor's degree at Amherst College (1949) and a master's (1952) and doctorate (1955) at the University of Michigan, where he worked as Research Associate in the Museum of Zoology (1955–1957). He moved to Yale as Assistant Curator, Ornithology and Assistant Professor, Zoology (1957–1962). He was Curator, Division of Birds, Department of Zoology, Smithsonian (1962–1965) and Chairman of Vertebrate Zoology (1965–1967). He was at the University of Kansas thereafter.

Philippi

Water Frog sp. *Telmatobius philippii* Cuevas & Formas, 2002

Rodulfo Amando Philippi (1905–1969) was a Chilean zoologist and the grandson of another naturalist – German-born Rodolfo Amando Philippi (1808–1904). He co-wrote *Las Aves de Chile* (1946).

Picado

La Estrella Salamander *Nototriton picadoi* **Stejneger**, 1911
Picado's Bromeliad Treefrog *Isthmohyla picadoi* **Dunn**, 1937

Dr Clodomiro Picado Twight (1887–1944), also known as Clorito Picado, was a leading Costa Rican botanist, zoologist and toxicologist. The Sorbonne awarded his doctorate (1913). He was internationally recognized for his studies of venomous snakes and his development of anti-venom serums. The scientific research institute Instituto Clodomiro Picado, Costa Rica, is named after him, as is a reptile.

Pickering

Pickering's Hyla *Hylodes pickeringii* **Holbrook**, 1839 NCR
[JrSYN *Pseudacris crucifer*]

Dr Charles F. Pickering (1805–1878) was an ornithologist and ethnologist who was a friend of Audubon. He explored the White Mountains, New Hampshire (1825). He qualified as a physician at Harvard (1826), set up practice in Philadelphia (1827) and was Librarian of the Philadelphia Academy of Natural Sciences (1828–1833) and, thereafter, Curator. He was on the US Exploring Expedition (1838–1842) as Naturalist. He helped Holbrook with the publication of *North American Herpetology* (1842). He went abroad to study ethnology (1844), traveling from Egypt to Zanzibar and then on to India. He wrote *The Races of Man and their Geographical Distribution* (1847). A bird and a reptile are named after him.

Pickersgill

Pickersgill's Reed Frog *Hyperolius pickersgilli* Raw, 1982
[Alt. Avoca Reed Frog]
Pickersgill's Banana Frog *Afrixalus delicatus* Pickersgill, 1984
[Alt. Delicate Leaf-folding Frog]

Martin Pickersgill (b.1956) is an English herpetologist whose main interest is amphibians. He collected in southern Africa (1997). He wrote *The taxonomy and ethology of the* Afrixalus stuhlmanni *complex (Anura: Hyperoliidae)* (2005).

Pierotti

Lesser Chini Frog *Chacophrys pierottii* **Vellard**, 1948

Serafin Angel Pierotti was an Argentine botanist and herpetologist. He collected botanical specimens in Argentina and Bolivia (1944–1948). He worked as a collector and taxidermist at the Instituto Lillo with Cei (q.v.) and they co-wrote *Notas batracológicas y biogeográficas Argentinas* (1957).

Pierre

Pierre's Cricket Frog *Fejervarya pierrei* **Dubois**, 1975

Pierre Dubois is, presumably, related to the author, but we have been unable to find anything about him: the original description gives his name as the dedicatee, but no more information.

Pillai

Pillai's Tropical Frog *Micrixalus nudis* Pillai, 1978
[Alt. Naked Torrent Frog]
Pillai's Night Frog *Nyctibatrachus pillaii* **Biju** et al., 2011

Dr R. Sridharan Pillai worked for the Zoological Survey of India and is a member of the Department of Zoology, University College, Trivandrum, India. He co-wrote *Gymnophiona (Amphibia) of India: A Taxonomic Study* (1999).

Pinango

Pinango Stubfoot Toad *Atelopus pinangoi* **Rivero**, 1982

This amphibian is named after the type locality, Piñango, Mérida, Venezuela.

Pinchon, A

Pinchon's Salamander *Batrachuperus pinchonii* **David**, 1872

Annet-Théophile Pinchon (1814–1891) was a French Roman Catholic missionary in China. He was ordained as a priest (1839) in La Société des Mission Étrangères. He was appointed titular Bishop of Polemonium (1859) and Vicar Apostolic, Northwest Szechwen (1861). Père David met him (1871) and named the amphibian after him, saying in the description (free translation of the French): 'I have dedicated this remarkable amphibian to Monseigneur Pinchon, apostolic vicar of upper Szechwan, under the name *Dermodactyliis pinchonii,* in memory of the service he has rendered to my expedition, and because he gave me the first information about this species.'

Pinchon, R

Grand Cafe Robber Frog *Eleutherodactylus pinchoni* **Schwartz**, 1967

Professor Père Robert Pinchon (1913–1980) of Community of the Fathers of the Holy Spirit was ordained a priest (1939) in France, where he worked as a parish priest until sailing (1945) for Martinique, where he was made Professor of Biology, Seminaire College, Fort-de-France. He was a man of many interests including archaeology, and he convened (1961) the first meeting of the International Association for Caribbean Archaeology to compare notes on early Amerindian cultures. From his base in Martinique he travelled in the Caribbean and amassed diverse collections of fishes, crustaceans, butterflies, spiders, scorpions, snakes etc. He was a great help to naturalists in their fieldwork and Schwartz commemorated him for it. There is a reserve in Martinique named La Réserve du Père Pinchon after him.

Pine

Mexican Pine Woods Treefrog *Exerodonta pinorum* **Taylor**, 1937

Pinorum means 'of the pines' and refers to the habitat in Mexico where the holotype was collected.

Ping

Salamander subgenus *Pingia* **Chang**, 1935
Ping's Toothed Toad *Oreolalax pingii* **Liu**, 1943
[Alt. Chinese Lazy Toad]

Dr Chih Ping (1886–1965), whose doctorate was awarded (1918) by Cornell, became a zoology investigator at Wistar Institute, Philadelphia, before being appointed Professor of Zoology, College of Agriculture, National Southeastern University,

Nanking (1921). He was (1935) Director of the biological laboratory at Nanking (Nanjing). A bird is named after him.

Pinto

Pinto's River Frog *Phrynobatrachus pintoi* Hillers, Zimkus & Rödel, 2008

Pinto was the nickname of Sidy Mohamed Diawara (d.2006). He was a great contributor to field research and nature conservation in Guinea.

Pinto, OM

Brown Bromeliad Frog *Crossodactylodes pintoi* **Cochran**, 1938

Olivério Mario de Oliveira de Pinto (1896–1981) was a Brazilian ornithologist. He was Head, Zoological Museum, University of São Paulo (1940s) and led expeditions under its auspices. He wrote *Catalogo das Aves do Brasil* (1938). Five birds are named after him.

Pisano

Argentine Water Frog *Telmatobius pisanoi* **Laurent**, 1977

Professor Dr A. Pisano is a biologist at the Institute of Zoology, National University of Tucumán, Argentina. He was (1959) at the Universidad Nacional de Cuyo, Mendoza. He co-wrote *Submicroscopical aspects of Telmatobius hauthali schreiteri spermatids* (1968).

Pittier

Rocket Frog sp. *Allobates pittieri* **La Marca**, Manzanilla & Mijares-Urrutia, 2004

Henri François Pittier (1857–1950) was a Swiss-born botanist who moved to Costa Rica (1887). He collected plant specimens in Central America and Venezuela. The location where this frog occurs, the Henri Pittier National Park in Venezuela, is named after him, as are a bird and a mammal.

Plancy

Eastern Golden Frog *Pelophylax plancyi* **Lataste**, 1880

[Alt. Peking Frog, Beijing Gold-striped Pond Frog]

Victor Collin de Plancy (1853–1922) was a diplomat and traveller in China, Korea and Siam (now Thailand). While other European nations' diplomats were interfering in the politics of Korea, de Plancy, who was the head of the French mission there for 15 years, was interested in the language and culture. He made studies of both, catalogued as many Korean books as he could, and also donated a number of Buddhist paintings to the Guimet Museum. The Bibliothèque Nationale de France purchased his impressive collection of books, which included xylographic editions of Korean books and maps. A mammal is named after him.

Plaumann

Nova Teutonia White-lipped Frog *Leptodactylus plaumanni* **Ahl**, 1936

Friedrich 'Fritz' Plaumann (1902–1994) was a German-born entomologist who moved to Brazil (1924). He collected in southern Brazil, including Santa Catarina province where the frog holotype was discovered. The entomological museum at Seara, Santa Catarina, Brazil is named after him.

Pleione

Pleione's Torrent Frog *Taudactylus pleione* Czechura, 1986

[Alt. Kroombit Tinker Frog]

In Greek mythology Pleione was a nymph, the mother of the seven Pleiades.

Pleske

Pleske's High Altitude Frog *Nanorana pleskei* **Günther**, 1896

[Alt. Plateau Frog, Tibetan Frog]

Fedor Dimitrievich Pleske (1858–1932) was a zoologist, geographer and ethnographer. From childhood he collected birds and insects in the different provinces of European Russia. He graduated (1882) from St Petersburg University. He became a Fellow of the Russian Imperial Academy of Science, St Petersburg, and was appointed (1886) Scientific Keeper, Department of Ornithology, Zoological Museum, Imperial Academy of Sciences. After the Russian Revolution (1917) he was an active member of the USSR Zoological Museum. He wrote *Ornithological Fauna of Imperial Russia* (1891). Five birds and a reptile are named after him.

Plimpton

Plimpton's Dainty Frog *Cacosternum plimptoni* **Channing** et al., 2005

George Ames Plimpton (1927–2003) was an American sports journalist, actor, editor and ornithologist. He went to Harvard (1944), graduating in 1950 because of military service (1945–1948). After Harvard he went to England, taking both bachelor's and master's degrees (1952) at Cambridge. He joined *The Paris Review*, a magazine largely funded by the CIA. His classmate at Harvard was Robert Kennedy and Plimpton was one of those who jumped on Sirhan Sirhan when he shot Kennedy

(1968). Plimpton undertook assignments 'as an amateur' in the world of the professional, including going three rounds with Sugar Ray Robinson. Channing and his co-authors chose to honour him in the frog's binomial as he played an important role in generating support for the Tanzania Wildlife Fund and the African Rainforest Conservancy.

Poelz

Poelz's Worm Salamander *Oedipina poelzi* **Brame**, 1963

Dietrich Pölz of Radbuel, Germany, was an old friend of the describer. We know nothing more about him.

Poeppig

Grey Toad *Rhinella poeppigii* **Tschudi**, 1845

Professor Eduard Friedrich Poeppig (1798–1868) was a German naturalist, collector and explorer. He studied medicine and natural science at Leipzig University, leaving to undertake an expedition to Cuba and the USA (1823–1826). He then went to Chile, and was in South America until 1832. When he returned to Germany he became Professor of Zoology, Leipzig University. He wrote *Reise nach Chili, Peru, und auf dem Amazonen-Flusse* (1835). A mammal, a bird and a reptile are named after him.

Poiret

Poiret's Newt *Pleurodeles poireti* Gervais, 1835
[Alt. Algerian Ribbed Newt]

Professor Abbé Jean Louis Marie Poiret (1755–1834) was a French priest, botanist and explorer. King Louis XVI sent him to Algeria (1785) to study the flora. He travelled in Algeria (1785–1786) and wrote *Voyage en Barbarie. Animaux Amphibies de l'Algérie* (1789). After the French Revolution he became Professor of Natural History, Écoles Centrales, Aisne.

Pokomchi

Rio Sananja Spikethumb Frog *Plectrohyla pokomchi* **Duellman** & **Campbell**, 1984

The Pokomchi are an indigenous people of Guatemala, where this frog is found.

Poll

Tshimbulu Reed Frog *Hyperolius polli* **Laurent**, 1943

Dr Max Fernand Leon Poll (1908–1991) was a Belgian ichthyologist and 'connoisseur of the fish fauna'. He worked in the Congo and in the Musée Royal du Congo Belge, Tervuren, and was Professor at the Université Libre de Bruxelles. He led expeditions to Lake Tanganyika (1946–1947) and a Congo River survey (1953). Many fishes and a mammal are named after him.

Polunin

Polunin's Spiny Frog *Nanorana polunini* **Smith**, 1951
[Alt. Langtang Frog]

Oleg Vladimirovitch Polunin (1914–1985) was an English traveller and botanist who taught botany at Charterhouse, Surrey for over 30 years. The Oleg Polunin Memorial Fund, which provides scholarships, was established in his memory. He co-wrote *Flowers of the Himalaya* (1985). He collected the frog holotype (1949).

Polymnia

Sierra Juarez Robber Frog *Craugastor polymniae* **Campbell**, Lamar & Hillis, 1989

The describers say that Polymnia was one of the sirens who lured sailors to their death, but they may have muddled up their Greek myths. Polymnia, or Polyhymnia, was the Muse of lyric poetry and the mother of Orpheus. Her name is not usually found among those of the sirens.

Pombal

Pombal's Toad *Rhinella pombali* Baldissera, **Caramaschi** & **Haddad**, 2004
Treefrog sp. *Hypsiboas pombali* Caramaschi, Pimenta & **Feio**, 2004
Saddleback Toad sp. *Brachycephalus pombali* Alves et al., 2006

José P. Pombal Jr is a herpetologist who is an associate professor and Curator of Amphibians at the Museu Nacional, Rio de Janeiro, Brazil. He has written with at least one of the describers: with Haddad et al. *Breeding behavior of the Pumpkin Toadlet,* Brachycephalus ephippium *(Brachycephalidae)* (1994).

Poole

Poole's Robber Frog *Eleutherodactylus poolei* **Cochran,** 1938

Arthur J. Poole was a zoologist at the Smithsonian. He collected the type specimen (1928). He was Treasurer, American Society of Mammalogists (1921–1930). He was a member, Biological Society of Washington. He co-wrote *Catalog of the Type Specimens of Mammals in the United States National Museum, including the Biological Surveys Collection* (1942).

Pope

Pope's Salamander *Leurognathus marmorata intermedia* Pope, 1928 NCR
[JrSYN *Desmognathus marmoratus*]
Pope's Lazy Toad *Oreolalax popei* **Liu**, 1947
[Alt. Baoxing Toothed Toad]

Clifford Hillhouse Pope (1899–1974) was an American herpetologist. He spent several years in China (1920s) as a member of expeditions organised by the AMNH, New York. He became a fluent speaker of Chinese. He was Assistant Curator of Herpetology, American Museum (1928–1935) but was sacked after a disagreement with the Director, Gladwyn Kingsley Noble. To support himself, unemployed in the Great Depression, he wrote a number of popular books on herpetology. He was (1941–1954) Curator of Reptiles, Field Museum, Chicago. He took early retirement and went to live in California where he continued to write. He wrote *Reptiles of China* (1935). His wife, Sarah H. Pope was also a herpetologist. Four reptiles are named after him.

Poppy

Poppy's Shrub Frog *Pseudophilautus poppiae* Meegaskumbura & Manamendra-Arachchi, 2005

Poppy Valentina Meyer (b. 2003) was only two years old when this frog was named after her as a compliment to her parents, George A. Meyer and Maria Semple, who support the Global Amphibian Assessment.

Porras

Pijol Salamander *Bolitoglossa porrasorum* **McCranie** & Wilson, 1995

Jorge Porras Zuniga and his son Jorge Porras Orellana were honoured by the describers as they '...have assisted us in every way imaginable in the pursuit of our work in Honduras.' It was noted at the time of the description that the elder Porras had died.

Porter

Toad sp. *Incilius porteri* **Mendelson** et al., 2005

[Syn. *Bufo porteri*, *Cranopsis porteri*]

Kenneth Raymond Porter was a herpetologist who collected the toad holotype in Honduras (1964). He received his bachelor's degree from the University of Wyoming, and a master's from Oregon State College, before undertaking his PhD in zoology at the University of Texas (1962). He was a first lieutenant in the US Army (1954–1957). He wrote a number of scientific papers (1950s–1970s), mostly on North American amphibians, as well as the textbook *Herpetology* (1972). He undertook fieldwork in Mexico and Honduras, collecting there several times (1963–1965).

Posada

Glass Frog sp. *Nymphargus posadae* **Ruiz-Carranza** & **Lynch**, 1995

Dr Andrés Posada Arango (1839–1923) was a Colombian physician, zoologist, botanist, educationalist and promoter of natural sciences. He graduated (1859) in Bogotá and served as an army surgeon (1860–1862) in the army of the Confederation of Granada (long gone, and now part of a number of South American countries, including Colombia). He went to Europe and stayed in Paris (1868–1872). He became the first Professor of Physical and Natural Sciences, Universidad de Antioquia, Colombia. Later in life he held many important appointments, but a fire (1921) destroyed his home and most of the papers recording his life's work.

Power

Power's Rain Frog *Breviceps poweri* **Parker**, 1934

Power's Toad *Amietophrynus poweri* **Hewitt**, 1935

[Alt. Western Olive Toad]

Power's Reed Frog *Hyperolius poweri* **Loveridge,** 1938

John Hyacinth Power (1884–1964) was born in Ireland and emigrated to South Africa (1904) to teach at a Christian Brothers College in Kimberley. He became (1920) Principal, South African School of Mines. He was Director, Alexander McGregor Memorial Museum, Kimberley (1947–1958). He collected plants and reptiles as well as amphibians. He wrote *Some Notes on the Colouration of Reptiles and Amphibians found near Kimberley, C.P.* (1918).

Poynton

South African frog genus *Poyntonia* **Channing** & Boycott, 1989

Toad genus *Poyntonophrynus* **Frost** et al., 2006

Poynton's Dainty Frog *Cacosternum poyntoni* Lambiris, 1988

[Alt. Poynton's Caco, Nursery Metal Frog]

Poynton's Forest Toad *Nectophrynoides poyntoni* Menegon, Salvidio & Loader, 2004

Professor Dr John Charles Poynton (b.1931) is a South African herpetologist who is Emeritus Professor, Biology, University of Natal, Pietermaritzburg, South Africa. He is a research associate at the Natural History Museum, London. He wrote *The Amphibia of Southern Africa: A faunal study* (1964). His other main interest is in psychical research: he was President, Society for Psychical Research, London (2004–2007) and has since acted as the Society's Honorary Secretary.

Praderio

Rocket Frog sp. *Anomaloglossus praderioi* **La Marca**, 1997

María José Praderio is a Venezuelan herpetologist who graduated from the Universidad Simón

Bolívar, Caracas (1985). He was at the Museo de Historia Natural La Salle, Caracas (1990) and (2004) was on staff of the Laboratory of Biogeography, Universidad de los Andes, Mérida. He co-wrote with La Marca and Esqueda, from the same department, *A new species of high tepuy lizard of the genus* Riolama *(Squamata: Gymnophthalmidae) from Marahuaca mountain, Amazonas state, Venezuela.*

Pratt

Pratt's Rocket Frog *Colostethus pratti* **Boulenger**, 1899
Pratt's Treefrog *Litoria pratti* Boulenger, 1911

Antwerp Edgar Pratt was one of a family of great travellers and explorers: Antwerp Edgar (c.1850–c.1920), after whom both these amphibians are named, and his sons, Felix and Charles, and a cousin, Joseph, all explored – sometimes alone and sometimes together. Antwerp and Felix collected lepidoptera in the Colombian Andes in the 1890s, and later spent two years in New Guinea. While travelling (1891) to Tibet, Antwerp visited Tatsienlu, China, and there met two famous naturalists, both of whom were Lazarite missionaries – Bishop Felix Biet and Père Jean André Soulie. He wrote *To the Snows of Tibet through China* (1892). Three mammals and two reptiles are named after him.

Preuss

Buea Smalltongue Toad *Werneria preussi* Matschie, 1893

Professor Paul Preuss (1861–1926), although born in Poland, was a German naturalist, botanist and horticulturist. He collected in West Africa (1886–1898 and 1910) and in New Guinea (1903). He was a member of a military expedition to explore the hinterland of Cameroon – a German colony at that time (1888–1891). While storming a native village the troop commander was killed and the second-in-command severely wounded. Preuss took command, leading the remaining troops back to the coast. He constructed the botanical gardens of Victoria (Limbe), Cameroon (1901). Six birds, a reptile and three mammals are named after him.

Price

El Centro Caecilian *Parvicaecilia pricei* **Dunn**, 1944

Wilfred Devereux Price was Dunn's host when he visited El Centro, Colombia.

Prince Charles

Prince Charles' Stream Treefrog *Hyloscirtus princecharlesi* **Coloma** et al., 2012

Charles Philip Arthur George Windsor, Prince of Wales (b.1948) is heir to the British throne and a noted conservationist of rainforests, for which the describers honoured him.

Protche

Rochebrune's Reed Frog *Hyperolius protchei* **Rochebrune**, 1885

Rochebrune's very brief description gives no etymology. We think that this frog was named after a man by the name of Protche who was a taxidermist, and who collected natural history specimens in the Congo (1874) and other areas, including Cabinda, for the Muséum National d'Histoire Naturelle, Paris.

Puebla

Puebla Frog *Lithobates pueblae* **Zweifel**, 1955
[Syn. *Rana pueblae*]

The holotype was acquired in Puebla state, Mexico.

Pugh

Pugh's Frog *Philoria pughi* Knowles et al., 2004
[Alt. Southern Mountain Frog]

Dailan Pugh (b.1955) is an Australian artist and conservationist who has done much to help preserve forest habitats, and (2003) was awarded the Order of Australia for services to forest conservation.

Pujol

Pujol's Grassland Frog *Ptychadena pujoli* **Lamotte** & Ohler, 1997

Raymond Pujol is a French agronomist and ethnologist. He is Professor Emeritus of Ethno-Zoology at the Muséum National d'Histoire Naturelle, Paris, where he was originally employed in the entomological department before working in West Africa in association with the senior describer. He led an expedition to Sérédou, Guinea (1957–1958). He co-wrote *Dictionnaire raisonné de biologie* (2003). A bird is named after him.

Pulido

Treefrog sp. *Hypsiboas pulidoi* **Rivero**, 1968

Juan R. Pulido (b.1926) was a Venezuelan entomological technician who collected with Rivero (1968) on Monte Duida, Venezuela, where the holotype was discovered.

Purcell

Purcell's Ghost Frog *Heleophryne purcelli* Sclater, 1898
[Alt. Purcell's Torrent Frog, Cape Ghost Frog]

Dr William Frederick Purcell (1866–1919) was a South African entomologist and naturalist, particularly interested in spiders and scorpions. He was born in England but partly educated in Germany:

the University of Berlin awarded his doctorate (1895). He worked at the South African Museum (1896–1905) and collected the holotype. A reptile is named after him.

Pyburn

Pyburn's Robber Frog *Pristimantis savagei* Pyburn & **Lynch**, 1981

Pyburn's Pancake Frog *Otophryne pyburni* **Campbell** & **Clarke**, 1998

Professor Dr William Frank 'Billy' Pyburn (1927–2007) founded the specimen collection (1956), University of Texas, Arlington, which became the Amphibian and Reptile Diversity Research Centre. He made expeditions to the jungles of Colombia (1970s). These trips were not for the faint-hearted, as he would climb into his motor vehicle in Arlington and drive all the way to Colombia! He was also a very good artist who painted birds. Three reptiles are named after him.

Q

Quang

Quang's Tree Frog *Gracixalus quangi* Rowley et al., 2011

Professor Quang Xuan Hoang is a herpetologist at Vinh University, Vietnam.

Quecchi

Las Palmas Spikethumb Frog *Plectrohyla quecchi* **Stuart**, 1942

The Quecchi are an indigenous people of Guatemala.

Quelch

Roraima Bush Toad *Oreophrynella quelchii* **Boulenger**, 1885

John Joseph Quelch (b.1854) was a zoologist who was on the Challenger expedition (1872–1876) and contributed to some of its reports. He collected in British Guiana (Guyana) and Venezuela. He was curator at the British Guiana Museum (Demerera) (1886) and went on an expedition to Mount Roraima (1898). He wrote *The Boa Constrictors of British Guiana* (1898).

Quoy

Treefrog sp. *Hyla quoyi* Bory de Saint-Vincent, 1828 NCR

[*Nomen oblitum* - taxon now called *Hypsiboas prasinus* **Burmeister,** 1856]

Chilean Toad sp. *Peltocephalus quoyi* **Tschudi**, 1838 NCR

[*Nomen oblitum* - taxon now called *Calyptocephalella gayi* **Duméril** & **Bibron**, 1841]

Jean René Constant Quoy (1790–1869) was a French zoologist and naval surgeon who was on a number of voyages of discovery (1826–1829). He was on board the 'Uranie' when she was wrecked but continued his journey on the rescuing vessel, which the expedition bought and re-named 'Physicienne'. Many taxa, including five birds and a reptile, are named after him.

Quyet

Rough-coated Tree Frog *Gracixalus quyeti* Nguyen et al., 2008

[Syn. *Philautus quyeti*]

Le Khac Quyet is a biologist who works with the Flora and Fauna International Vietnam programme. He is studying for his doctorate at Hanoi University of Science, Vietnam National University.

R

Rabb

Guatemalan Bromeliad Salamander *Dendrotriton rabbi* **Lynch** & **Wake**, 1975

Rabbs' Fringe-limbed Treefrog *Ecnomiohyla rabborum* **Mendelson** et al., 2008

Dr George Bernard Rabb (b.1930) is a zoologist and herpetologist who was Director, Brookfield Zoo and President (1976–2003), Chicago Zoological Society, which he originally joined (1956) as Curator of Research – and which presents an annual George B Rabb Conservation Award. The College, Charleston, South Carolina awarded his bachelor's degree in biology; his master's degree and doctorate in zoology were awarded by the University of Michigan. He is Chairman of the Illinois State Museum Board. The treefrog is named after both George and his wife, Mary S, Rabb, also a herpetologist. Together they wrote *Additional observations on breeding behavior of the Surinam Toad,* Pipa pipa (1963).

Rabe

Venezuelan Tiny Caecilian *Microcaecilia rabei* **Roze** & Solano, 1963

G. Rabe accompanied Julian Steyermark on botanical expeditions to southern Venezuela (1961). The etymology says no more and we have been unable to identify him further.

Rabor

Rabor's Forest Frog *Platymantis rabori* Brown et al., 1997

Dr Dioscoro S. [Joe] Rabor (1911–1996) was the pre-eminent Philippine ornithologist, mammalogist, herpetologist, ichthyologist and conservationist of the twentieth century. He graduated from the University of the Philippines (1934) and took his doctorate at Yale (1958). He led over 50 expeditions (1935–1977), normally accompanied by his wife, two sons and four daughters, all of whom became physicians. He produced the most thorough documentation of the birds and mammals of the Philippines and wrote *Philippine Birds and Mammals: A project of the U. P. Science Education Center* (1977) and *Philippine Reptiles and Amphibians* (1981). Four mammals, two reptiles and five birds are named after him.

Radde

Radde's Toad *Pseudepidalea raddei* **Strauch**, 1876 [Alt. Mongolian Toad]

Gustav Ferdinand Richard Radde (1831–1903) was a Prussian naturalist and apothecary. Born in Danzig in Prussia (now Gdansk in Poland), he moved (1852) to Russia. He was on numerous expeditions through Eastern Siberia, the Crimea, the Caucasus, Trans-Caucasus and other regions of Russia, and also Iran and Turkey. He settled (1863) in Georgia and founded the Caucasian Museum, Tbilisi (1867). Towards the end of the nineteenth century, he made two further journeys, accompanying Russian Imperial family members: to India and Japan (1895) and to North Africa (1897). He wrote *Reisen im Süden von Ost-Sibirien in den Jahren 1855–1859* (2-vols. 1862–1863). Seven birds, two mammals and three reptiles are named after him.

Rafinesque

Salamander subgenus *Rafinus* **Dubois** & Raffaëlli, 2009

Professor Constantine Samuel Rafinesque-Schmaltz (1783–1840) was born in Constantinople to a French father and German mother. He was sent to live in Tuscany to escape the turmoil of the French Revolution. His father was a merchant who died in Philadelphia (1793), leaving the family very badly off. Despite being unable to attend a university, Rafinesque was a highly gifted individual, accomplished as a botanist, geologist, historian, poet, philosopher, philologist, economist, merchant, manufacturer, professor, architect, author and editor. He was apprenticed (1802) to a merchant house in Philadelphia, and for the next two years he roamed the fields and woods and made collections of plants and animals. He was in Sicily (1805–1815) as Secretary to the US Consul and carried on a lucrative trade in commodities. He scoured the island for plants and collected previously unrecorded fishes from the stalls of the Palermo market. He sailed for New York (1815) but was shipwrecked in Long Island Sound, losing all his unpublished manuscripts and collections. He sailed down the Ohio River (1818) and conducted a comprehensive survey of the fish species there, published as *Ichthyologia Ohiensis* (1820). He visited Henderson, Kentucky, and stayed for eight days with John James Audubon. He was Professor of Botany and Natural Science, University of Transylvania (Lexington, Kentucky) (1819–1826). He returned to Philadelphia (1826) with 40 crates of specimens. He had a remarkable gift for inventing scientific names, some 6,700 in botany alone. He died in poverty but was later re-interred in Lexington. A mammal is named after him.

Ragazzi

Ragazzi's Tree Frog *Leptopelis ragazzii* **Boulenger**, 1896

Dr Vincenzo Ragazzi (1856–1929) of the Modena Natural History Society explored and collected in Ethiopia. He was a physician and was posted to the Italian research station (1884), Let Marefia, Ethiopia, later being its Director for several years. He was on good terms with the Emperor Menelik and accompanied his military expeditions to Harrar (1886–1887) to make geographical surveys. Menelik selected (1887) Ragazzi as his emissary to Italy. Two reptiles and a bird are named after him.

Ramage

Paraiba Robber Frog *Ischnocnema ramagii* **Boulenger**, 1888

George Albert Ramage (1858–1937) was a zoologist who collected amphibians and birds in St Lucia (1889) and Dominica (1915). He collected the holotype of this frog at Igaraçu, Pernambuco, Brazil.

Ramaswami

Ramaswami's Bush Frog *Micrixalus swamianus* **Rao**, 1937
[Alt. Rao's Bush Frog, Swami's Bubble-nest Frog]

Ramaswami's Caecilian *Gegeneophis ramaswamii* **Taylor**, 1964

Dr L. S. Ramaswami was an Indian biologist and herpetologist at the Department of Zoology, Reproductive Physiology Section, University of Rajasthan, Jaipur, where he was a professor.

Ramírez

Ramírez's Cochran Frog *Cochranella ramirezi* **Ruiz-Carranza** & **Lynch**, 1991

Ramírez's Rocket Frog *Colostethus ramirezi* **Rivero** & **Serna**, 2000

Dr Fabio Ramírez was in charge of the 'Las Orquídeas' National Park in Colombia.

Ramos

Ramos' Rocket Frog *Hyloxalus ramosi* **Silverstone**, 1971

Ramos' Mushroomtongue Salamander *Bolitoglossa ramosi* **Brame** & **Wake**, 1972

Professor Jorge Eduardo Ramos-Pérez (b.1945) is a Colombian botanist who was a colleague of Silverstone, describer of the rocket frog, at the herbarium of the Universidad del Valle, Cali, Colombia. He worked as a technician (1968–1977), was the herbarium Director (1984–1996), being succeeded by Silverstone, with whom he wrote *Floristic Exploration and Phytogeography at the Carro del Torra, Choco, Colombia* (1993).

Rand

Boraceia Robber Frog *Ischnocnema randorum* **Heyer**, 1985

Rand's Dwarf Frog *Engystomops randi* Ron, **Cannatella** & **Coloma**, 2004

Dr Austin Stanley Rand (1932–2005) was a herpetologist at the Smithsonian. His father was the famous Canadian ornithologist Austen Loomer Rand (1905–1982). His mother was a herpetologist, as was his wife, Patricia Grubbs Rand, whom he met when they catalogued salamanders together at the Smithsonian as a student summer job. Harvard awarded his doctorate (1961). He was an assistant, Division of Amphibians and Reptiles, Field Museum, Chicago (1950). He served in the US Army in Germany (1955–1957). He worked at the Smithsonian's Tropical Research Institute, Panama (1964–1997) and after retiring still visited Panama annually. The robber frog is named after both him and his wife. A reptile is named after him.

Ranger

Ranger's Toad *Amietophrynus rangeri* **Hewitt,** 1935
[Alt. Raucous Toad]

Gordon Arthur Ranger (b.1903) was a South African farmer, self-taught naturalist, ornithologist and collector of botanical specimens for the National Herbarium, Pretoria, and the museums at Grahamstown (where Hewitt worked) and King William's Town.

Rank

Rank's Snouted Treefrog *Scinax ranki* Andrade & **Cardoso**, 1987

Abilio Rank, according to the etymology, contributed much to the study of the biology of amphibians in the region of São Bento do Sul, Santa Catarina, Brazil.

Ranwella

Ranwella's Spined Tree Frog *Polypedates ranwellai* Wickramasinghe, Munindradasa & Fernando, 2012

Dr Sanjeewa Ranwella (1965–2003) was a Sri Lankan physician, zoologist and herpetologist. He, his wife and their young son were all drowned in a tragic boating accident on Lake Bolgoda. His brother, Dimuth Ranwella, is a Sri Lankan wildlife photographer who has provided photographs to illustrate articles written by the senior describer, Mendis Wickramasinghe. A reptile is named after him.

Rao

Rao's Bush Frog *Micrixalus swamianus* Rao, 1937
[Alt. Ramaswami's Bush Frog]
Rao's Lateral-eyed Frog *Nyctibatrachus sylvaticus* Rao, 1937
[Alt. Forest Night Frog]
Rao's Pale Brown Toad *Duttaphrynus brevirostris* Rao, 1937

Professor C R Narayan Rao (1882–1960) was an Indian herpetologist and zoologist. He was educated at Madras Christian College and, after gaining his diploma, became a teacher in Coimbatore and Ernakulam. He moved to Central College, Bangalore, University of Mysore, where he was Head, Department of Zoology, which he created, until his retirement (1909–1937). He co-founded (1932) and was first Editor of *Current Science*. He was influential in the decision that led to the foundation of the Indian Academy of Sciences. His speciality was frogs and their taxonomy. (See also **Narayan**)

Rathbun

Texas Blind Salamander *Eurycea rathbuni* **Stejneger**, 1896
[Syn. *Typhlomolge rathbuni*]

Dr Richard Rathbun (1852–1918) was an American zoologist, ichthyologist, geologist and museum administrator. He worked (1873–1875) as a voluntary assistant to Baird at the US Fish Commission, whilst being Assistant, Zoology, Boston Society of Natural History (1874–1875). He was Assistant Geologist, Geological Commission of Brazil (1875). He was Scientific Assistant, US Fish Commission (1878) and was Assistant, Zoology at Yale (1879–1880). He received a master's degree in science from Indiana University (1883) and an honorary Doctorate of Science from Bowdoin University (1894). He was Curator, Marine Invertebrates, AMNH (1880–1914). He was a member of the joint US/British Fisheries Commission (1892–1896). He became Assistant Secretary, Smithsonian Institution and (1899–1918) was in charge of the US National Museum.

Rattray

Rattray's Frog *Anhydrophryne rattrayi* **Hewitt**, 1919
[Alt. Hogsback Frog]

Dr George J. Rattray (1872–1941) was a British botanist who graduated at Aberdeen University (1894), taught for a few years in South Africa, returned to Aberdeen to take a BSc degree and (1904–1931) was headmaster of Selborne College, East London, South Africa. He was awarded a doctorate by Aberdeen (1910) for a thesis on cycads and wrote *Notes on the pollination of some South African Cycads* (1913). He discovered this frog and collected the holotype.

Raven, RJ

Raven's Brood Frog *Pseudophryne raveni* Ingram & **Corben**, 1994
[Alt. Copper-backed Brood Frog, Copper-backed Toadlet]

Dr Robert John Raven is an arachnologist at the Queensland Museum, Australia, where he is the Senior Curator (Arachnids). He did post-doctoral work as a CSIRO Fellow at the AMNH, New York, and at CSIRO Division of Entomology, Canberra. He was Chairman of the XII International Congress of Arachnology (Queensland Museum, Brisbane, 1992).

Raven, WA

Volcano Reed Frog *Hyperolius raveni* **Ahl**, 1931
[Probably a synonym of *Hyperolius kivuensis*]

Dr Werner Alborus von Raven was a German physician, traveller and explorer who was a member of the Duke of Mecklenburg's 1907–1908 expedition to Lake Edward and Lake Kivu. He was the expedition's medical officer and bacteriologist. He was a member of the Institute for Infectious Diseases – his major interest was research into sleeping sickness – but was unfortunately injured by a wounded buffalo (1908). He was resident physician in charge of a sleeping sickness clinic at Mount Klutow, Togo (Benin), German West Africa (1908–1910). He was a councillor in the German government (Weimar Republic), Berlin (1921).

Ravi

Ravi's Bush Frog *Raorchestes ravii* Zachariah et al., 2011

Ravi Chandran (fl.2007, d.<2011) was an amateur naturalist, a shopkeeper from Wayanad, who collected the holotype (2007). He was one of a number of dedicated amateur naturalists who regularly accompanied Dr Zachariah and other professional on their collecting trips. The etymology reads: 'This species is named in memory of the late Ravi Chandran, an enthusiastic nature lover, who collected the species and accompanied in most of our field expeditions.'

Rebecca

Rebecca's Marsupial Frog *Gastrotheca rebeccae* **Duellman** & **Trueb**, 1988

Dr Rebecca A. Pyles is Associate Professor, Department of Biological Sciences, East Tennessee State University, Johnson City. She joined the faculty (1991) as Assistant Professor, becoming Associate

Professor (1997). The University of Missouri, Kansas City awarded her bachelor's degree; both her master's and doctorate (1988) were awarded by the University of Kansas. She taught at the Universities of Kansas and Tulsa (1988–1991). To quote the etymology: '...she assiduously applied her knowledge of multivariate analytical methods and computational techniques to our research, especially that dealing with Gastrotheca. We take this opportunity to express our gratitude and indebtedness to her for her remarkable patience and forbearance of our markedly more limited grasp of these tools.' One might say that she blinded them with science! She co-wrote, with Duellman, *A new marsupial frog (Hylidae, Gastrotheca) from the Andes of Ecuador* (1980).

Rees

Rees' Toad *Amietophrynus reesi* **Poynton,** 1977

Magombero Reed Frog *Hyperolius reesi* **Schiøtz**, 1982

Allen Rees was a principal game warden for the Tanzanian Wildlife Department, and collected in the Rufiji area of Tanganyika (Tanzania) (1960s). He wrote *A check list of the mammals and amphibia of Ulanga District, Tanganyika* (1964).

Refugi

Lubok Antu Bubble-nest Frog *Philautus refugii* **Inger** & Stuebing, 1996

The Latin for a refuge is *refugium*, and this binomial means 'of the refuge'. It refers to the nature reserve in Sarawak where the holotype was taken.

Regis

Royal Ghost Frog *Heleophryne regis* **Hewitt**, 1910

[Alt. Rex's Ghost Frog, Southern Ghost Frog]

George Rex (1765–1839) was a British entrepreneur who founded the town of Knysna in the Western Cape, South Africa. He held a number of important positions in the early days of the Cape Colony, including being Registrar of Courts Martial (1797–1801) and Advocate for the Crown (1798). He owned slaves and ships and was a timber exporter and trader. In Latin, 'rex' means 'king' and the genitive *regis* means 'of the king': in Rex's day, and subsequently, it was believed that he was an illegitimate son of King George III. The story was de-bunked and found to have no basis in fact as a result of DNA testing (2004).

Reiche

Reiche's Squeaker *Arthroleptis reichei* **Nieden**, 1911

[Alt. Poroto Screeching Frog]

Dr Karl Friedrich Reiche (1860–1929) was a German botanist. He took his doctorate at the University of Leipzig, and was a professor at the University of Dresden (1886–1889). He worked at the National Museum of Chile, Santiago de Chile, being Director, Botany Department (1896–1911). He held a professorship in Mexico (1912–1924), then lived in Monaco (1924–1926), went back briefly to Mexico, and returned to Germany (1928). A reptile is named after him.

Reichle

Treefrog sp. *Dendropsophus reichlei* Moravec et al., 2008

Robber Frog sp. *Pristimantis reichlei* **Padial** & **De La Riva**, 2009

Steffen Reichle (b.1968) is a German herpetologist who is the science co-ordinator of the Nature Conservancy's Southern Andes Conservation Programme, in relation to Chile, Peru and Bolivia.

Reinhardt

Reinhardt's Caecilian *Mimosiphonops reinhardti* Wilkinson & **Nussbaum**, 1992

Dr Johann Theodor Reinhardt (1816–1882) was a Danish zoologist and herpetologist who collected the holotype (1878) and studied caecilian systematics – for which he was honoured 110 years after his death. He was on board the corvette 'Galathea' which circumnavigated the globe (1845–1847), and made two collecting expeditions to Brazil in later years. He was Director of the Kongelige Naturhistoriske Museum, Copenhagen. He wrote *Om svommeblaeren hos Familien Gymnotini* (1852). A bird and five reptiles are named after him.

Reinwardt

Reinwardt's Flying Frog *Rhacophorus reinwardtii* **Schlegel**, 1840

[Alt. Reinwardt's Gliding Frog, Reinwardt's Tree Frog]

Dr Caspar Georg Carl Reinwardt (1773–1854) was a German-born Dutch ornithologist who collected in Java (1817–1822). He studied botany, chemistry, mathematics, classical and modern languages, and history in Amsterdam. He gained a doctorate in natural philosophy and medical science (1801) and became Professor, Chemistry and Natural History, University of Harderwijk, where he was responsible for the University's botanical garden. Louis Napoleon made him director of his menagerie and he became Professor of Natural History in Amsterdam. He was in Java (1816–1821), responsible for everything concerning agriculture, arts and sciences, and contributed greatly to education and public health. He founded the Botanical Gardens, Buitenzorg (Bogor). He laid the foundation for two important museum collections in Leiden – the National Museum of Natural History and the National Museum of Ethnography. While collecting he was meticulous and detailed in documenting his specimens. He was very concerned with systematic development of collections and their documentation, and he was always happy to extend opportunities for research to others. Among other taxa, many plants and eight birds are named after him.

Renjifo

Robber Frog sp. *Pristimantis renjiforum* **Lynch**, 2000

Dr Juan Manuel Renjifo (b.1948) is a biologist, herpetologist and wildlife photographer at the Instituto Nacional de Salud, Bogotá, in which city he is also Associate Professor of Biology, Universidad del Magdalena. A reptile is also named after him. The plural binomial *renjiforum* also includes Camila and Patricia Renjifo.

Restrepo

Robber Frog sp. *Pristimantis restrepoi* **Lynch**, 1996

Father Sergio Restrepo Jaramillo (1939–1989) was a Colombian Jesuit priest and botanist who cultivated orchids. He was a great promoter of human rights and was assassinated in Medellín, Colombia, by two apparently paramilitary hitmen.

Reverberi

Laguna Raimunda Frog *Atelognathus reverberii* **Cei**, 1969

Dr Oscar Valentin Reverberi is an Argentine geologist and mineralogist who became Director of the Argentinian Servicio Minero Nacional. Cei wrote: 'The new species which shall be described herein are named *Telmatobius reverberii* in honor of Dr. Oscar Reverberi, who studied the geology of the region…'

Rex

Rex's Ghost Frog *Heleophryne regis* **Hewitt**, 1910
[Alt. Royal Ghost Frog, Southern Ghost Frog]

(See under **Regis**)

Rhodes

Rhodes' Robber Frog *Eleutherodactylus rhodesi* **Schwartz**, 1980

S. Craig Rhodes was a field companion of the describer, Albert Schwartz, and collected with him on Haiti (1979), when the type specimen was discovered.

Ria

Stefania Treefrog *Stefania riae* **Duellman** & **Hoogmoed**, 1984

Ria Hoogmoed-Verschoor is the wife of the junior describer.

Richard

Richard's Salamander *Nototriton richardi* **Taylor**, 1949

(See **Taylor, RC**)

Richards

Richards' Big-eyed Treefrog *Litoria oktediensis* Richards & Johnston, 1993
Richards' Mawatta Frog *Hylophorbus richardsi* **Günther**, 2001
Richards' Treefrog *Litoria richardsi* Dennis & Cunningham, 2006

Dr Stephen J. Richards is a herpetologist working as Honorary Research Associate, Vertebrates Department, South Australian Museum. He wrote *A new species of torrent-dwelling frog (Anura: Hylidae:* Litoria*) from the mountains of Indonesian New Guinea (West Papua)* (2001). Dennis's etymology says the treefrog is named 'For Stephen Richards in recognition of his immense contribution to understanding the herpetofauna, its diversity and its conservation, on the island of New Guinea.'

Richmond, CW

Richmond's Robber Frog *Eleutherodactylus richmondi* **Stejneger**, 1904

Charles Wallace Richmond (1868–1932) was an American ornithologist who specialized in nomenclature and bibliography. His first job, aged about 13, was as a page boy at the House of Representatives in Washington DC. He was on the US Geological Survey of Montana (1888) and later worked for the US Department of Agriculture as an 'ornithological clerk'. The next job he was able to get was with the Smithsonian as a night watchman. He was, however, soon promoted, eventually becoming Assistant Curator of Birds (1894). He created *The Richmond Index to the Genera and Species of Birds*. It was his life's work (1889–1929), culminating in 70,000 file cards. A mammal and thirteen birds are named after him.

Richmond, ND

Ravine Salamander *Plethodon richmondi* **Netting** & Mittleman, 1938

Neil Dwight Richmond (1912–1992) was a herpetologist and zoologist who was an instructor in zoology at Syracuse University (1935–1938), and

Marshall College, West Virginia (1938–1939). He worked for the US Fish & Wildlife Service, Pennsylvania (1947–1948). He became Curator, Amphibians and Reptiles, Carnegie Museum (1951–1972) and made four collecting expeditions to the Bahamas. A reptile is named after him.

Rickett

Chinese Sucker Frog *Amolops ricketti* **Boulenger**, 1899

Charles Boughey Rickett (1851–1943) was a British banker and amateur naturalist, particularly an ornithologist. He left records of his observations in Scotland and Cornwall (1901–1907). He collected in China and other parts of Asia, including Japan, the Straits Settlements (now Malaysia), Java and India. Throughout his 33-year career he worked for the Hong Kong & Shanghai Banking Corporation. Nine birds and a mammal are named after him.

Ricord

Ricord's Frog *Eleutherodactylus ricordii* **Duméril** & **Bibron**, 1841

Alexandre Ricord (1798–1876) was a French naval surgeon who qualified as a physician in Paris (1824), and became a corresponding member of the French Academy of Medicine (1838). He collected in Latin America (1826–1834). A bird and two reptiles are named after him.

Riggenbach

Riggenbach's Reed Frog *Hyperolius riggenbachi* **Nieden**, 1910

Fritz Wilhelm Riggenbach (1864–1944) was a zoologist and collector employed on German Central African expeditions, led by Adolf Friedrich, Duke of Mecklenburg, in the first two decades of the twentieth century. A mammal, a reptile and eight birds are named after him.

Rilett

Rilett's Mushroomtongue Salamander *Bolitoglossa riletti* Holman, 1964

Dr Robert Omar Rilett was an American biologist and entomologist who became (1958) Head, Department of Biological Sciences, Illinois State University, retiring as Emeritus Professor (1974). The University administers an Omar Rilett Scholarship fund in his memory. The University of Wisconsin awarded his doctorate (1946). He taught (1957) entomology at the State University of New York, Buffalo.

Rimarum

Gorzula's Treefrog *Tepuihyla rimarum* **Ayarzaguena, Señaris** & **Gorzula**, 1993

Rimarum is Latin for 'of the crevices' and we assume the binomial is descriptive of this amphibian's hiding places.

Rita

Rita's Cochran Frog *Cochranella ritae* **Lutz**, 1952

Gertrud Rita Kloss was a parasitologist, mentioned by the describer Bertha Lutz as her assistant naturalist at the Museum Nacional, Rio de Janeiro, Brazil. In the early 1960s she joined the staff of the Department of Zoology, Secretary of Agriculture, São Paulo State, which became the Museu de Zoologia, São Paulo (1969). She wrote *Alimentação das aves silvestres* (1950).

Rita Rasquin

Strabomantid frog sp. *Noblella ritarasquinae* **Köhler**, 2000

[Syn. *Phyllonastes ritarasquinae*]

Rita Rasquin of Krefeld, Germany is a supporter of BIOPAT and this frog was named '...in recognition of her efforts in supporting taxonomic research and nature conservation.'

Rivas

Robber Frog sp. *Pristimantis rivasi* Barrio-Amorós, Rojas-Runjaic & Barros, 2010

Gilson Rivas Fuenmayor is a Venezuelan herpetologist and was one of the original collectors of this frog species. After working for ten years at the Museo de Historia Natural La Salle, Caracas, where he was curator of the herpetology department, he is now an associate researcher at the Museo de Biología, Universidad del Zulia (Venezuela). His main interests are the systematics of South American lizards and colubrid snakes.

Rivero

Rivero's Poison Frog *Mannophryne riveroi* **Donoso-Barros**, 1965

Rivero's Toad *Rhinella beebei* **Gallardo**, 1965

Rivero's Amazon Treefrog *Dendropsophus riveroi* **Cochran** & **Goin**, 1970

Rivero's Tiny Treefrog *Dendropsophus minusculus* Rivero, 1971

Rivero's White-lipped Frog *Leptodactylus riveroi* **Heyer** & **Pyburn**, 1983

Rivero's Cochran Frog *Cochranella riveroi* **Ayarzgüena**, 1992

Rivero's Robber Frog *Pristimantis riveroi* **Lynch** & **La Marca**, 1993

Yuruaní-tepui Treefrog *Stefania riveroi* **Señaris,** Ayarzagüena & **Gorzula**, 1997

Plains Coqui Frog *Eleutherodactylus juanariveroi* Rios-López & **Thomas**, 2007

Dr Juan Arturo Rivero-Quintero (b.1923) is a Puerto Rican herpetologist. He gained his bachelor's degree (1945) from the University of Puerto Rico, and both his master's (1951) and doctorate (1953) from Harvard. He is the founder of the Dr Juan A Rivero Zoo, University of Puerto Rico, which he joined (1945), becoming Professor of Biology (1958). He was a research associate at Harvard. For some years he hosted a popular science program on local television. A reptile is named after him.

Rivet

Despax's Robber Frog *Pristimantis riveti* **Despax**, 1911

Dr Paul Rivet (1876–1958) originally trained as a physician. He was on the second French Geodesic mission (1901) to Ecuador, staying until 1906. After returning to Paris he worked at the Muséum National d'Histoire Naturelle and was one of the founders of the Institut d'Ethnologie (1926). He became Director, Muséum National d'Histoire Naturelle (1928) and founded the Musée de l'Homme (1937). He was in Colombia, where he founded the Anthropological Institute (1942–1945). Despax published an article (1911) on the collection of reptiles and amphibians that Rivet had made in Ecuador. A reptile and a bird are named after him.

Robert Inger

Hejiang Spiny Frog *Quasipaa robertingeri* Wu & **Zhao**, 1995 NCR

[JrSYN *Quasipaa boulengeri*]

Flying Frog sp. *Rhacophorus robertingeri* **Orlov** et al., 2012

(See **Inger**)

Roberts

Roberts' False Brook Salamander *Pseudoeurycea robertsi* **Taylor**, 1939

Roberts' Treefrog *Plectrohyla robertsorum* Taylor, 1940

Dr Howard Radclyffe Roberts (1906–1982) was an entomologist who became Director, Academy of Natural Sciences, Philadelphia (1947–1972) and thousands of his specimens are now housed there. He had collected whilst serving in the Medical Entomological Department, US Army, with the Malaria Survey Unit in the Philippines and New Guinea. During the 1940s he co-wrote *The Mosquito Atlas*, a standard reference work for medical servicemen describing mosquitoes that transmit malaria. The treefrog is named after both him and his wife, who both helped collect the type series in Mexico. A bird is named after him.

Robinson, AC

Central Ranges Toadlet *Pseudophryne robinsoni* Donnellan, Mahony & Bertozzi, 2012

Dr Anthony C. 'Tony' Robinson was formerly Manager of Biological Surveys and Monitoring at the South Australian Department for the Environment and Natural Resources. He was chiefly responsible for the initiation and sustained management of a statewide biodiversity survey conducted over 25 years from the early 1980s. He has written or co-written many reports on these surveys.

Robinson, DC

Robinson's Web-footed Salamander *Bolitoglossa robinsoni* Bolaños & **Wake**, 2009

Dr Douglas C. Robinson (1936–1991) was a herpetologist in Costa Rica, where he lived for the last 25 years of his life. He joined the Universidad de Costa Rica (1966), becoming full Professor (1970). Texas A&M University awarded his doctorate (1968). He founded the Museum of Zoology, Universidad de Costa Rica, and was its director (1966–1969 & 1984–1987). One of his major interests was the study of sea turtles and he wrote *Management of olive ridley sea turtles (*Lepidochelys olivacea*) nesting at Playas Nancite and Ostional, Costa Rica* (1991).

Robinson, GA

Paradise Toad *Vandijkophrynus robinsoni* **Branch** & Braack, 1996

Dr Gilbert A. Robinson was Chief Executive of South Africa's National Parks Board.

Robinson, HC

Robinson's Flying Frog *Rhacophorus robinsoni* **Boulenger**, 1903

[Alt. Robinson's Tree Frog]

Robinson's Grainy Frog *Kalophrynus robinsoni* **Smith**, 1922

Herbert Christopher Robinson (1874–1929) was a British zoologist and ornithologist. He went to Switzerland for his health (1894) and visited Queensland (1896). He was assistant to Dr Henry Ogg Forbes at Liverpool Museum (1897–1900), and in later years sent many specimens to that museum. He spent the rest of his life in the tropics, in the Malay Peninsula (1900–1902) and as Curator, Federated Malay States Museum, Selangor (1903–1926). Cecil Boden Kloss (q.v.) joined him in exploring the Indo-Malay region (1908). He wrote *The Birds of the Malay Peninsula*

(1927). Amongst other taxa, four mammals, two reptiles and twenty birds are named after him.

Robinson, MAC

Treefrog sp. *Litoria robinsonae* Oliver, Stuart-Fox & **Richards**, 2008

Margaret Ann Clare Robinson was a judge in the Family Court of Australia in whose memory a donation was made for research into Melanesian zoology.

Robledo

Giant Glass Frog sp. *Centrolene robledoi* **Ruiz-Carranza** & **Lynch**, 1995

Dr Emilio Robledo Correa (1875–1961) was a physician, surgeon, historian, ethnographer, botanist and naturalist who was Rector of the Universidad de Antioquia, Colombia.

Rochebrune

Rochebrune's Reed Frog *Hyperolius protchei* Rochebrune, 1885

Dr Alphonse Trémeau de Rochebrune (1836–1912) was a French military doctor who took a zoological doctorate (1874) and became a botanist, malacologist and ichthyologist. He was in Senegal (1875–1883) and wrote *Fauna de Sénégal* (1883). He worked at the Muséum National d'Histoire Naturelle, Paris (1884–1911).

Rodríguez

Rodríguez's Amazon Treefrog *Tepuihyla rodriguezi* **Rivero**, 1968

Dr Gilberto Domingo Rodríguez Ramírez (1929–2004) was South American President of the Crustacean Society (1982–1995). The Universidad Central de Venezuela awarded his bachelor's degree (1955). The University of Miami awarded his degree in marine biology (1959). After visiting marine laboratories in the United Kingdom and Denmark, he co-founded (1959) the Oceanographic Institute, Universidad de Oriente, Cumaná. He joined the newly founded Instituto Venezolana de Investigaciones Científicas, near Caracas (1960). He studied for his doctorate at the University of Wales, Swansea (1968–1970). He published more than 100 scientific papers. On a lighter note, he was a great cinema buff, enjoying the dark comedies of Ealing Studios, especially *The Ladykillers* and Frank Capra's classic *Arsenic and Old Lace*.

Roeschmann

Roeschmann's Treefrog *Hyla roeschmanni* De Grys, 1938 NCR
[JrSYN *Hypsiboas raniceps*]

Dr Hermann Röschmann collected in Bolivia (1902) for the Naturhistorisches Museum, Hamburg. He may be the same man as the Hamburg physician who was a prominent venereologist in the 1920s and 1930s, and who died in 1950.

Rogério

Snouted Treefrog sp. *Scinax rogerioi* Pugliese, Baeta & **Pombal**, 2009

Dr Rogério P Bastos is a herpetologist at the Universidade Federal de Goiás, Brazil. He co-wrote *The Character and Dynamics of the Cerrado Herpetofauna* (2002).

Rogers, FW

Rogers' Three-fingered Frog *Hoplophryne rogersi* **Barbour** & **Loveridge**, 1928
[Alt. Usambara Banana Frog]

F. W. Rogers, who had been seconded there from Kew, was Custodian of the Amani Institute, Tanganyika (Tanzania) at the time of the describers' visit. They noted that his local knowledge and assistance were great assets to them.

Rogers, JS

Bahamas Flathead Frog *Eleutherodactylus rogersi* **Goin**, 1955

Dr James Speed Rogers (1892–1955) was an American biologist and entomologist who was Director, Museum of Zoology, University of Michigan (1946–1955). He had been a professor at Grinnell College, Iowa, and was (1923) at Gainsville, Florida forming a Department of Biology and Geology for the University of Florida. He co-wrote *Man and the biological world* (1952). Goin was a pupil of Rogers and wrote: 'I take pleasure in naming this new form for Prof. J. Speed Rogers, who has constantly encouraged me in my studies of *Eleutherodactylus*.'

Rohde

Rohde's (Leaf) Frog *Phyllomedusa rohdei* **Mertens**, 1926
[Alt. Mertens' Leaf Frog]

Dr H. Rohde was a ship's doctor who made a large collection of Brazilian herpetofauna which he sent to Mertens at the Senckenberg Institute, Frankfurt-am-Main.

Rolanda

Rolanda's Burrowing Frog *Sphaerotheca rolandae* **Dubois**, 1983
[Alt. Marble Sand Frog]

Rolande Roux-Estève (see **Roux, R**).

Román

Robber Frog sp. *Pristimantis romanorum* Yánez-Muñoz et al., 2010

Ángel Román, Guido Román, Hernando Román and José L. 'Tun' Román C. all work at the Ecuadorian National Museum of Natural History and Biology as taxidemists and palaeontologists. They have contributed significantly to the museum's collections of mammals, birds and herpetology.

Romer

Romer's Tree Frog *Liuixalus romeri* **Smith**, 1953

John D. Romer (1920–1982) was a herpetologist who worked at the Urban Services Department, Central Government Offices, Hong Kong. He discovered (1952) this frog, which is found on a few of Hong Kong's smaller islands. He co-wrote *Origin and Taxonomic Status of the Salamander* Cynops shataukokensis (1980).

Ron

Ron's Caecilian *Rhinatrema ron* Wilkinson & Gower, 2010

Dr Ron Archie Nussbaum (see **Nussbaum**).

Ronald

Ronald's Robber Frog *Eleutherodactylus ronaldi* **Schwartz**, 1960

Ronald Francis Klinikowski (see **Klinikowski**).

Rookmaaker

Flores Cross Frog *Oreophryne rookmaakeri* **Mertens**, 1927

Hendrik Roelof Rookmaaker (1887–1945) was a Dutch colonial administrator (1911–1937) who was Assistant Resident on Flores (now in Indonesia). He supplied Mertens with two Komodo Dragons for the Frankfurt Zoo. A bird is named after him.

Roraima

Roraima Treefrog *Stefania roraimae* **Duellman** & **Hoogmoed**, 1984

This treefrog is named after the type locality, Mount Roraima in Guyana.

Rosado

Rosado's Robber Frog *Pristimantis rosadoi* **Flores**, 1988

José P. O. Rosado joined the Harvard Museum of Comparative Zoology (1977) and became Collection Manager, Herpetology Department. A reptile is named after him.

Rose

Rose's Ghost Frog *Heleophryne rosei* **Hewitt**, 1925
[Alt. Table Mountain Ghost Frog]
Rose's Toad *Capensibufo rosei* Hewitt, 1926
[Alt. Rose's Mountain Toadlet]
Rose's Rain Frog *Breviceps rosei* **Power**, 1926
[Alt. Sand Rain Frog, Rose's Short-headed Frog]

Dr Walter Rose (d.1964) was a South African physician living in the Cape, and an early member of the Herpetological Association of Africa. He co-wrote with Hewitt *Description of a new species of* Xenopus *from the Cape Peninsula* (1927). It was said of him: 'Skilled professional men are not rare and many of them have hobbies, but there must be very few who have made such a notable thing of their hobby as Mr. Rose'. He wrote *Reptiles and Amphibians of Southern Africa* (1950).

Rosenberg

Rosenberg's Gladiator Treefrog *Hypsiboas rosenbergi* **Boulenger**, 1898

William Frederick Henry Rosenberg (1868–1957), who described himself as 'a traveller-naturalist', was a natural history dealer who had a shop in London (1930–1948). He collected in South America, including Ecuador (1896–1898) where the treefrog holotype was taken. Eight birds are named after him.

Ross Allen

Ross Allen's Treefrog *Dendropsophus rossalleni* **Goin**, 1959

Ensil Ross Allen (1908–1981) was an American herpetologist. He studied natural history at Stetson University, Florida (1927–1929) but dropped out to pursue taxidermy and to extend his live reptile collection. He opened a taxidermy shop (1929) in a new tourist area of Florida and founded the Ross Allen Reptile Institute. This was an 'attraction', with alligator-wresting and Ross making a show of milking rattlesnakes for their venom. He went on a collecting expedition to Honduras (1937). He was responsible for milking 73,990 poisonous snakes to make anti-venom for the armed forces (1941–1945). He collected in Colombia (1947) and in Cuba (1948). The Florida Department of Agriculture gave him an award in recognition of his conservation work (1965). He founded and was President of the International Crocodile Society.

Rostand, E

Rostand's Papuan Treefrog *Litoria rostandi* Kraus, 2007
[Probably a junior synonym of *Litoria hilli*]

Edmond Eugène Alexis Rostand (1869–1918) was a French playwright and poet whose best known work is *Cyrano de Bergerac*. The etymology states that he is honoured 'in commemoration of his interest in long-snouted life forms.'

Rostand, J

Rostand's Paa Frog *Nanorana rostandi* **Dubois**, 1974
[Alt. Dubois' Frog]

Jean Rostand (1894–1977) was a French biologist and philosopher who spent 23 years studying the incidence and causes of deformities in frogs. He studied amphibian embryology and parthenogenesis. He was actively opposed to nuclear proliferation and the death penalty, and held agnostic convictions. His father was Edmond Rostand (above). He is noted for the saying: 'Kill one man and you are a murderer. Kill millions of men, and you are a conqueror. Kill them all, and you are a god.'

Roth

Roth's Treefrog *Litoria rothii* De Vis, 1884
[Alt. Northern Laughing Tree Frog]

Henry Ling Roth (1866–1925) was an English anthropologist who worked in Australia. He was educated in England and Germany. He was in Russia (1876–1877) before moving to Australia and settling in Queensland (1878). He wrote *The Aborigines of Tasmania* (1890). He collected the treefrog holotype.

Roux, J

Roux's Puddle Frog *Phrynobatrachus rouxi* **Nieden**, 1913

Dr Jean Roux (1876–1939) was a Swiss zoologist who gained his doctorate at Geneva University (1899) and initially focused his studies on protozoa. He turned towards herpetology on becoming Curator, Natural History Museum, Basel (1902–1930). He was in New Guinea and Australia (1907–1908), and in New Caledonia and the Loyalty Islands (1911–1912) with Fritz Sarasin. Six reptiles and a bird are named after him.

Roux, R

Madagascar Climbing Frog *Anodonthyla rouxae* **Guibé**, 1974

Mme Rolande Roux-Estève is a French herpetologist and ichthyologist who was Curator of Reptiles, Muséum National d'Histoire Naturelle, Paris. In the early 1950s she described fish species which Jacques Cousteau had collected in the Red Sea and off Aldabra during his expeditions with the 'Calypso'. She wrote *Révision systématique des Typhlopidae d'Afrique Reptilia-Serpentes* (1974). Two reptiles are named after her (see **Rolanda**).

Roy

Roy's Robber Frog *Pristimantis royi* Morales, 2007

Dr Roy Wallace McDiarmid (see **McDiarmid**).

Roze

Roze's Robber Frog *Pristimantis rozei* **Rivero**, 1961

Dr Janis Arnolds Roze (b.1926) is a herpetologist who was born in Latvia, and is currently Emeritus Professor of Biology, City College and Graduate School, City University of New York. He was formerly associated with the AMNH, New York, and was a founder of the International Center for Integrative Studies. He wrote *On the Synonymy and Holotypes of the Coral Snakes* (1966). Six reptiles are named after him.

Rueda

Robber Frog sp. *Pristimantis ruedai* **Ruiz-Carranza, Lynch** & **Ardila-Robayo**, 1997
Glass Frog sp. *Hyalinobatrachium ruedai* Ruiz-Carranza & Lynch, 1998
Glass Frog sp. *Nymphargus vicenteruedai* Velásquez-Álvarez et al., 2007

José Vicente Rueda-Almonacid (b.1954) is a Colombian herpetologist, who works for Conservation International. He co-wrote *Libro rojo de los anfibios de Colombia* (2004). A reptile is named after him.

Ruiz

Ruiz's Robber Frog *Strabomantis ruizi* **Lynch**, 1981
Ruiz's Rocket Frog *Hyloxalus ruizi* Lynch, 1982
Ruiz's Marsupial Frog *Gastrotheca ruizi* **Duellman** & **Burrowes**, 1986
Ruiz's Cochran Frog *Nymphargus ruizi* Lynch, 1993
Stubfoot Toad sp. *Atelopus petriruizi* **Ardila-Robayo**, 1999
Toad sp. *Rhinella ruizi* **Grant**, 2000
Backpack Frog sp. *Cryptobatrachus pedroruizi* Lynch, 2008

Professor Pedro Miguel Ruiz-Carranza (1932–1998) was a Colombian herpetologist and Curator of Herpetology, Instituto de Ciencias Naturales, Museo de Historia Natural, Bogotá. He co-wrote *Two New Species of* Hyla *from the Andes of Central Colombia and their relationships to other small Andean Hyla* (1957). A reptile is named after him.

Ruly(rana)

Glass Frog genus *Rulyrana* Guayasamin et al., 2009

This name combines the first letters of the surnames of Pedro Miguel **Ruiz-Carranza** and John Douglas **Lynch** (see their entries for biographical details) with the Latin *rana* ('frog'). Additionally, the describers say that 'Ruly' is the nickname of Ecuadorean herpetologist Martin Bustamante, who is also recognized for his work on amphibian conservation.

Rumboll

Salta Toad *Rhinella rumbolli* **Carrizo**, 1992

Maurice A. E. Rumboll (b.1940) is an Argentine naturalist and ornithologist. He has worked at the Museo Argentino de Ciencias Naturales, Buenos Aires, and with the US Fish & Wildlife Service. He co-authored *Collins Illustrated Checklist to the Birds of southern South America and Antarctica.* He acts as a wildlife guide on cruise ships, accompanies wildlife tours, and has made over 60 trips to Antarctica.

Runewsweek

Maharira Madagascar Frog *Gephyromantis runewsweeki* Vences & **De la Riva**, 2007

This frog is named after the Russian edition of *Newsweek*, which magazine is amongst those giving support to research through the BIOPAT scheme.

Rüppell

Rüppell's Big-eyed Treefrog *Litoria rueppelli* **Boettger**, 1895

Wilhelm Peter Eduard Simon Rüppell (1794–1884) was a German naturalist and explorer, and a collector in the broadest sense: coins and manuscripts as well as natural history specimens. He went to Egypt and ascended the Nile to Aswan (1817) and later made two extended expeditions to eastern Africa – Sudan (1821–1827) and Ethiopia (1830–1834). He brought back large zoological and ethnographical collections. He wrote *Reise in Abyssynien* (1838–1840). Eleven birds, two reptiles and five mammals are named after him.

Ruschi

Ruschi's Treefrog *Dendropsophus ruschii* **Weygoldt** & **Peixoto**, 1987

Augusto 'Gutti' Ruschi (1915–1986) was a Brazilian naturalist who dedicated his life to conservation. He was considered to be a world authority on hummingbirds but is probably even more famous for his second love: orchids. He studied law and agronomy, so was largely self-taught in natural history. He learned English, French, Latin and German in order to read standard botanical texts. He lived on Santa Lucia Reserve, Santa Teresa, in a nineteenth- century house built by his grandfather. He transformed his house into the 'Professor Mello Leitão Biology Museum' and, before he died, he asked the Pro-Memória Foundation to bury him in the Reserve. It was thought that touching poisonous frogs during years of research caused the liver disease from which he died. However, the doctor who treated him said his problems were caused by an overdose of anti-malarial drugs. He wrote over 400 scientific articles and book. A mammal is named after him.

Russel

Makamunu Assumbo Caecilian *Idiocranium russeli* **Parker**, 1936

Parker gives no etymology in his description, but it appears that W. M. Russell (1911–1979) is the person intended. He liked to be known as 'Gerald' Russell. He was a cryptozoologist who was born in the USA and educated in England at Cambridge, and served in the Royal Navy in the Second World War as an American volunteer. He went on a number of expeditions: Tibet (1933–1934 and 1936–1937) in search of the giant panda, and Nepal (1954 & 1958) searching for the yeti. He took part in the Percy Sladen Zoological Expedition to the British Cameroons (1932), after which Ivan T. Sanderson (whom the caecilian holotype was brought from) wrote that it was W. M. Russell 'who collected much of the herpetological material'. Parker apparently made an error in not spelling the binomial with a double-L (*russelli*).

Russell

Russell's Toadlet *Uperoleia russelli* **Loveridge**, 1933

Alfred Robert Eric Russell lived and farmed near Landor Station, Western Australia, on the confluence of the Gascoyne River with the Aurillia creek, near where the toadlet holotype was acquired. He served in both World Wars, in France as a lieutenant in the Australian Imperial Force (1916–1918), and as a captain in the Australian Army (1941–1944).

Rutenberg

Rutenberg's Reed Frog *Heterixalus rutenbergi* **Boettger**, 1881

Diedrich Christian Rutenberg (1851–1878) was a German explorer and collector, who studied medicine and natural history at Jena. He was in Dalmatia and Montenegro (1872). He went to South Africa (1877), collecting botanical specimens in Natal and Transvaal, before moving on to Mauritius and Madagascar, where he was murdered. A bird is named after him.

Ruth

Ruth's Robber Frog *Eleutherodactylus ruthae* **Noble**, 1923

Mrs Ruth Crosby Noble (1897–1988) was the describer's wife. She discovered this species and collected the first series of specimens. She wrote *The Nature of the Beast* (1944), a book on animal behaviour based on her husband's research.

Ruthven

Tukeit Hill Frog *Allophryne ruthveni* **Gaige**, 1926

Tungara Frog *Eupemphix ruthveni* **Netting**, 1930 NCR [JrSYN *Engystomops pustulosus*]

Ruthven's Robber Frog *Pristimantis ruthveni* **Lynch** & **Ruiz-Carranza**, 1985

Rocket Frog sp. *Colostethus ruthveni* **Kaplan**, 1997

Backpack Frog sp. *Cryptobatrachus ruthveni* Lynch, 2008

Dr Alexander Grant Ruthven (1882–1971) was an American herpetologist who was awarded his doctorate in zoology by the University of Michigan (1906), and worked at the University's Museum of Zoology (1906–1929), first as Curator, then Director (1913) and Professor of Zoology (1915). He was Director of the University of Michigan (1921–1951) and Emeritus Professor thereafter. He was President of the American Society of Ichthyologists and Herpetologists (1962). He co-wrote *Herpetology of Michigan* (1912). Eleven reptiles are named after him.

Ryabov

Ryabov's Bug-eyed Frog *Theloderma ryabovi* **Orlov** et al., 2006

Sergei A. Ryabov is a Russian herpetologist who was Director of the Tula Exotarium, which at one time possessed perhaps the world's biggest (living) snake collection. He works with the senior author and together they described the snake species *Boiga tanahjampeana* (2002).

S

Sabin

Guest House Guinea Frog *Aphantophryne sabini* **Zweifel** & **Parker**, 1989
Sabin's Glass Frog *Centrolene sabini* Catenazzi et al., 2012

Andrew E. Sabin (b.1958) is the President of Sabin Metal Corporation, a philanthropist and environmentalist who is said to be 'an individual with a passion for amphibians'. He has served on the environmental law advisory boards at Columbia Law School and UCLA School of Law, through his role as treasurer of the Evan Frankel Foundation, which made grants of more than $17million during its 15-year existence. Mr Sabin is also a member of the board of visitors of the M D Anderson Cancer Center at the University of Texas (2005). To provide more professional support and involve his family in charitable giving, he established the Andrew Sabin Family Foundation (2007). He is also President of the South Fork Natural History Society. He participated in the expedition to New Guinea which collected the *Aphantophryne* holotype, the eponymous hostelry being the 'Myola Guest House', Northern Province, Papua New Guinea.

Sabrina

Sierra Juarez Treefrog *Plectrohyla sabrina* **Caldwell**, 1974

The etymology says that 'sabrina' is Latin for 'river nymph' and only refers to the animal's aquatic habitat.

Saha

Sahai Bush Frog *Raorchestes sahai* Sarkar & Ray, 2006
[Syn. *Philautus sahai*]

Dr Subhendu Sekhar Saha is an Indian naturalist who was an assistant zoologist with the Zoological Survey of India. He surveyed (1980 & 1985) in the state of Arunachal Pradesh, whence came the frog holotype, writing a report proposing a biosphere reserve there.

Sakurai

Japanese Brown Stream Frog *Rana sakuraii* **Matsui** & Matsui, 1990

Atsushi Sakurai (b.1946) is a Japanese wildlife photographer and ichthyologist. He co-wrote *Aquarium Fish of the World* (1992).

Sala

Portezuelo Frog *Atelognathus salai* **Cei,** 1984

Osvaldo Rubén Sala (b.1944) is an Argentine politician and engineer who was Director of the Centro Nacional Patagónico, Puerto Madryn, Chubut, Argentina (1979–1983). He was twice elected mayor of the city of Puerto Madryn, and was also Senator for the province of Chubut. He was honoured in the frog's binomial '...in recognition of his important aid to our recent Patagonian research.'

Sall

Treefrog sp. *Dendropsophus salli* Jungfer, **Reichle** & Piskurek, 2010

John Sall (b.1948) is an American businessman and developer of software, as well as a supporter of the BIOPAT program. He was honoured in the binomial 'for his continuous generous contributions to forest conservation'.

Salmin

Salmin's Frog *Limnodynastes salmini* **Steindachner**, 1867
[Alt. Salmon-striped Frog, Steindachner's Frog]

C. L. Salmin was a dealer in natural history objects in Hamburg. The Leiden Museum had dealings with him (1863–1875).

Salser

Timbo Disc Frog *Synapturanus salseri* **Pyburn**, 1975

Jay K. Salser Jr, who collected the holotype with Pyburn (1974), is an ethnographer at the Instituto Linquístico de Verano, Bogotá, Colombia. He has a particular interest in the linguistics of native languages in Colombia. He wrote *The Cubeo gourd blackening process* (1974).

Salvaje

Copan Treefrog *Ecnomiohyla salvaje* Wilson, **McCranie** & Williams, 1985

Salvaje is the Spanish spelling to allow the pronunciation of 'Savage'. See Jay Mathers **Savage**, after whom this frog is named.

Salvator

Paraiso Escuerzo *Odontophrynus salvatori* **Caramaschi**, 1996

Salvador Monteiro co-edited the book *Amazon: Flora Fauna* (1993). Caramaschi contributed the section on amphibians and reptiles.

Salvin

Salvin's Mushroomtongue Salamander *Bolitoglossa salvinii* **Gray**, 1868

Osbert Salvin (1835–1898) was an English naturalist. He made the first of several visits to Guatemala (1857). He was the first European to record observing a Resplendent Quetzal, pronouncing it 'unequalled for splendour among the birds of the New World' ...and promptly shot it. Salvin co-authored the 40-volume *Biologia Centrali-Americana* (1879), a near-complete catalogue of Middle American species. Thirty-eight birds, four reptiles and two mammals are named after him.

Sanborn

Sanborn's Treefrog *Dendropsophus sanborni* **Schmidt**, 1944

Sanborn's Water Frog *Telmatobius sanborni* Schmidt, 1954

Dr Colin Campbell Sanborn (1897–1962) was an American ornithologist and mammalogist. He was Assistant Curator, Ornithology Section, Field Museum, Chicago (1922–1924) and subsequently became Curator of Mammals. He wrote *New mammals from Guatemala and Honduras* (1935). Six mammals and three birds are named after him.

Sánchez

Sánchez's Giant Glass Frog *Centrolene sanchezi* **Ruiz-Carranza** & **Lynch**, 1991

Ricardo Sánchez collected in the field with Lynch on a number of expeditions.

Sander

Strabomantid frog sp. *Oreobates sanderi* **Padial**, **Reichle** & **De la Riva**, 2005

Holger Sanders is a German who gives support via BIOPAT to taxonomic research and conservation in Bolivia. (It is curious that the binomial is *sanderi*, not *sandersi* – which would have been more accurate)

Sanderson

Sanderson's Hook Frog *Phrynobatrachus sandersoni* **Parker**, 1935

Sanderson's Rainfrog *Craugastor sandersoni* **Schmidt**, 1941

Ivan Terence Sanderson (1911–1973) was a British naturalist, explorer, author, traveller and television commentator. His father was a wealthy Scotch whisky distiller and so he was not short of funds. He started collecting animals (1924) and travelled the world (1927–1929) collecting for the Natural History Museum, London, before going to Cambridge where he was awarded a bachelor's degree in zoology and subsequently a master's in geology and botany. He led the Percy Sladen Expedition (1932–1933) to Cameroon and spent the rest of the 1930s travelling and collecting. The specimens were sent to the NHM where Parker worked on the collection. During the Second World War he served in British Naval Intelligence (1940–1945) and continued to work for British Intelligence until 1957. He moved to the USA in 1950s, took US citizenship and made a career as a television naturalist. He was a great admirer of Charles Fort (the chronicler of bizarre phenomena) and wrote extensively on UFOs and cryptozoology – particularly the Abominable Snowman, on which he published a memorable book (1961). He founded the Society for the Investigation of the Unexplained (1965).

Sandra

Bright-eyed Frog sp. *Boophis sandrae* **Glaw** et al., 2010

Sandra Nieto Román is the wife of David Vieites, one of the describers and original collectors of this species in Madagascar.

San José

Anori Stubfoot Toad *Atelopus sanjosei* **Rivero** & **Serna**, 1989

This frog is named to celebrate a hundred years (1890–1990) of the College San José de Medellín, Colombia. It coincides with the centenary of the Community of La Salle in Colombia.

San Martín

San Martín's Redbelly Toad *Melanophryniscus sanmartini* **Klappenbach**, 1968

San Martín's Rocket Frog *Allobates sanmartini* **Rivero, Langone** & Prigioni, 1986

Pablo Rubens San Martín (1934–1969) of the Department of Entomology, Universidad de la República, Montevideo, Uruguay, had a principal interest in scorpions but also collected herpetological specimens on his travels. He collected the holotype of the rocket frog in Venezuela (April 1957) during the Uruguayan Orinoco Expedition, but the species was not described until almost 30 years later.

Sarasin

Müller's Shrub Frog *Pseudophilautus sarasinorum* **Müller**, 1887

Paul Benedikt Sarasin (1856–1929) and Karl Friederich 'Fritz' Sarasin (1859–1942) were cousins. They were Swiss zoologists, explorers and collectors who wrote *Reisen in Celebes* (1905). Five birds, five reptiles and a mammal are named after them.

Sartor

Taylor's Yellow Treefrog *Dendropsophus sartori* **Smith**, 1951

Caecilian sp. *Minascaecilia sartoria* Wake & **Campbell**, 1983 NCR

[JrSYN *Gymnopis syntrema*]

Edward Harrison Taylor (1889–1978) was a collector. Both binomials are one of those clever uses of language much beloved of academics. The etymology for the treefrog says: 'From the Latin *sartor*, meaning tailor; here used loosely in reference to Edward H. Taylor, who was responsible for accumulation of all known specimens of the present subspecies.'. The etymology for the caecilian makes the same point. (See **Taylor, EH**)

Sato

Sato's Oriental Salamander *Hynobius abei* Sato, 1934

[Alt. Abe's Salamander]

Ikio Sato (1902–1945) was a Japanese herpetologist. He took his early degrees at the University of Hiroshima and then became a naturalist in Nagano prefecture. He returned to Hiroshima, gaining his doctorate (1941) and becoming Assistant Professor (1943). He was still there when the atom bomb exploded (August 1945). His collection and his manuscripts were all destroyed and he died of radiation poisoning five days later. He researched arachnids and Asian salamanders. He wrote *A Monograph of the Tailed Batrachians of Japan* (1943).

Sauer

Paka Bubble-nest Frog *Philautus saueri* Malkmus & Riede, 1996

Strabomantid frog sp. *Bryophryne hanssaueri* **Lehr** & Catenazzi, 2009

Hans Sauer supports the BIOPAT initiative to find new sources of finance to help taxonomical research and biosystematics.

Saul

Santa Cecilia Rocket Frog *Hyloxalus sauli* **Edwards**, 1974

Dr F. William Saul (d.1992) was a physician who defended conventional medicine against what he saw as quackery in his book *Pink Pills for Pale People* (1949). He was also a naturalist who made many contributions to the Museum of Natural History, University of Kansas and actively supported the University's work at Santa Cecilia, Ecuador. The University administers a fund named after him.

Sauron

Tree Frog sp. *Litoria sauroni* **Richards** & Oliver, 2006

The describers are presumably fans of Tolkien, as Sauron is the chief villain of *The Lord of the Rings*.

Sauter

Sauter's Brown Frog *Rana sauteri* **Boulenger**, 1909

[Alt. Taiwan Groove-toed Frog]

Dr Hans Sauter (1871–1943) was a German entomologist who also became interested in herpetology. He studied biology at the Universities of Munich and Tübingen. He was in Formosa (Taiwan), then under Japanese occupation, collecting insects (1902–1904). He was in Tokyo (1905) before returning to Taiwan for the rest of his life. He worked for a British trading company but entomology was his passion. Although Japan and Germany were enemies in the First World War he kept his job and continued collecting, albeit under observation. He was said to be the first person to offer private piano lessons in Taiwan, and gave German and English lessons. Three reptiles are also named after him.

Sauvage

Painted-belly Leaf Frog *Phyllomedusa sauvagii* **Boulenger**, 1882

[Alt. Waxy Monkey Frog]

Dr Henri-Émile Sauvage (1844–1917) was a French palaeontologist, herpetologist and ichthyologist. He wrote *Note sur les Geckotiens de la Nouvelle-Calédonie* (1878). Two reptiles are named after him.

Savage

Savage's Mushroomtongue Salamander *Bolitoglossa savagei* **Brame** & **Wake**, 1963

Savage's Brook Frog *Duellmanohyla lythrodes* Savage, 1968

Rio Changena Robber Frog *Craugastor* jota* **Lynch**, 1980

Pyburn's Robber Frog *Pristimantis savagei* **Pyburn** & Lynch, 1981

Glass Frog sp. *Centrolenella savagei* Harding, 1991 NCR

[JrSYN *Ikakogi tayrona*]

Savage's Glass Frog *Centrolene savagei* **Ruiz-Carranza** & Lynch, 1991

Savage's Worm Salamander *Oedipina savagei* Garcia-Paris & Wake, 2000

Savage's Thin-toed Frog *Leptodactylus savagei* **Heyer**, 2005

Dr Jay Mathers Savage (b.1928) is an American herpetologist. He is Emeritus Professor of Biology at the University of Miami and Adjunct Professor at San Diego State University. He took all his degrees at Stanford. He was a member of the faculty at the

University of Southern California before taking up his post in Florida. Among his many publications is *The Amphibians and Reptiles of Costa Rica: A Herpetofauna Between Two Continents, Between Two Seas* (2002). Five reptiles are also named after him.

* Lynch explains that *Craugastor jota* is named after Savage as *jota* is the letter 'J' in Spanish, and thus very appropriate for someone called Jay.

Savigny

Savigny's Treefrog *Hyla savignyi* Audouin, 1827
[Alt. Lemon-yellow Treefrog]

Marie Jules Cesar Lelorgne de Savigny (1778–1851) was a French zoologist and artist. He studied medicine but, under Geoffroy Saint-Hilaire's influence, turned to zoology. He was in Egypt during Napoleon's occupation (1798–1800) and undertook several expeditions, including one to Lake Manzala, where he studied the birds, sending specimens to Saint-Hilaire. The two men eventually quarrelled so badly that Saint-Hilaire prevented Savigny from becoming professor at the Muséum National d'Histoire Naturelle. Illness and near-blindness plagued his later years. He wrote *Description d'Egypte, ou Recueil des observations et des recherches qui ont été faites en Egypte pendant l'expédition de l'armée française,* (1798–1801). Three birds and two reptiles are named after him.

Sawaya

Banhado Frog *Scythrophrys sawayae* **Cochran**, 1953

Professor Dr Paulo Sawaya (1903–2003) was a Brazilian physician, zoologist and teacher. He qualified in medicine (1928) and took up a post as Assistant Professor (1934) at the University of São Paulo. He soon became Head of the newly formed Department of General and Animal Physiology, as well as establishing a marine laboratory and a zoo. The original etymology for this frog states: 'Through the kind interest of Dr. Paulo Sawaya of the University of São Paulo, the United States National Museum has received a number of frogs collected by him and his students in the states of São Paulo and Paraná, Brazil.'

Sazima

Sazima's Treefrog *Scinax pinima* **Bokermann** & Sazima, 1973
Sazima's Leaf Frog *Phasmahyla jandaia* Bokermann & Sazima, 1978
[Alt. Parakeet Leaf Frog]
Zagaia Treefrog *Bokermannohyla sazimai* **Cardoso** & Andrade, 1982
Campinas Tree Toad *Hylodes sazimai* **Haddad** & **Pombal**, 1995

Professor Dr Ivan Sazima is a Brazilian ichthyologist and herpetologist. He is now retired from the Department of Zoology and Museum of Natural History, University of Campinas, Brazil. His three degrees - bachelor's in biological sciences (1971), master's (1975), and doctorate (1980) in zoology - were all awarded by the University of São Paulo. He co-wrote *Snakes of the Brazilian Atlantic Forest. An illustrated field guide for the Serra do Mar range* (2004). A reptile is named after him.

Scala

La Escalera Treefrog *Stefania scalae* **Rivero**, 1970

'Scala' is Latin for 'ladder' and 'Escalera' is the Spanish equivalent. This frog's holotype was acquired near La Escalera in Venezuela; the binomial reflects this.

Scheepstra

Angguruk Fanged Frog *Xenorhina scheepstrai* **Blum** & **Menzies**, 1989

G. Scheepstra collected the holotype (1979).

Scheffler

Scheffler's Puddle Frog *Phrynobatrachus scheffleri* **Nieden**, 1911

Georg Scheffler (d.1910) was a German collector in East Africa at the end of the nineteenth century and the beginning of the twentieth. Many plants, a bird and a reptile are named after him.

Schenck

Blackchin Red Salamander *Pseudotriton ruber schencki* **Brimley**, 1912

Dr Carl Alwin Schenck (1868–1955) was a forester who was born in Germany. He studied at Tübingen and Giessen Universities and received his doctorate (1895). He was invited (1898) to found and run Biltmore Forest School on George Vanderbilt's Biltmore Estate, North Carolina. The school closed down (1913) and Schenck returned to Germany and served in the German army (1914–1915) on the Russian front, where he was badly wounded. He lived in Germany throughout the Second World War, afterwards travelling throughout the world giving lectures on scientific forestry. He wrote *Forestry for Kentucky* (1899).

Scheville

Scheville's Barred Frog *Mixophyes schevilli* **Loveridge**, 1933
[Alt. Northern Barred Frog]

William Edward Schevill (1906–1994) was a cetologist and biologist who was Librarian of the

Museum of Comparative Zoology, Harvard, and Assistant Curator of Invertebrate Palaeontology (1938). He was at the Woods Hole Oceanographic Institution, Massachusetts (1950s) when he and his wife, the zoologist Barbara Lawrence, made the first hydrophone recordings of cetaceans in the wild. A reptile is named after him.

Schiefenhoevel

Munggona Fanged Frog *Xenorhina schiefenhoeveli* **Blum** & **Menzies**, 1989

Dr Wulf Schiefenhövel (b.1943) is a German physician and anthropologist who became Professor of Medical Psychology and Ethnomedicine, Ludwig Maximilians University, Munich (1991). He qualified in medicine (1970) at Universität Erlangen-Nürnberg. Since 1965 he has carried out anthropological studies in the field in Melanesia and Indonesia, leading to him co-founding ethno-medicine in Germany.

Schilf

Schilf's Madagascar Frog *Gephyromantis schilfi* **Glaw** & Vences, 2000

[Syn. *Mantidactylus schilfi*]

Dr Wolfgang Schilf (b.1950) taught microbiology at Beuth Hochschule für Technik, Berlin, where he is a professor. His mother, Mrs Margot Schilf of Augsburg, organized (presumably through support for BIOPAT) that this frog be named after Wolfgang as a fiftieth birthday present.

Schilluk

Schilluk Ridged Frog *Ptychadena schillukorum* **Werner**, 1908

The Schilluk are a Nilotic people in the Sudan.

Schiøtz

Schiotz's Reed Frog *Hyperolius laurenti* Schiøtz, 1967

Apouh Night Frog *Astylosternus schioetzi* **Amiet**, 1978

Acha Tugi Long-fingered Frog *Cardioglossa schioetzi* Amiet, 1982

Running Frog sp. *Kassina schioetzi* Rödel et al., 2002

Dr Arne Schiøtz (b.1932) is a Danish herpetologist who is an authority on African tree frogs. He received a master's degree in zoology (1959) and a doctorate (1967). He was Curator, Herpetology, Copenhagen Zoo (1959–1964) and Director, Danish Aquarium (1964–1996). He was on leave from the Aquarium (1980–1983) as Director of Conservation, WWF-International and (1990–1993) as Senior Technical Advisor to the Government of Bhutan. He contributed much of the comprehensive collection of treefrogs in the Zoological Museum, Copenhagen. He wrote *Treefrogs of Africa* (1999). (See also **Arne**)

Schirch

Rio Mutum Frog *Dasypops schirchi* **Miranda-Ribeiro**, 1924

Santo Smooth Horned Frog *Proceratophrys schirchi* Miranda-Ribeiro, 1937

Dr Paulo F. Schirch was a Brazilian biologist who was an expert on flatworms. He made a collection of frogs and gave them to the describer. He wrote *Excitantes estimuladores e inhibitores como principio physiologico cellular em endocrinologia* (1938).

Schlaginhaufen

Wandolleck's Land Frog *Liophryne schlaginhaufeni* **Wandolleck**, 1911

Otto Schlaginhaufen (1879–1973) was Professor of Anthropology and Ethnology at the University of Zurich (1917–1950) and Director of the University's Anthropological Institute (1917–1951). He researched in New Guinea and Melanesia. He edited (1924 onwards) *Bulletin of the Swiss Society for Anthropology and Ethnology*.

Schlegel

Schlegel's Flying Frog *Rhacophorus schlegelii* **Günther**, 1858

Schlegel's Snouted Frog *Xenorhina oxycephala* Schlegel, 1858

Professor Hermann Schlegel (1804–1884) was a German zoologist, the first person to use trinomials to describe separate races of animals. In his youth C. L. Brehm, the father of Alfred Brehm, tutored him. He studied in Vienna (1824–1825), was recruited by Temminck as a preparator for the museum at Leiden (1825) and succeeded Temminck as Director there (1858). Although primarily an ornithologist, he wrote extensively on herpetology. He co-wrote *Fauna Japonica* (1845–1850). Many other taxa are named after him including sixteen birds, ten reptiles and two mammals.

Schlüter

Schlueter's Wart Frog *Fejervarya schlueteri* **Werner**, 1893

Wilhelm Schlüter (1828–1919) founded the eponymous firm of natural history dealers in Halle (1853). His father Friedrich was an entomologist and malacologist. His brother Julius emigrated to Brazil and was one of a network of collectors, including the Geisler brothers in Australia and

New Guinea, who kept the company supplied with specimens. The company was pre-eminent in the trade and until 1914 dealt with all major museums and private collectors. Werner's description specifically mentions the company as having supplied the specimens. Wilhelm's sons, both of whom had studied natural sciences, later ran the company: Wilhelm 'Willy' Schlüter Jr (1866–1938) and Curt (1881–1944). After the Second Word War Schlüter relocated from the Russian Zone (later East Germany) to West Germany. Two reptiles and four birds are named after Wilhelm senior.

Schmacker

Schmacker's Bubble-nest Frog *Philautus schmackeri* **Böttger**, 1892

Schmacker's Frog *Odorrana schmackeri* Böttger, 1892 [Alt. Piebald Odorous Frog]

Philipp Bernhard Schmacker (1852–1896) was a German amateur malacologist. As he came from a trading family, he was sent to Hong Kong (1872) and worked in Canton and Shanghai. He collected molluscs in China and other parts of the Far East. He and Boettger jointly presented a paper to the Malacological Society (1894). He died in Yokohama and left his collection to the city of Bremen. Three birds are named after him.

Schmarda

Schmarda's Shrub Frog *Pseudophilautus schmarda* **Kelaart**, 1854

Dr Ludwig Karl Schmarda (1819–1908) was an Austrian physician, naturalist and traveller who went round the world (1853–1857). After qualifying as a physician in Vienna (1843) he became an army surgeon. He taught at the Joanneum, Graz (1848–1850). He was Professor of Natural History, Karl-Franzens-Universität, Graz (1850–1852) and Professor of Zoology and Director of the Zoological Museum, Univerzita Karlova v Praze (Charles University, Prague) (1852–1862). He was Professor of Zoology, Universität Wien (1862–1883). After retirement he travelled in Spain and North Africa (1884–1887). He wrote *Die Geographische Verbreitung der Thiere* (1853).

Schmid

Sierra Maestra Blotched Frog *Eleutherodactylus michaelschmidi* **Díaz**, Cádiz & Navarro, 2007

Michael Schmid is Professor of Human Genetics, University of Würzburg, Germany. He is also involved in herpetology as he studies the chromosome of amphibians. He co-wrote *Chromosome Banding of six Dendrobatid Frogs (Colostethus, Mannophryne)* (2003)

Schmidt, EO

Schmidt's Marsupial Frog *Gastrotheca splendens* Schmidt, 1857

Schmidt's Stubfoot Toad *Atelopus pachydermus* Schmidt, 1857

Dr Eduard Oscar Schmidt (1823–1886) was a German zoologist who originally studied mathematics at Halle and Berlin. He received his doctorate from the University of Jena (1847) and was appointed Professor of Zoology at the University of Cracow (1855). He also taught at the University of Graz (from 1857) and the University of Strasbourg (from 1872). He was an early proponent of Darwin and his theories. He wrote *Lehrbuch der Zoologie* (1854).

Schmidt, KP

Schmidt's Robber Frog *Eleutherodactylus schmidti* **Noble**, 1923

Schmidt's Uruguay Treefrog *Scinax uruguayus* Schmidt, 1944

Schmidt's Lazy Toad *Oreolalax schmidti* **Liu**, 1947

Chiala Mountain Salamander *Batrachuperus karlschmidti* Liu, 1950

Schmidts' Mountain Brook Frog *Duellmanohyla schmidtorum* **Stuart**, 1954

Schmidt's Forest Frog *Hydrolaetare schmidti* **Cochran** & **Goin**, 1959

Schmidt's Pygmy Salamander *Thorius schmidti* Gehlbach, 1959

Schmidt's Spinythumb Frog *Crossodactylus schmidti* **Gallardo**, 1961

Schmidt's Wrinkled Ground Frog *Platymantis schmidti* **Brown** & **Tyler**, 1968

Schmidt's Snouted Frog *Mertensophryne schmidti* **Grandison**, 1972

Karl Patterson Schmidt (1890–1957) was an American herpetologist. He graduated from Cornell (1916), then worked as a scientific assistant in herpetology at the AMNH, New York, until 1922. He was Assistant Curator of the newly founded Department of Amphibians and Reptiles, Field Museum, Chicago (1922–1940), then Curator of Zoology there (1941–1955), becoming Emeritus Curator in the latter year. He undertook many expeditions, including Santo Domingo (1916), Puerto Rico (1919), Central America (1923), Brazil (1926) and Guatemala (1933). He edited the journal *Copeia* (1937–1949). He was an avid collector of herpetological specimens and wrote books on them; he co-wrote *Field Book of Snakes of the US and Canada* (1941). He was bitten by a juvenile boomslang he had thought unable to inject a fatal dose of venom.

He went home to his wife and received no medical treatment. He kept a careful note of the development of the symptoms he experienced until he died. Sixteen reptiles and a mammal are named after him. (See **Karl**).

The plural binomial of the mountain brook frog (*schmidtorum*) honours both Karl and his brother, Frank, who collected with him, especially in Guatemala.

Schneider, C

Schneider's Banana Frog *Afrixalus schneideri* **Boettger**, 1889

Carl Schneider was a naturalist and natural history dealer at Eschweiler bei Aachen, Germany. He corresponded with Boulenger at the BMNH regarding the sale of specimens (1889).

Schneider, CJ

Schneider's Shrub Frog *Pseudophilautus schneideri* Meegaskumbura & Manamendra-Arachchi, 2011

Christopher J. Schneider is a biologist at Boston University, USA, where he is currently (2012) Associate Professor and Acting Director of the Center for Ecology and Conservation Biology.

Schneider, G

Schneider's Toad *Rhinella schneideri* **Werner**, 1894

Gustav Schneider (1834–1900) was a conservator at Basel (1859–1875) but appears to have left to start up his own business as a natural history dealer, as by 1879 he was regularly supplying museums such as the Senckenberg Institute and herpetologists like Peters in Berlin. Werner wrote of this toad that 'Diese Art wurde mir von Herrn Naturalienhändler G. Schneider in Basel zur Bestimmung eingesandt...' and named it after him. A bird is named after him.

Schneider, JGT

Schneider's Dwarf Toad *Duttaphrynus scaber* Schneider, 1799

Johann Gottlob Theaenus Schneider (1750–1822) was a German scholar in the days when scholars were expected to be polymaths conversant with everything from the natural sciences to dead languages. He corrected, expanded and republished Bloch's work on fish: *Systema Ichthyologiae iconibus CX illustratum* (1801). He also published editions of works by classical authors, such as Aelian's *De Natura Animalium*. Six reptiles and a mammal are named after him.

Schouteden

Screeching Frog genus *Schoutedenella* **De Witte**, 1921 NCR

[Now included in *Arthroleptis*]

Kunungu Reed Frog *Hyperolius schoutedeni* **Laurent**, 1943

Henri Eugene Alphonse Hubert Schouteden (1881–1972) was a Belgian zoologist who undertook many expeditions to the Congo. Thirteen birds, three reptiles and two mammals are named after him.

Schreiber

Schreiber's Smooth Newt *Triturus vulgaris schreiberi* **Wolterstorff**, 1914 NCR

[JrSYN *Lissotriton graecus*]

Dr Egid Schreiber (1836–1913) was a zoologist who was Professor at the State High School in Salzburg, Austria. His most notable publication was *Herpetologia Europaea* (1875). Two reptiles are also named after him.

Schreiter

La Rioja Water Frog *Telmatobius schreiteri* **Vellard**, 1946

Carlos Rodolfo Schreiter (1877–1942) was a German-born Argentine naturalist and entomologist who was also interested in archaeology and ethnology. He settled in San Miguel de Tucumán and joined (1914) the group working for Dr Miguel Lillo to concentrate on the collections of insects and reptiles. He was the first Director of Instituto Lillo and from 1915 he became Curator, Natural History Museum, National University of Tucumán. He wrote *Megalithic monuments and paintings in the upper valleys of the province of Tucumán* (1928).

Schrenck

Schrenck's Newt *Salamandrella tridactyla* Nikolskii, 1906

[Use of *Salamandrella schrenckii* for this taxon is apparently incorrect]

Leopold Ivanovich von Schrenck (sometimes transcribed as Schenk or Shrenk) (1826–1894) was a Russo-German zoologist, geographer and ethnographer who was Director (1879) of the Imperial Academy of Sciences, St Petersburg. He explored the Amur River and Sakhalin Island (1854–1856), publishing the results in *Reisen und Forschungen im Amur-Lande in den Jahren 1854–1856* (four vols. 1860–1900). He believed that mammoths which he found preserved in the permafrost must have died recently, and thought they were subterranean animals that ate earth! Among other taxa, two birds and a reptile are named after him.

Schubart

Schubart's Humming Frog *Chiasmocleis schubarti* **Bokermann,** 1952

Schubart's Rondonia Treefrog *Dendropsophus schubarti* Bokermann, 1963

Dr Otto Schubart (1900–1962) was a German biologist who emigrated to Brazil (1934). He was regarded as the grand old man of Brazilian diplopodology. He worked at the Museum in São Paulo, where a street is named after him as is a reptile.

Schubert

Schubert's Stefania Treefrog *Stefania schuberti* **Señaris, Ayarzagüena** & **Gorzula,** 1997

Dr Carlos Schubert Paetow (1938–1994) was born in Germany but arrived in Venezuela as a young child. He was a geologist who worked variously for Shell, the Ministry of Mines and Hydrocarbons, and for the state-owned oil company PDVSA. The University of Arizona awarded his bachelor's degree (1961); both his master's (1963) and doctorate (1967) were from Rice University, Houston. He did much research on the mountain Auyan-tepui, where the treefrog holotype was acquired. He co-wrote *Geomorphology of the Gran Sabana, Guyana Shield, southeastern Venezuela* (1990).

Schubo

Bright-eyed Frog sp. *Boophis schuboeae* **Glaw** & Vences, 2002

Dr Anna Schubö is a lecturer at Department of Psychology, Ludwig-Maximillian University, Munich, Germany. She and Dr Werner Schubö gave generous support to the describers.

Schubotz

Schubotz's Squeaker *Arthroleptis schubotzi* **Nieden,** 1911

[Alt. Burundi Screeching Frog]

Central Grassland Frog *Rana schubotzi* **Sternfeld,** 1917 NCR

[JrSYN *Ptychadena tellinii*]

Dr Johann G. Hermann Schubotz (1881–1955) was a German zoologist. He was in East Africa (1907–1908) with the Duke of Mecklenburg's Expedition, and acted as overall editor of *Ergebnisse der deutschen Zentral-Afrika-Expedition 1907/08 unter Fuehrung Ad. Friedrichs, Herzog zu Mecklenburg,* in which Sternfeld, describer of the grassland frog, wrote the section on reptiles. Schubotz travelled from Ubangi to the Nile (1910–1911). He became Professor at the Natural History Museum, Hamburg (1914). A reptile and ten birds are named after him.

Schulte

Schulte's Robber Frog *Pristimantis schultei* **Duellman,** 1990

Rainer Schulte is a German biologist and herpetologist who graduated from Hohenheim University. He has lived in Peru since 1982 as an independent researcher. He works at the Instituto de Investigación Biológica de las Cordilleras Orientales (INIBICO), Tarapoto, Peru and co-operates closely with Duellman, who described this frog.

Schultze

Schultze's Stream Frog *Ptychohyla leonhardschultzei* **Ahl,** 1934

Dr Leonhard Schultze-Jena (1872–1955) was a geographer, zoologist, botanist, philologist and ethnographer. He was Professor of Geography at the University of Marburg (1913–1937). He led a number of expeditions to New Guinea early in the twentieth century. He was in Mexico and Guatemala studying Mayan and Aztec culture and languages, and published translations and dictionaries of them (1930s). Four reptiles are named after him.

Schwartz

Schwartz's Caribbean Toad *Peltophryne cataulaciceps* Schwartz, 1959

Schwartz's Robber Frog *Eleutherodactylus schwartzi* **Thomas,** 1966

Dr Albert Schwartz (1923–1992) was an American biologist and entomologist. He was Professor Emeritus of Biology, Miami-Dade Community College, and was associated with the Florida Museum of Natural History, the National Museum of Natural History, and the Museo Nacional de Historia Natural, Santo Domingo. He was a specialist in the fauna of the West Indies, writing extensively on amphibians, reptiles and lepidoptera, including *The Butterflies of Hispaniola* (1989). Seven reptiles, two mammals and a bird are named after him.

Scortecci

Scortecci's Toad *Duttaphrynus scorteccii* **Balletto** & Cherchi, 1970

Dr Giuseppe Scortecci (1898–1973) was an Italian zoologist and herpetologist. He took his doctorate at the University of Florence (1921) and joined the staff of the Institute of Comparative Anatomy there. He became Professor of Zoology, University of Genoa (1942). Before the Second World War he explored in the Sahara, Italian Somaliland (Somalia) and Ethiopia. He produced some 50 publications on herpetology, particularly about the fauna of desert regions. Twelve reptiles and a bird are named after him.

Scovazzi

Moroccan Painted Frog *Discoglossus scovazzi* Camerano, 1878

Commendatore Scovazzi was Italian Consul in Morocco, and an amateur herpetologist who collected reptile and amphibian specimens.

Scrocchi

Andalgala Water Frog *Telmatobius scrocchii* **Laurent** & **Lavilla**, 1986

Dr Gustavo José Scrocchi Manfrini (b.1956) is an Argentine herpetologist who works at the Fundación Miguel Lillo and is Secretary of the Argentine Herpetological Association. His doctorate (1956) is in biology and zoology. He has written widely on South American herpetology, such as co-writing *A new species of the genus* Atractus *from the northwestern Argentina (Serpentes, Colubridae)* (1991). Two reptiles are named after him.

Seabra, A

Quilombo Reed Frog *Hyperolius seabrai* Ferreira, 1906

Dr Anthero Frederico de Seabra (1874–1952) was a Portuguese zoologist at the University of Coimbra. He wrote *Catalogue systematique des vertebros du Portugal* (1911). A mammal is named after him.

Seabra, C

Tukeit Hill Frog *Sphoenohyla seabrai* **Bokermann**, 1958 NCR

[JrSYN *Allophryne ruthveni*]

Dr Carlos Alberto Campos Seabra was a Brazilian physician, biologist and entomologist at the National Museum, Rio de Janeiro (1920–1958).

Seegobin

Seegobin's Tepui Toad *Oreophrynella seegobini* Kok, 2009

Giuliano 'Kinky' Seegobin of Menzies Landing, Guyana, is a collector who has been on a number of expeditions to the Kaieteur National Park.

Señaris

Señaris' Treefrog *Tepuihyla galani* **Ayarzaguena**, Señaris & **Gorzula**, 1993

Dr Josefa Celsa Señaris is a biologist and herpetologist at the Fundación La Salle de Ciencias Naturales, Caracas, Venezuela, which she joined (1986) as Assistant, becoming Assistant Director (1994) and Director (2004), Natural History Museum La Salle. The Universidad Central de Venezuela awarded her baccalaureate (1990) and the University of Santiago de Compostela, Spain her doctorate (2001). She is a representative member in Venezuela for 'Conservation International'.

Sengupta

Dicroglossid frog sp. *Fejervarya sengupti* Purkayastha & **Matsui**, 2012

Dr Saibal Sengupta is an Indian herpetologist who teaches at Arya Vidyapeeth College, Assam, where he is head of the zoology department. The etymology contains the words '...in honour of Dr. Saibal Sengupta, the first author's mentor and one of the pioneer workers in the field of herpetology in northeastern India after the independence of India.'

Sequoyah

Sequoyah Slimy Salamander *Plethodon sequoyah* Highton, 1989

Sequoyah, also known as George Gist (c.1770–1843) was the son of a white father and a Cherokee mother. He devised the Cherokee syllabary, which made reading and writing the Cherokee language possible.

Séraphin

Gaboon Caecilian *Geotrypetes seraphini* **Duméril,** 1859

Séraphin Braconnier was a French naturalist who was employed in the Herpetology and Ichthyology Department, Muséum National d'Histoire Naturelle, Paris for over 25 years. He collected in New Caledonia and sold a number of specimens to the Berlin Museum. Three reptiles are named after him.

Serchhipi

Serchhip's Horned Toad *Xenophrys serchhipii* Mathew & Sen, 2007

This toad is named after the type locality in India – 'Serchhip, Agriculture Guest House compound' – and the common name would be better given as the Serchhip Horned Toad.

Serna

Serna's Robber Frog *Pristimantis dorsopictus* **Rivero** & Serna, 1988

Stubfoot Toad sp. *Atelopus sernai* **Ruiz-Carranza** & Osomo-Muñoz, 1994

Brother Marco Antonio Serna Díaz (1936–1991) was a biologist and naturalist who was Professor of Ornithological Biology at La Salle University, Colombia, and in charge of its museum (1970–1991). A bird is named after him.

Seshachar

Seshachar's Bush Frog *Raorchestes charius* **Rao**, 1937 [Alt. Chari's Bush Frog]

Seshachar's Caecilian *Gegeneophis seshachari* Ravichandran, Gower & Wilkinson, 2003

Professor Bagepalli Ramachandrachar Seshachar (1908–1994) was Professor of Zoology at Delhi University. He began his career as a demonstrator in zoology at the University of Mysore (1926) while studying under Rao. He spent 34 years in the Department of Zoology, Bangalore University. After retiring (1960) he was invited by Rao, then Vice-chancellor, to head the Zoology Department at Delhi University. He was President of the Indian Academy of Science (1971–1972). The etymology says that the species is named for Seshachar 'in recognition of his pioneering studies of the cytogenetics, reproductive biology, and natural history of Indian caecilians'. Seshachar conducted several successful and influential projects on caecilians with L. S. Ramaswami, for whom Taylor (1964) named another species of *Gegeneophis*. (See **Chari**)

Shasta

Shasta Salamander *Hydromantes shastae* Gorman & **Camp**, 1953

This species is named after Mount Shasta in California, near which the holotype was collected.

Sheldrick

Sheldrick's Reed Frog *Hyperolius sheldricki* Duff-MacKay & **Schiøtz**, 1971

David Leslie William Sheldrick (1919–1977) was an outstanding warden in Kenya (1948–1976), where he established Tsavo National Park. Born in Egypt, he was taken to Kenya (1919) where his parents settled. He was sent to school in England (1925), only seeing his parents again 11 years later and failing to recognize his mother at the railway station! He commanded a battalion of the King's African Rifles in Burma in the Second World War. Until 1948 he worked for a safari company and acted as guide for the Aga Khan, until he joined Kenya National Parks. The David Sheldrick Wildlife Trust was founded in his memory.

Shelkovnikov

Shelkovnikov's Tree Frog *Hyla arborea schelkownikowi* Cernov, 1926

[Taxonomic status uncertain. Synonym of *Hyla (arborea) orientalis*?]

Alexander Bebutovich Shelkovnikov (1870–1933) was a Russian zoologist who founded (1922) the Herbarium of Armenia, which later became part of the Institute of Botany of the Armenian National Academy of Sciences. He worked for a time as an official in the Armenian agricultural administration. His collections, mainly of zoological and botanical specimens from Georgia, Armenia and Azerbaijan, are held across parts of the old Soviet Union, particularly in Yerevan, St Petersburg, and Moscow. He was a graduate of a military school in St Petersburg, so had no formal zoological training, but was a gifted amateur who contributed much to the knowledge of the natural history of the Caucasus and Persia (Iran). He spent part of his career as the unofficial Curator of Scientific Collections at Moscow State University. Among other taxa, two mammals are named after him.

Shelly

Shelly's Rainforest Frog *Cophixalus shellyi* **Zweifel,** 1956

Father O. Shelly SVD was appointed a missionary in New Guinea (1949) when he left his job as Professor of Mathematics, St Vincent's Mission Seminary, Philadelphia. He made several collections of amphibians and reptiles in northeast New Guinea, which he sent to the AMNH, New York.

Sherman

Red-legged Salamander *Plethodon shermani* **Stejneger**, 1906

Professor Franklin Sherman Jr (1877–1947) was an entomologist and zoologist at Clemson University, South Carolina, where he was curator of the University's arthropod collection (1926–1947).

Shih

Wushan Salamander *Liua shihi* **Liu**, 1950

Pei-nan Shih collected the holotype (1938) in eastern Szechwan, China. We know nothing more about him.

Shin

Spiny-flanked Frog *Quasipaa shini* **Ahl**, 1930

Professor Dr Shin, of Sun Yat-sen University, collected reptiles and amphibians in the mountainous regions of Guangxi (Kwangsi) Province, China. As well as the frog he found the Chinese Crocodile Lizard, *Shinisaurus crocodilurus*, which honours him in the genus name.

Shiv

Shiv's Rhinatrema *Rhinatrema shiv* Gower et al., 2010

Shivnarine 'Shiv' Chanderpaul (b.1974) is the most capped and highest run-scoring Guyanese cricketer, and was International Cricketer of the Year 2008.

Shola

Shola's Wrinkled Frog *Nyctibatrachus sholai* Radhakrishnan, Dinesh & Ravichandran, 2007
[Alt. Eravikulam Night Frog]

This frog is named after its habitat and the type locality: Kolukkumalai shola forest, Eravikulam National Park, Kerala State, India. 'Sholas' are patches of stunted evergreen forest found in valleys amid montane grasslands.

Sholigar

Karnataka Rice Frog *Microhyla sholigari* **Dutta** & Ray, 2000

This frog is named after the Soliga tribal people who inhabit the area of southern India where the holotype was collected.

Shompen

Shompen Frog *Limnonectes shompenorum* **Das**, 1996

The Shompen are an indigenous tribe of Great Nicobar Island, where the frog holotype was collected.

Shreve

Shreve's Sarayacu Treefrog *Dendropsophus sarayacuensis* Shreve, 1935
Shreve's Rocket Frog *Anomaloglossus shrevei* **Rivero**, 1961
Shreve's Robber Frog *Eleutherodactylus heminota* Shreve & **Williams**, 1963
St. Vincent Piping Frog *Pristimantis shrevei* **Schwartz**, 1967

Benjamin Shreve (1908–1985) was an American herpetologist at the Harvard Museum of Comparative Zoology. He co-wrote articles such as *Concerning some Bahamian Reptiles* (1935). Five reptiles are named after him.

Shyamrup

Shyamrup's Bush Frog *Chiromantis shyamrupus* Chanda & **Ghosh**, 1989
[Alt. Hornbill Bubble-nest Frog]

Dr Shyamrup Biswas led the survey team to the area in Arunachal Pradesh, India where the holotype was collected. He wrote *Occurrence of* Draco blanfordi *Boulenger* (Sauria: Agamidae) *in Assam India* (1967). A spider is also named after him.

Siblesz

La Escalera Treefrog *Hypsiboas sibleszi* **Rivero**, 1972

Gerardo Siblesz of Caracas, Venezuela, was a long-term friend and co-operator of the describer in studying amphibians.

Siebold

Siebold's Giant Salamander *Andrias japonicus* Temminck, 1836
[Alt. Japanese Giant Salamander; *Megalobatrachus sieboldi* is a junior synonym]

Dr Philipp Franz Balthasar von Siebold (1796–1866) was a German physician, biologist and traveller, and a medical officer to the Dutch East Indian army in Batavia and at the Dutch trading post, Dejima Island, Nagasaki, Japan. He taught Western medicine and treated Japanese patients, accepting ethnographic and art objects as payment, and established a boarding school in the outskirts of Nagasaki. Using local Japanese agents he collected in the interior (1823–1829). With the connivance of the Imperial librarian and astronomer, he copied a map of the northern regions of Japan, so upsetting the Government that all his known Japanese contacts were imprisoned, his house was searched and many possessions confiscated. He packed all of his manuscripts, maps and books in a large lead-lined chest, which was then hidden. He was banished from Japan (1829) and was forced to leave behind his young Japanese mistress and a two-year-old daughter (shades *of Madame Butterfly*), returned to Holland, prepared his Japanese materials for publication, and was appointed by the King to advise on Japanese affairs (1831). The Japanese ban was lifted (1859) and he became (1861) chief negotiator for all European nations who were trying to establish trade links. However, his mission was a failure and he was pensioned off (1863). His eldest son served the Japanese government as an interpreter and adviser for external and financial affairs (1870–1911). He wrote much on Japan including *Fauna Japonica – aves* (1844). In Nagasaki the Siebold Memorial Museum was founded to honour his contributions to the modernization of Japan. He sent specimens of the giant salamander to Leiden. Two reptiles and three birds are named after him.

Siegfried

Siegfried's Rainforest Frog *Albericus siegfriedi* **Menzies**, 1999

Siegfried is the eponymous hero of the third opera in Wagner's ring cycle, *Der Ring des Nibelungen*. He is based on the figure of Sigurd in Norse mythology. This species is one of several in the genus that are fancifully named after Wagnerian characters.

Siemers

Red-spotted Argentina Frog *Argenteohyla siemersi* **Mertens**, 1937

Siemers collected the holotype in Buenos Aires province, Argentina, but we know nothing more about him.

Silverstone

Rocket Frog genus *Silverstoneia* **Grant** et al., 2006
Silverstone's Mushroomtongue Salamander *Bolitoglossa silverstonei* **Brame** & **Wake**, 1972
Silverstone's Rocket Frog *Colostethus imbricolus* Silverstone, 1975
Silverstone's Poison Frog *Ameerega silverstonei* **Myers** & **Daly**, 1979
Robber Frog sp. *Pristimantis silverstonei* **Lynch** & **Ruiz-Carranza**, 1996

Professor Dr Philip Arthur Silverstone-Sopkin (b.1939) is an American botanist and herpetologist. His doctorate was awarded by the University of Southern California, Los Angeles. He works at the Department of Biology, Universidad del Valle, Cali, Colombia, where he has been Director of the University's Herbarium since 1996. He wrote *A Revision of the Poison-Arrow Frogs of the Genus* Phyllobates *Bibron in Sagra (Family Dendrobatidae)* (1976).

Simmons

Simmons' Big-headed Frog *Oreobates simmonsi* **Lynch**, 1974
Simmons' Treefrog *Hyloscirtus simmonsi* **Duellman**, 1989

John Edward Simmons (b.1951) was Collections Manager, Division of Herpetology, Natural History Museum & Biodiversity Research Center, University of Kansas (1981–2007). He left the University to work as an independent museum consultant (setting up the consulting company 'Museologica'). He was awarded an MA in historical administration and museum studies by the University of Kansas (1988). He processed all the photographs for one of Duellman's publications in 1984, and co-wrote with Duellman *Two new species of dendrobatid frogs* (1988). Among his many other publications is *Herpetological Collecting and Collections Management* (2002). Simmons and Bruce MacBryde collected the holotype of the big-headed frog in Ecuador (1972). Lynch wrote in his etymology: 'This species is named for John E. Simmons who during a one year tenure in Amazonian Ecuador collected extensive ecologic material on amphibians as well as made an important collection on the Cordillera del Condor.'

Simón (Bolívar)

Simón's Robber Frog *Pristimantis simonbolivari* **Wiens** & **Coloma**, 1992

Simón José Antonio de la Santísima Trinidad Bolívar y Palacios Ponte Blanco aka Simón Bolívar (1783–1830) is one of the most significant figures in South American history, and needs no biography here!

Simons

Paramo Robber Frog *Pristimantis simonsii* **Boulenger**, 1900
Sucre Water Frog *Telmatobius simonsi* **Parker**, 1940

Perry O. Simons (1869–1901) was an American who collected in the Neotropics. He was collecting herpetofauna in Peru (1899–1900) and birds in Bolivia (1901). Later, while crossing the Andes, he was murdered by his lone guide. (A man by the name of F. A. Simons was collecting in Colombia a decade earlier and we are curious to know whether the two were related) There are more than a dozen holotypes in the Natural History Museum, London, which Simons collected. The water frog is a good example of how a collector's work can lie around in a museum unattended for decades. When Parker examined the specimens, c.1939/1940, he found that each still had the collector's label on them: 'The Andes, PO Simons 2800m'. Seven birds, a mammal and four reptiles are named after him.

Simpson

Robber Frog sp. *Pristimantis sirnigeli* Yánez-Muñoz et al., 2010
Simpson's Plump Toad *Osornophryne simpsoni* Páez-Moscoso, Guayasamin & Yánez-Muñoz, 2011

Dr Sir Nigel Simpson is a British scientist who worked in the chemicals industry. He is also an ornithologist and a conservationist with particular interest in protecting the Andean cloud forests of Ecuador. He is a collaborator of two of the most important conservation NGOs in Ecuador, EcoMinga Foundation and Jocotoco Foundation.

Sir Nigel

Robber Frog sp. *Pristimantis sirnigeli* Yánez-Muñoz et al., 2010

Dr Sir Nigel Simpson (above)

Sisyphus

Obree Rainforest Frog *Cophixalus sisyphus* Kraus & **Allison**, 2006

In Greek mythology, Sisyphus was the founder of Corinth. He was condemned by the gods to spend all eternity in futile labour: forever pushing a rock up a hill only to have it repeatedly roll back down to the bottom. The describers wrote: 'The name alludes to the seeming task involved in sorting out the taxonomy of the species described herein.' It seems that amphibian taxonomy can get you down at times!

Sjostedt

Four-digit Toad *Didynamipus sjostedti* **Andersson**, 1903

Bror Yngve Sjöstedt (1866–1948) was a Swedish entomologist and ornithologist. He was in Cameroon collecting for the Uppsala University Zoological Department (1890–1891) and for the Swedish Natural History Museum, which he joined (1897). He travelled for the Museum to the USA and Canada to visit entomological stations and study their methods. He went on the Swedish Zoological Expedition to Mount Kilimanjaro (1905–1906). His publications include *Zur ornithologie Kameruns, Wissenschaftliche ergenbisse der schwedischen expedition nach dem Kilimanjaro.* Four birds and a reptile are named after him.

Skuk

Beaked Toad sp. *Dendrophryniscus skuki* **Caramaschi**, 2012

[Syn. *Rhinella skuki*]

Dr Gabriel 'Gabo' Skuk (1962–2011) was a Uruguayan herpetologist. The Universidad de La República Oriental del Úruguay awarded his bachelor's degree (1986) and the Universidade de São Paulo, Brazil his master's (1994) and doctorate (1999). He was a professor at the Universidade Federal de Alagoas, Maceió, Brazil. He greatly enjoyed diving and submarine hunting but was drowned at a beach north of Maceió.

Slater

Slater's Callulops Frog *Callulops slateri* **Loveridge**, 1955

[Syn. *Metamagnusia slateri*]

Kenneth R. Slater (1923–1999) was a zoologist and herpetologist. He worked in New Guinea (1950s) and collected and described (1956) the Papuan Taipan, *Oxyuranus scutellatus canni*. In 1954 he collected the frog holotype near Port Moresby. He later worked for the Animal Industry Branch, Northern Territory Administration, Alice Springs (1960s). He wrote *A Guide to the Dangerous Snakes of Papua* (1968). A repile is named after him.

Sligo

Sligo's Giant Salamander *Megalobatrachus sligoi* **Boulenger**, 1924 NCR

[Presumed to be a junior synonym of *Andrias davidianus*]

George Ulick Browne, Marquess of Sligo (1856–1935) was visiting Hong Kong when he saw a specimen of giant salamander living in the botanical gardens there. He used his influence to get the local authorities to send the creature to London Zoo, where it was exhibited until well into the 1930s. (Boulenger believed this specimen to be different from the 'typical' Chinese Giant Salamander (*Andrias davidianus*) but nowadays the taxon is regarded as synonymous with the latter)

Sloane

Sloane's Froglet *Crinia sloanei* **Littlejohn**, 1958

Ian Fyfe Sloane (1903–1986) was an Australian grazier and environmentalist. His grandfather settled in the area now known as 'Savernake Station', Savernake, New South Wales (1862) and Ian took it over (1926). He recorded with his camera the change in agriculture from horse power to modern machinery. He wrote *The Mulwala Aborigines* (1968).

Smith, A

Smith's Frog *Phrynobatrachus natalensis* Smith, 1849

[Alt. Natal Puddle Frog]

Smith's Reed Frog *Hyperolius tuberilinguis* Smith, 1849

[Alt. Tinker Reed Frog]

Dr Sir Andrew Smith (1797–1872) was a Scottish surgeon and zoologist who joined the Army Medical Service (1819) after graduating from Edinburgh University. He was famous for his scrupulous accuracy. He was in Cape Colony, South Africa (1820–1837), and was the first Superintendent of the South African Museum of Natural History, Cape Town (1825). He visited Namaqualand to discover more about its inhabitants (1828), publishing a paper on the history and lives of 'bushmen' (1831). He led the first scientific expedition into the South African interior (1834–1836). He returned to Britain, becoming Principle Medical Officer at Fort Pitt, Chatham (1841), and later Director General of the Army Medical Services (1853) – a post that included organizing medical services in the Crimean War (an enquiry cleared him of charges of inefficiency and incompetence instigated by *The Times* newspaper). He wrote *Illustrations of the Zoology of South Africa* (28 parts, 1838–1850). Nine reptiles, four mammals and twelve birds are named after him.

Smith, Barton

Barton's Robber Frog *Eleutherodactylus bartonsmithi* **Schwartz**, 1960

(See under **Barton**)

Smith, CC

Dwarf American Toad *Anaxyrus americanus charlesmithi* Bragg, 1954

Dr Charles Clinton Smith (1910–1966) was Research Professor of Biology, Arkansas College, which he joined (1957). Previously he had been a colleague and friend of the toad's describer, Arthur Bragg, at Pawhuska, Oklahoma. They co-wrote *Observations on the Ecology and Natural History of Anura IV. The Ecological Distribution of Toads in Oklahoma* (1943).

Smith, HH

Dwarf Mexican Treefrog *Tlalocohyla smithii* **Boulenger**, 1902

Herbert Huntingdon Smith (1851–1919) was an American naturalist who mainly worked on the natural history of Brazil. He wrote *Brazil, the Amazons and the coast* (1879) and *Do Rio de Janeiro a Cuyabá: Notas de um naturalista* (posthumous publication, 1922). He first went to Brazil in 1870 on the Morgan expedition and returned to stay (1874–1876), spending a year exploring the Amazon. Back in the USA he began working for Scribner's Magazine, writing on Brazil and frequently returning there. He married Amelia Woolwirth (1880), also a naturalist. They lived in Brazil until 1886, travelling widely and visiting Paraguay but spending most time at Chapada dos Guimarães, where intensive collecting (especially of insects) resulted in the discovery of many new species. After a few months in Rio de Janeiro, they returned to the USA. In 1889 Smith collected in Mexico for Godman, the results appearing in *Biologia Centrali-Americanum.* He was then commissioned by the Royal Society to collect in the West Indies (1889–1895). He became a curator at the Carnegie Museum. He collected in Colombia (1898–1902) before resuming curatorship of the Carnegie Museum on his return. Tragically, while collecting snails along a railway, the deaf naturalist was fatally hit by a train.

Smith, HM

Smith's Pygmy Robber Frog *Craugastor hobartsmith* **Taylor**, 1937

Smith's False Brook Salamander *Pseudoeurycea smithi* Taylor, 1939

Smith's Minute Salamander *Thorius smithi* **Hanken** & **Wake**, 1994

Dr Hobart Muir Smith (né Frederick William Stouffer) (b.1912) was adopted (1916) and his names were then changed. His bachelor's degree (1932), master's (1933) and doctorate (1936) were awarded by Kansas State University. He worked for different institutions including the Field Museum, Chicago and the Smithsonian (1937–1941), and undertook a number of field trips to Mexico, collecting over 20,000 specimens. He was Professor of Zoology, University of Rochester, New York (1941–1945), then returned to Kansas (1946). He became Associate Professor, Wildlife Management, Texas A&M University (1946). He was Professor of Zoology, University of Illinois (1947–1968). After retiring he became Professor of Biology, Colorado University, Boulder, becoming Chairman, Department of Ecology and Evolutionary Biology (1972). He retired again (1983), as Emeritus Professor. He was still writing and researching in his nineties. He has become the most published herpetologist ever. His first major publication was *Handbook of Lizards, Lizards of the US and Canada* (1945). Eleven reptiles are named after him.

Smith, HW

Smith's Squat Frog *Calluella smithi* **Barbour** & **Noble**, 1916

Professor Dr Harrison Willard Smith (1872–1947) was a teacher of physics at the Massachusetts Institute of Technology. He graduated in 1894 and by 1897 was an associate professor at MIT. He was one of a party of scientists who travelled to Sumatra (1901) to observe a total eclipse of the sun. In 1910 he was in the Malay Archipelago. He moved to Tahiti and took up botany. The frog holotype was part of a collection of reptiles and amphibians from Sarawak that Smith gave to the Museum of Comparative Zoology. He wrote an article (1919) in National Geographic: *Sarawak: The Land of the White Rajahs.* He also took many photographs there during several trips (1909, 1912, 1915 & 1919).

Smith, MA

Smith's Asian Frog *Chiromantis laevis* Smith, 1924

Malcolm Smith's Shrub Frog *Pseudophilautus malcolmsmithi* **Ahl**, 1927

Smith's Litter Frog *Leptobrachium smithi* **Matsui**, Nabhitabhata & **Panha**, 1999

Dr Malcolm Arthur Smith (1875–1958) was an English physician and herpetologist. He practised medicine in Bangkok (1902–1924), including five years as Court Physician. He visited French Indo-China (1918). He was President of the British Herpetological Society (1949–1954). He published *A Physician at the Court of Siam* (1947). Seven reptiles are named after him.

Snethlage

Utinga Surinam Toad *Pipa snethlageae* **Müller**, 1914

Dr Henriette Mathilde Maria Elizabeth Emilia Snethlage (1868–1929) was a German ornithologist and former assistant in zoology at the Berlin Museum. She collected in the Amazon rainforests (1905–1929), having been recommended by Reichenow to the Goeldi Museum. She succeeded Goeldi and was head of the museum's zoological section (1914–1929), except for suspension (1917–1918) when Brazil entered the First World War against Germany. She was the first woman scientist to direct a Brazilian Museum and to work in Amazonia. She wrote *Catalogo das Aves Amazonicas* (1914) as well as works on local languages. Two reptiles, a mammal and seven birds are named after her.

Soares

Santa Cruz Dwarf Frog *Physalaemus soaresi* **Izecksohn**, 1965
Picos Treefrog *Dendropsophus soaresi* **Caramaschi** & **Jim**, 1983

Professor Benedicto Abilio Monteiro Soares (1914–1985) was a biologist, entomologist and arachnologist at the Zoology Department, São Paulo Zoo. He collected with Izecksohn.

Sommer

Massif du Nord Whistling Frog *Eleutherodactylus sommeri* **Schwartz**, 1977

William W. Sommer was an entomologist at the Allyn Museum of Entomology, Sarasota, Florida, now part of the University of Florida. He and Schwartz collected together (1978) in Haiti. He published *A new species of* Atlantea *(nymphalidae) from Hispaniola, West Indies* (1980). A reptile is named after him.

Sonan

Sonan's Salamander *Hynobius sonani* Maki, 1922
[Alt. Taichu Salamander]

Jinhaku Sonan (1892–1984) was a Japanese entomologist working in Formosa (now Taiwan) in the first half of the twentieth century, at which time it was occupied by Japan. He served in the Imperial Japanese Navy (1912–1916) before returning to Formosa. After the Second World War he moved back to Japan, working in agriculture, forestry and the tea industry.

Sonja

Ivory Coast Wart Frog *Acanthixalus sonjae* Rödel et al., 2003

Sonja Wolters (b.1974) is a biologist and behavioural zoologist. Westfälische Wilhelms-Universität Münster awarded her master's degree. She carried out research on primates in the Ivory Coast (2000) and co-ordinated a conservation project in North Vietnam before joining West African Primate Conservation Action, Accra, Ghana (2005). She returned to Germany (2008). She collected the first specimen of this frog species.

Sooy

Cordillera Talamanca Salamander *Bolitoglossa sooyorum* Vial, 1963

Mr Kirk Van Sooy and Mrs Jean L. Van Sooy were close friends of the describer, James Vial.

Soriano

Cloud Forest Stubfoot Toad *Atelopus sorianoi* **La Marca**, 1983

Dr Pascual José Soriano is a Venezuelan zoologist at the Institute of Environmental Science and Ecology, University of the Andes, Mérida. A mammal is also named after him.

Sornoza

Glass Frog sp. *Teratohyla sornozai* Cisneros-Heredia et al., 2009

Francisco 'Pancho' Sornoza-Molina is an Ecuadorian ornithologist and conservationist who worked at the Museo Ecuatoriano de Ciencias Naturales and is now President of Fundación Jocotoco.

Sos

Barton Springs Salamander *Eurycea sosorum* Chippindale, Price & Hillis, 1993

This salamander is named after a conservation group called 'Save Our Springs'.

Sosa

Toad sp. *Metaphryniscus sosai* **Señaris, Ayarzagüena** & **Gorzula**, 1994

Miss Maricela Sosa (1966–1992) was a biologist of the Department of Biology, Faculty of Science, Universidad de los Andes, Mérida, Venezuela. She published (1991) her degree thesis, *Ecological relationships between bat* Glossophaga longirostris *and columnar cacti in the dry pocket of Lagunillas, Mérida, Venezuela*. A reptile is named after her.

Sowerby

Narrow-mouthed Frog sp. *Microhyla sowerbyi* **Stejneger,** 1924 NCR
[JrSYN *Microhyla butleri*]

Arthur de Carle Sowerby (1885–1954) was a zoologist, naturalist, explorer and artist who was born in China, where his father was a Baptist missionary. He went to Bristol University but only stayed a short time before returning to China, where he began to collect specimens for the Natural History Museum in Tai-yuan Fu. He collected mammals (1907) for the BMNH during an expedition to the Ordos Desert in Mongolia and (1908) was part of the Clark Expedition to Shansi and Kansu provinces. Jointly with Robert Sterling Clark he wrote an account *Through Shên Kan, the Account of the Clark Expedition in North China 1908–09* (1912). Clark was a very wealthy man and he financed a number of collecting trips for Sowerby. There was a revolution in China (1911) and Sowerby led an expedition to evacuate foreign missionaries from Shensi and Sianfu provinces. During the First World War, Sowerby was a technical officer in the Chinese Labour Corps and saw service in France. After the war Sowerby settled in Shanghai and established *The China Journal of Science and Arts*, which he edited until the Japanese occupied Shanghai during the Second World War. The Japanese army in Shanghai interned him for the duration but, despite that, he appears to have been able to go on writing and publishing as evidenced by *Birds recorded from or known to inhabit the Shanghai* area (1943). He moved to the USA (1949) and lived the rest of his life in Washington DC, spending his time in genealogical research which resulted in a family history, *The Sowerby Saga*. Three birds and a reptile are named after him.

Spartacus

Tree Frog sp. *Litoria spartacus* **Richards** & Oliver, 2006

Spartacus (ca. 120–69 B.C.) was about 51 when he either died in battle or was one of those crucified along the Via Appia – quite an old man for his era and occupation. Life for a slave-gladiator must normally have been short and brutal. Plutarch, the Roman historian, reported Spartacus as being cultured and intelligent and more like a Greek than a Thracian, as tradition states he was. For reasons unknown he was enslaved and trained as a gladiator at Capua, and in 73 B.C. he and some companions escaped and started a slave revolt that spread widely until he had a large army of ex-slaves, plus women and children - perhaps as many as 120,000 people at the height of his revolt. It caused the Roman authorities immense trouble, as he routed a number of armies and crushed several legions. Finally Crassus defeated him at the battle of Silarus, and the survivors of that battle were caught and defeated by Pompey the Great. A mammal is named after him.

Spatz

Spatz's Reed Frog *Hyperolius spatzi* **Ahl**, 1931

Paul W. H. Spatz (1865–1942) was an explorer, collector and dealer in natural history specimens in northwest Africa. He lived in Gabès (Tunisia) (1893–1928) and covered most countries from Tunisia to the Gambia. After returning home he set up a company in Diemitz (Germany). Seven birds are named after him.

Speer

Poison Frog sp. *Mannophryne speeri* **La Marca**, 2009

Jason W. Speer is Vice-President and General Manager of Quality Float Works Inc. of Illinois. He was a generous donor towards research leading to the description of this species.

Spencer

Spencer's Burrowing Frog *Platyplectrum spenceri* **Parker**, 1940

[Alt. Centralian Burrowing Frog; Syn. *Opisthodon spenceri*]

Spencer's Tree Frog *Litoria spenceri* **Dubois**, 1984

Sir Walter Baldwin Spencer (1860–1929) was an English explorer, biologist, anthropologist and patron of the arts. He went to Australia (1887) to become the first Professor of Biology, University of Melbourne, and Director of the Natural History Museum, Melbourne. He was photographer and zoologist on the Horn Expedition, the first major scientific expedition to the centre of Australia (1894). In 1919 he resigned his professorship and became (1920) vice-president of the trustees of the public library of Victoria. He died at Navarin Island, Tierra del Fuego, where he had gone to undertake anthropological fieldwork. He mainly wrote on ethnology, such as *The Northern Tribes of Central Australia* (1904). Two reptiles and a mammal are named after him.

Spix

Spix's Horned Treefrog *Hemiphractus scutatus* Spix, 1824

Spix's Saddleback Toad *Brachycephalus ephippium* Spix, 1824

[Alt. Pumpkin Toadlet]

Spix's Snouted Treefrog *Scinax nebulosus* Spix, 1824

Spix's White-lipped Frog *Leptodactylus spixi* **Heyer**, 1983

Dr Johann Baptist Ritter Von Spix (1781–1826) was a German naturalist working in Brazil (1817–1820). He gained his doctorate (1800) aged only 19! He studied theology for three years in Würzburg, then medicine and natural sciences, qualifying as a physician (1807). King Maximilian of Bavaria awarded him a scholarship to study zoology in Paris (1808), then *the* centre for the natural sciences with renowned scientists such as Cuvier, Buffon, Lamarck and Etienne Geoffroy de Saint-Hilaire at the height of their reputations. The King appointed him Assistant to the Bavarian Royal Academy of Sciences with special responsibility for the natural history exhibits (1810). A group of academicians was invited (1816) to travel to Brazil, and King Maximilian agreed to two members of the Bavarian Academy of Sciences going along. Spix was in South America (1817–1820), returning with specimens of 85 species of mammal, 350 birds, 130 amphibians, 116 fishes, 2,700 insects and 6,500 plants, plus 57 species of living animals, mainly monkeys, parrots and curassows; this formed the basis for the Natural History Museum, Munich. The King awarded him a knighthood and a pension for life. He catalogued and published his findings despite extremely poor health contracted in Brazil. The report on the expedition was published in three volumes (1823–1831). Five reptiles, eight mammals and eight birds are named after him.

Sponga

Azulita Salamander *Bolitoglossa spongai* Barrio-Amorós & Fuentes-Ramos, 1999

Manuel Ángel González Sponga (1929–2009) was a Venezuelan biologist and arachnologist. He taught biology at a secondary school (1954–1964), and then (1964–1991) at Universidad Pedagógica Experimental Libertador, Caracas, from which he retired as Profesor Emeritus. He wrote *Escorpiones de Venezuela* (1997). His wife collected the type series – so perhaps the salamander should have been named after her!

Spurrell

Spurrell's Leaf Frog *Agalychnis spurrelli* **Boulenger**, 1913 [Alt. Gliding Leaf Frog]

Condoto Stubfoot Toad *Atelopus spurrelli* Boulenger, 1914

Professor Dr Herbert George Flaxman Spurrell (1877–1918) was a British physician and zoologist who collected in Ghana and Colombia. He wrote *Modern Man and his Forerunners, a short study of the human species living and extinct* (1917). He served in the Royal Army Medical Corps in the First World War as a captain. He died in Egypt of pneumonia. Three mammals, a bird and three reptiles are named after him.

Stadelman

Stadelman's Coqui *Craugastor stadelmani* **Schmidt**, 1936

Raymond Edward Stadelman (1907–1991) was the curator of a serpentarium at Tela, Honduras (1930s). He may also have been an agronomist since it was probably the same Stadelman who wrote the report *Maize cultivation in Northwestern Guatemala* (1940). Two reptiles are named after him.

Stafford

Stump-toed Frog sp. *Stumpffia staffordi* **Köhler** et al., 2010

Dr Peter Stafford (1961–2009) was a botanist and herpetologist at the BMNH, London (1983–2009). His master's degree (2003) in biodiversity conservation was awarded by Kent University, Canterbury and his doctorate (2006) by the School of Life Sciences, Oxford Brookes University. He was closely involved with the establishment and running of the Natural History Museum's field station in Belize, and co-wrote *A Guide to the Reptiles of Belize* (2000). He died of cancer.

Stan Lai

Stan Lai's Toad. *Rhinella stanlaii* **Lötters** & **Köhler**, 2000

Stan Lai (b.1969) was (2000) an investment banker with Salomon Brothers on Wall Street. After a stint at Bank of America Merrill Lynch, he became (2009) a Managing Director at Moelis & Company. His wife, Lara Lai, gave money to BIOPAT to have this toad named after Stan as a birthday present to him.

Stanley

Stanley's Warty Frog *Callulina stanleyi* Loader et al., 2010

William 'Bill' T. Stanley of the Field Museum, Chicago, is primarily a mammalogist. He has made numerous contributions to understanding the biological diversity of the Eastern Arc Mountains of Kenya and Tanzania. Humboldt State University awarded his bachelor's degree (1986) and his master's (1991).

Starrett

Starrett's Treefrog *Isthmohyla tica* Starrett, 1966

Limon Giant Glass Frog *Sachatamia ilex* **Savage**, 1967 [Syn. *Centrolene ilex*]

Starrett's Glass Frog *Hyalinobatrachium vireovittatum* Starrett & Savage, 1973

Dr Priscilla 'Holly' Latham Hollister Starrett (1929–1997) was an American herpetologist, at one time married to Dr Andrew Starrett (see **Andi**), on the staff of the University of Southern California, Los Angeles (1969–1992). The University of Connecticut awarded her bachelor's degree (1951) and her doctorate was from the University of Michigan (1969). She co-wrote *Notes on a collection of centrolenid frogs from the Colombian Chocó* (1980).

The binomial of the giant glass frog, *ilex*, is Latin for the 'holly' tree and refers to the name by which Dr Starrett was familiarly known.

Stauffer, J

Stauffer's Treefrog *Scinax staufferi* **Cope**, 1865

Jacob Stauffer (1808–1880) was a storekeeper, painter, printer, pharmacist, lawyer, botanist, biologist and photographer. He lived in Richland, Pennsylvania (1839–1858) and from 1858 at Lancaster, being librarian of Lancaster Athenaeum. He was a regular correspondent of Cope's. He wrote, published and printed *On the Parasitism of Comandra and Gerardia* (1853).

Stauffer, JH & R

Jondachi Treefrog *Hyloscirtus staufferorum* **Duellman** & **Coloma**, 1993

John H. Stauffer (a lawyer who graduated (1979) from the University of Kansas Law School) and Ruth Stauffer of Topeka, Kansas, were honoured in the binomial as they '...provided generous financial support to the Neotropical Biological Diversity Program of the Museum of Natural History at the University of Kansas.'

Stawiarsky

Tree Toad sp. *Dendrophryniscus stawiarskyi* **Izecksohn**, 1994

Professor Victor Stawiarsky taught biology at a Buenos Aires College (1929) and was at the Museu Nacional, Rio de Janeiro, Brazil (1930–1945). He collected the holotype (1948).

Stebbins

Tehachapi Slender Salamander *Batrachoseps stebbinsi* **Brame** & Murray, 1968

Sonoran Tiger Salamander *Ambystoma mavortium stebbinsi* Lowe, 1954

Dr Robert Cyril Stebbins (1915–2000) was an American zoologist. His master's degree (1941) and doctorate (1943) were both awarded by the University of California, Los Angeles, where he worked (1944–1958), becoming Associate Professor. He also worked in the California Museum of Vertebrate Zoology as Assistant Curator (1945), then Curator, Herpetology (1948). He became Professor of Zoology at the Museum of Vertebrate Zoology, University of California, Berkeley (1958); then Emeritus Professor after retiring (1978). He travelled in many parts of the world and was on the University of California's Scientific Expedition to Galápagos Islands (1964). He wrote *Amphibians and Reptiles of Western North America* (1954).

Stefani, L

Treefrog genus *Stefania* **Rivero**, 1968

Luis Stefani Raffucci was Vice-Chancellor Emeritus, University of Puerto Rico, Mayaguez. Rivero wrote: 'The genus is named after don Luis Stefani....In Mr. Stefani's stimulating attitude, science found the greatest admirer, the scientist a willing collaborator and the friend a helping hand, full of promises and hopes.'

Stefani, R

Stefani's Salamander *Speleomantes flavus* Stefani, 1969

[Alt. Monte Albo Cave Salamander]

Dr Renzo Stefani (1922–2007) was an Italian zoologist. He graduated in natural sciences at the Università degli Studi di Padova (1950). He joined the Institute of Zoology, Università degli Studi di Cagliari, as a lecturer (1952), and was Professor of Comparative Anatomy (1966–1994). He was an internationally famous entomologist, being a member of both Senckenberg Naturforsch Gesellschaft, Frankfurt, and of the New York Academy of Sciences. He wrote *Contributo alla conoscenza dei formicidi cavernicoli della Sardegna* (1970).

Steindachner

Steindachner's Dwarf Frog *Physalaemus ephippifer* Steindachner, 1864

Steindachner's Robber Frog *Ischnocnema guentheri* Steindachner, 1864

[Alt. Guenther's Frog]

Steindachner's Reed Frog *Hyperolius steindachneri* **Bocage**, 1866

Steindachner's Frog *Limnodynastes salmini* Steindachner, 1867

[Alt. Salmin's Frog, Salmon-striped Frog]

Steindachner's Frog *Ptychadena porosissima* Steindachner, 1867

[Alt. Grassland Ridged Frog, Striped Grass Frog]

Steindachner's Toad *Amietophrynus steindachneri* Pfeffer, 1893

Steindachner's River Frog *Phrynobatrachus steindachneri* **Nieden**, 1910

Franz Steindachner (1834–1919) was an Austrian herpetologist and ichthyologist. He originally planned to become a lawyer but became interested in fossil fishes. He worked at the Natural History Museum, Vienna (1860–1919), becoming a Curator (1861), Head of the Zoology Department (1874), and Director of the Museum (1898). Unlike many curators he travelled actively and collected in the Americas, including the Galápagos Islands and the Middle East. He wrote the herpetological sections of the report on the results of the circumnavigation by the Austrian frigate 'Novara'. Ten reptiles and two birds are named after him.

[Note that two different species have had the name 'Steindachner's Frog' applied to them.]

Steiner

Steiner's Shrub Frog *Pseudophilautus steineri* Meegaskumbura & Manamendra-Arachchi, 2005

Achim Steiner (b.1961) is a Brazilian-born, English-educated German environmentalist. Oxford University awarded his bachelor's degree and the University of London his master's. He worked for the International Union for Conservation of Nature and Natural Resources (IUCN) in Washington and Asia. He was Secretary-General of the World Commission on Dams, Cape Town (1998–2001). He was Director-General of IUCN (2001–2006). Since 2006 he has been Executive Director of the United Nations Environment Programme.

Steinitz

Treefrog sp. *Hyla heinzsteinitzi* Grach, Plesser & Werner, 2007

Dr Heinz Steinitz (1909–1971) was an Israeli marine biologist, ichthyologist and physician. His doctorate (1938) was the first in zoology awarded by the Hebrew University, Jerusalem, where he later became Associate Professor (1957) and Professor (1968). He took part in a number of expeditions, including the Lake Huleh Expeditions (1938–1940).

Stejneger

Stejneger's Caecilian *Grandisonia alternans* Stejneger, 1893

Stejneger's Robber Frog *Craugastor stejnegerianus* **Cope**, 1893

Cave Salamander sp. *Spelerpes stejnegeri* Eigenmann, 1901 NCR

[JrSYN *Eurycea longicauda*]

Stejneger's Caribbean Toad *Peltophryne longinasus* Stejneger, 1905

[Alt. Cuban Long-nosed Toad]

Stejneger's Narrow-mouthed Toad *Micryletta steinegeri* **Boulenger**, 1909

[Alt. Stejneger's Paddy Frog]

Salamander sp. *Ambystoma stejnegeri* **Ruthven**, 1912 NCR

[JrSYN *Ambystoma macrodactylum*]

Mindanao Horned Frog *Megophrys stejnegeri* **Taylor**, 1920

Stejneger's Oriental Salamander *Hynobius stejnegeri* **Dunn**, 1923

[Alt. Amber-colored Salamander]

Stejneger's Button Frog *Cycloramphus stejnegeri* **Noble**, 1924

Stejneger's Toad *Bufo stejnegeri* **Schmidt**, 1931

Dr Leonhard Hess Stejneger (1851–1943) was a herpetologist and ornithologist who grew up in Bergen, Norway (then belonging to Sweden). He studied philosophy and law at the University of Christiana (Oslo). He went on an expedition (1881) to the USA, to Bering Island, Kamchatka and Commander Island in the North Pacific, where he joined the Smithsonian collecting birds as Assistant Curator of Birds. He became Curator of Reptiles (1889), and was Head Curator of Biology (until 1911). He was laden with honours, such as Permanent Commander of the International Zoological Congress. He published extensively (1873–1936). Two mammals, sixteen reptiles and eleven birds are named after him.

Stellae

Sinimbu Treefrog *Hypsiboas stellae* Kwet, 2008

The treefrog is named, via the BIOPAT initiative, after Stellae Limited 'in recognition of its contribution to conserving biodiversity by aiding the reduction of greenhouse gases ...'

Stelzner

Redbelly Toad *Melanophryniscus stelzneri* Weyenbergh, 1875

Professor Dr Alfredo Wilhelm Stelzner (1840–1895) was a German (Saxon) scientist. He studied palaeontology and geology at the Dresden Polytechnic. In 1859 he studied mining at the Academy of Mining in Freiberg, being awarded his PhD in 1864. He became curator (1866–1870) there of the mineral and geological collections, as well as director of the library and education. He was invited (1870) by the German Director of the University of Córdoba, Argentina to head the newly formed National Academy of Science. He arrived in Argentina in 1871 with several notable German academics, to take up the position of Professor of Mineralogy. During his brief time there he made two long trips into the uncharted North and West of the country, and into Bolivia. However, he and his fellows returned to Germany in 1874 in dispute with the head of the institution. He became Professor at the Academy of Mining in Freiberg and there published on Argentina's geology.

Stéphan

Stéphan's Water Frog *Telmatobius stephani* **Laurent**, 1973

Dr Stéphan Roland Pierre Halloy is a Belgian botanist and alpine ecologist. He was brought up in Africa, USA and Argentina, and works for the New Zealand Institute for Crop and Food Research, Dunedin. After being a post-doctoral fellow at Otago University, New Zealand, he became an associate professor in phytogeography at the National University of Tucumán, Argentina. He and his brother Christian (q.v.), when they were teenagers, used to help their stepfather, Raimond Laurent, collect specimens; this particular species was discovered by them in the early 1970s in Catamarca province, Argentina. He co-wrote *Upward range extension of Andean anurans and chytridiomycosis to extreme elevations in response to tropical deglaciation.*

Stephen, G

Far Eastern Treefrog *Hyla stepheni* **Boulenger**, 1888 NCR

[JrSYN of *Hyla japonica*; possibly valid as a subspecies]

George Stephen of the Royal Navy collected the holotype (alive) at Port Hamilton, Korea, and presented it to the London Zoo.

Stephen, RE

Stephen's Rocket Frog *Anomaloglossus stepheni* Martins, 1989

[Syn. *Colostethus stepheni*]

Stephen R. Edwards recognised that the present species was different from one he had described (1974). (See **Edwards** for biographical detail)

Sternfeld

Sternfeld's River Frog *Phrynobatrachus sternfeldi* **Ahl**, 1924

Dr Richard Sternfeld (1884–1943) was a German herpetologist and zoologist. He was in charge of the herpetological section, Senckenberg Museum, Frankfurt, but was dismissed (1920). He wrote *Die Reptilien und Amphibien Mitteleuropas* (1912). He was murdered in the Holocaust. Three reptiles are named after him.

Stévart

Stévart's Egg Frog *Leptodactylodon stevarti* Rödel & Pauwels, 2003

Dr Tariq Stévart (b.1974) is a Belgian botanist who is Assistant Curator, Africa & Madagascar Department, Missouri Botanical Garden. His doctorate is from the Université Libre de Bruxelles. He and Pauwels collected the holotype together (2001).

Stewart

Stewart's Puddle Frog *Phrynobatrachus stewartae* **Poynton** & **Broadley**, 1985

Dr Margaret 'Meg' M. Stewart (1927–2006) was a biologist and herpetologist at the University of Albany (1956–1997), retiring as Professor Emeritus. She was awarded an honorary doctorate by the University of Mayaguez, Puerto Rico (1996). She wrote *Amphibians of Malawi* (1967).

Steyermark

Steyermark's Pancake Frog *Otophryne steyermarki* **Rivero**, 1968

Demonic Poison Frog *Minyobates steyermarki* Rivero, 1971

Dr Julian Alfred Steyermark (1909–1988) was an American botanist, explorer, taxonomist and plant collector who was Curator of the Missouri Botanical Gardens. His major works covered the flora of Missouri, Guatemala and Venezuela. His bachelor's degree (1929), his master's (1930) and his doctorate (1933) were all awarded by Washington University, St Louis. During his life he collected more than 132,000 plants in 26 countries, thus earning him an entry in the Guinness Book of Records! A reptile is named after him.

Sting

Sting's Treefrog *Dendropsophus stingi* **Kaplan**, 1994

[Alt. Kaplan's Garagoa Treefrog]

Sting [aka Gordon Matthew Thomas Sumner] (b.1951) is a British rock musician and founder member of the popular rock band 'The Police'. Winner of 16 Grammy awards, he has been nominated for an Oscar in best song category. After jobs as a bus conductor, construction labourer and tax officer, Gordon attended Northern Counties College of Education (later part of Northumbria University) (1971–1974), qualifying as a teacher. He taught at St Paul's First School, Cramlington (1975–1976). He was honoured in the treefrog's binomial in recognition of his 'commitment and efforts to save the rainforest'. He once performed wearing a black and yellow sweater with hooped stripes, which bandleader Gordon Solomon noted made him look like a bumblebee; thus Sumner became 'Sting'.

Stoliczka

Stoliczka's Frog *Limnonectes hascheanus* Stoliczka, 1870 [Alt. Hasche's Frog, Side-spotted Frog]

Dr Ferdinand Stoliczka (1838–1874) was a palaeontologist and zoologist born in Moravia (now in the Czech Republic). He was educated at Prague and the University of Vienna, after which he joined the Austrian Geological Survey (1861). A year later he joined the Geological Survey of India, and was assigned the task of documenting the Cretaceous fossils of southern India. He was on the Second Yarkand Mission (1873–1874) but died of spinal meningitis when 'returning loaded with the spoils and notes of nearly a year's research in one of the least-known parts of Central Asia'. He wrote many papers on Indian zoology (1863–1872). Among other taxa named after him are eight birds, nine reptiles and three mammals.

Stolzmann

Stolzmann's Horned Frog *Ceratophrys stolzmanni* **Steindachner**, 1882

Jean Stanislas Stolzmann (or Sztolcman) (1854–1928) was a Polish zoologist who worked as a collector in Peru (1875–1881). Oldfield Thomas described him as '...one of the best-known and most successful of Peruvian collectors...' After returning to Poland he became Director of the Branicki Museum. Among his publications is *On the Ornithological researches of M Jean Kalinowski in Central Peru* (1896). Nine birds, a reptile and a mammal are named after him.

Storm

Siskiyou Mountains Salamander *Plethodon stormi* Highton & **Brame**, 1965

Dr Robert Macleod Storm (b.1918) is Professor Emeritus of Ecology and Natural History, Oregon State University, where he worked (1939–1984, broken by service as a doctor in WW2). He is Canadian-born and US-educated. Both his master's degree (1941) and his doctorate (1948) were awarded by Oregon State College. He co-wrote *Amphibians and Reptiles of the Pacific Northwest* (1983).

Strachan

Pakistan Bullfrog *Sphaerotheca strachani* Murray, 1884 [Syn. *Tomopterna strachani*]

James Strachan was an engineer and architect who designed many important buildings in Karachi, including the Empress Market, in the last two decades of the nineteenth century. He sent the bullfrog holotype to Murray.

Straelen

Nagero Grassland Frog *Ptychadena straeleni* **Inger**, 1968

Professor Victor Émile van Straelen (1889–1964) was a Belgian geologist and naturalist. He became Director, Belgian Royal Institute of Natural Science, Brussels (1925). He travelled widely, especially in Indonesia and Belgian Congo (DRC) and became (1933) President, Institute of National Parks of the Congo. He was the first President (1959–1964) of the Darwin Foundation.

Strauch

Strauch's Spotted Newt *Neurergus strauchii* **Steindachner**, 1887

Dr Alexander Alexandrovich Strauch (1832–1893) was a Russian-German zoologist. He qualified as a physician in Estonia (1859), but was also a naturalist and herpetologist and his doctorate was in zoology. He was in Algeria (1859–1860). He became Director and Professor of the Zoological Museum, St Petersburg (1879), and Permanent Secretary of the Library of the Academy of Science (1890). He wrote *Essai d'une exploration de L'Algérie* (1862). Seven reptiles and a bird are named after him.

Strecker

Strecker's Chorus Frog *Pseudacris streckeri* Wright & **Wright**, 1933

John Kern Strecker (1875–1933) was a naturalist and herpetologist. He was also interested in conchology and folklore, being President of the American Folklore Society. He is regarded as the father of Texan herpetology, starting field surveys there (1895). He was Head Librarian, Baylor University, Waco, Texas (1919–1933) and Curator of the Museum (1903), which was renamed the Strecker Museum (1940). The John K Strecker Herpetological Society was formed (1964), but it lasted just two years. He wrote *Reptiles and Amphibians of Texas* (1915). A mammal and two reptiles are named after him.

Streiffeler

Cross Frog sp. *Oreophryne streiffeleri* **Günther** & **Richards**, 2011

Heinz Streiffeler (d.2011) was one of the senior author's oldest friends. He died when the paper describing the frog was nearly complete. He was a naturalist, actively engaged in conservation in the Pankow district of Berlin.

Strinati

Strinati's Cave Salamander *Speleomantes strinatii* Aellen, 1958

Dr Pierre Strinati was a Swiss biospeleologist, zoologist and herpetologist. He worked frequently with Aellen and they jointly wrote part of a report on a Swiss scientific expedition to Morocco (1950), and a guide to European caves. He wrote *Faune cavernicole de la Suisse* (1966), which was the thesis for his doctorate (1965), awarded by the Faculté des Sciences de Toulouse, France.

Stuart, LC

Stuart's Robber Frog *Craugastor stuarti* **Lynch**, 1967
Stuart's Worm Salamander *Oedipina stuarti* **Brame**, 1968
Stuart's Mushroomtongue Salamander *Bolitoglossa stuarti* Brame & **Wake**, 1969
Stuart's Moss Salamander *Nototriton stuarti* Wake & **Campbell**, 2000

Dr Laurence Cooper Stuart (1907–1983) of the Museum of Zoology, University of Michigan was an expert on Guatemalan herpetofauna. He collected (1938) in the Yucatán Peninsula. He co-wrote *A new* Hyla *from Guatemala.* (1934). Eight reptiles and a bird are named after him. (See **Pancho**)

Stuart, M

Stuart's Toad *Duttaphrynus stuarti* **Smith**, 1929

Dr Murray Stuart worked for the Geological Survey of India. He collected the holotype and made a small collection of fishes in the same area of northern Burma (Myanmar). He was also interested in fossil fuels and wrote *The Geology of Oil, Oil-Shale and Coal* (1926).

Stuart, SN

Stuart's Shrub Frog *Pseudophilautus stuarti* Meegaskumbura & Manamendra-Arachchi, 2005
Thin-toed Frog sp. *Leptodactylus simonstuarti* Angulo & Ichochea, 2010

Dr Simon Nicholas Stuart (b.1956) is a British conservation biologist. He was Acting Director General of IUCN (2000–2001), and Head of IUCN's Biodiversity Assessment Unit (2001–2006). He was elected Chair of the Species Survival Commission (2008).

Studer

Treefrog sp. *Dendropsophus studerae* Carvalho-e-Silva, Carvalho-e-Silva & **Izecksohn**, 2003

Anita Studer is a Swiss accountant, ornithologist, conservationist and ecologist who first arrived in Brazil (1976). She went to Quebrangulo (1980), where the treefrog holotype was collected, and decided that her mission was to save the rainforest. To that end she founded 'Nordesta' (1985), an association for reforestation, which since its inception has planted over three million trees. She runs the association from Geneva, where for most of the year she earns her living as an accountant.

Stuhlmann

Stuhlmann's Banana Frog *Afrixalus stuhlmanni* Pfeffer, 1893
[Alt. Stuhlmann's Spiny Reed Frog]

Professor Dr Franz Stuhlmann (1863–1928) was a zoologist, naturalist anthropologist and collector with the authorities in German East Africa (1888–1900). Stuhlmann travelled with Emin Pasha, after whose murder he and other survivors of a smallpox outbreak returned from Lake Albert with a large collection of specimens, plus cartographic material from which the first comprehensive map of German East Africa was made. About 1,800 years ago Ptolemy's map showed 'The Mountains of the Moon' (now Ruwenzori Mountains), snow-covered peaks that fed the Nile; Stuhlmann was the first European to confirm their existence. He wrote *Dr. Franz Stuhlmann: Mit Emin Pasha ins Herz von Africa* (1894). Five birds, a reptile and two mammals are named after him.

Stumpff

Stump-toed Frog genus *Stumpffia* **Böttger**, 1881

Anton Stumpff was a traveller and collector who assisted Ebenau (q.v.) on his visit to Nossi-Bé Island, Madagascar. Stumpff seems to have been the head of a factory on that island (1874–1890), but details of his life are hard to find. Böttger published *Diagnoses Reptilium et Batrachiorum novorum ab ill. Antonio Stumpff in insula Nossi Bé Madagascariensi lectorum* (1881). Four reptiles are named after him.

Sudell

Sudell's Frog *Neobatrachus sudelli* **Lamb**, 1911
[Alt. Spotted Burrowing Frog, Eastern Metal-eyed Frog]

Miss Jane Ann Sudell (b.1880) of Warwick, Queensland, collected the frog holotype. She was the younger half-sister of the describer, Joseph Lamb. (The binomial ought to be in the feminine, *sudellae*.)

Sumichrast

Sumichrast's Treefrog *Exerodonta sumichrasti* **Brocchi**, 1879

Adrien Jean Louis François de Sumichrast (1828–1882) was a naturalist who travelled with Henri de Saussure in the West Indies, USA and Mexico (1854–1856). Saussure took their considerable collection back to Geneva (1856), but Sumichrast stayed on in Mexico until his death. He used various parts of his name, depending upon whom he was dealing with. He may have taken Mexican nationality, though Mexican sources record him as a Swiss, called A. L. François Sumichrast. To the Smithsonian, which employed him for an expedition in Mexico, he was Professor François Sumichrast, a Frenchman. Nine birds, three reptiles and three mammals are named after him.

Summers

Summers' Poison Frog *Ranitomeya summersi* Brown et al., 2008

Dr Kyle Summers (b.1958) is an evolutionary biologist, and a professor at East Carolina University. His bachelor's degree was awarded by the University of California, Santa Cruz, and his doctorate by the University of Michigan. He has written or co-written a number of papers such as *A key ecological trait drives the evolution of biparental care and monogamy in a Peruvian poison frog* (2010).

Sung

Sung's Metacarpal-tubercled Toad *Leptolalax sungi* Lathrop et al.,1998

See **Cao Van Sung**

Supachai

Supachai's Caecilian *Ichthyophis supachaii* **Taylor**, 1960
[Alt. Hillstream Caecilian]

Professor Dr Supachai Vanijvadhana was secretary-general, Chulalongkorn University, Bangkok. He was interested in Thai fauna and started the train of events (and helped with arranging funding, through Fulbright grants) that led to Taylor's reptile-collecting expeditions in Thailand (1957–1958 & 1959–1960).

Susagua

Robber Frog sp. *Pristimantis susaguae* **Rueda-Almonacis, Lynch** & Galvis-Peñuela, 2003

This frog is named after the Rio Susagua, Colombia.

Susan

Susan's Forest Treefrog *Leptopelis susanae* **Largen**, 1977

Susan is the wife of the describer, Malcolm J. Largen. In 2001 he wrote a paper cataloguing all of the then known amphibians of Ethiopia, of which he said: 'this paper is dedicated with love and gratitude to my wife, Susan, without whose support and encouragement during the past 30 years it would perhaps never have been written.'

Susatama

Susatama's Glass Frog *Rulyrana susatamai* **Ruiz-Carranza** & **Lynch**, 1995
[Syn. *Cochranella susatamai*]

Gustavo Susatama has been a member of a number of expeditions with Lynch. He appears to have been something of a general factotum, and was described in 1988 as the 'expedition driver'.

Swan

Dharan Bullfrog *Sphaerotheca swani* **Myers** & Leviton, 1956
[Possibly a junior synonym of *Sphaerotheca breviceps*]

Professor Lawrence 'Larry' Wesley Swan (1922–1999) was a biologist, all-round naturalist and mountaineer, the son of Methodist missionaries in Darjeeling, who became a professor at the Department of Biology, San Francisco State College, California. His somewhat odd sense of humour led to his declaring himself to be the Raja of his own autonomous state: the Kingdom of Cooch Nahai (meaning 'Kingdom of Absolutely Nothing'). This 'state' took the extinct Dodo as its national symbol. Swan was on the 1954 California Himalayan Expedition to Makalu, Eastern Nepal, and co-wrote *The herpetology of Nepal: A history, check list, and zoogeographical analysis of the herpetofauna* (1962).

Swanhild

Southern Highlands Rainforest Frog *Albericus swanhildae* **Menzies**, 1999

In the Volsunga saga, the main source material for Wagner's *Der Ring des Nibelungen*, Svanhild was the beautiful daughter of Sigurd and Gudrun. Several frogs of the genus *Albericus* have been fancifully named after Wagnerian characters.

Swanson

Dale Swanson's Poison Frog *Ranitomeya daleswansoni* **Rueda-Almonacid** et al., 2006

Doris Swanson's Poison Frog *Ranitomeya dorisswansoni* Rueda-Almonacid et al., 2006

Dale Swanson (1927–2003) was an American conservationist and a generous contributor to the cause. His wife, Doris Swanson (d. 2010), was also a committed conservationist. He left school (1943) to work in Kaiser Corporation's shipyards and served in the US Navy in the Second World War. He took an engineering degree at Montana State College (1950). He met his wife, a nursing student, at the same college (1948). He invented a method of filtering impurities from molten aluminium (1954) and founded Pyrotek, a company with plants in

20 countries. The couple became millionaires and made a huge gift of $3.5million to their alma mater.

Swift

Microhylid frog sp. *Paedophryne swiftorum* Rittmeyer et al., 2012

John Swift and his family made generous contributions that enabled the establishment of the Kamiali Biological Station, Papua New Guinea, where the species was collected.

Swinhoe

Swinhoe's Brown Frog *Odorrana swinhoana* **Boulenger**, 1903

Robert Swinhoe (1836–1877) was born in Calcutta, India, and educated at the University of London (1852). He joined the China Consular Corps (1854). Before leaving for Hong Kong he deposited a small collection of British birds' nests and eggs with the British Museum. In China as a diplomat, he had a great opportunity as a naturalist to explore a vast area never previously open to any collector. He discovered new species at the rate of about one per month throughout 19 years. Although the majority of his discoveries were birds, he is also associated with many other Chinese taxa. He first returned to London (1862), taking part of his vast collection of specimens to meetings of the Zoological Societies in England, France and Holland. He was surprised at having to allow someone else to name the 200-plus new bird species he had discovered, but related: 'I have been blamed by some naturalists for allowing Mr. Gould to reap the fruits of my labours, in having the privilege of describing most of my novelties. I must briefly state, in explanation, that I returned to England elated with the fine new species I had discovered, and was particularly anxious that they should comprise one entire part of Mr. Gould's fine work on the Birds of Asia, still in progress. On an interview with Mr. Gould, I found that the only way to achieve this was to consent to his describing the entire series to be figured, as he would include none in the part but novelties, which he should himself name and describe. I somewhat reluctantly complied; but as he has done me the honour to name the most important species after me, I suppose I have no right to complain...' Four mammals, four reptiles and twenty-three birds are named after him.

Swynnerton

Swynnerton's Reed Frog *Hyperolius swynnertoni* **Fitzsimons**, 1941

Charles Francis Massy Swynnerton (1877–1938) was principally an entomologist. He was born in India and worked in Africa, becoming the first game warden in Tanganyika (now Tanzania) (1919–1929). He was head of tsetse research in East Africa (1929–1938). Among his publications is *On the birds of Gazaland, Southern Rhodesia* (1907). He was killed in an air crash. A reptile, a mammal and four birds are named after him.

Symington

Symington's Robber Frog *Eleutherodactylus symingtoni* **Schwartz**, 1957

Kenneth A. Symington was a British-Cuban chemical engineer and civic leader. He was the last National Commissioner for the Boy Scouts in Cuba. He was also a speleologist. He co-wrote *Caverns of St Tomas* (1955), which is the site where the frog holotype was acquired.

T

Tabasara

Tabasara Rain Frog *Craugastor tabasarae* **Savage** et al., 2004

Named after the Serranía de Tabasará, a mountain range in Panama.

Tagban

Litter Frog sp. *Leptobrachium tagbanorum* Brown et al., 2010

The Tagban are an indigenous tribe living on Palawan Island, Philippines.

Tago

Tago's Brown Frog *Rana tagoi* Okada, 1928

Katsuya Tago was a Japanese zoologist and herpetologist who was active in the first half of the twentieth century, and worked in the Zoology Department, Tokyo University Museum. He wrote *Notes on the habits and life history of* Megalobatrachus japonicus (1927).

Takeda

Hokuriku Salamander *Hynobius takedai* **Matsui** & Miyazaki, 1984

Toshio Takeda is a former headmaster, Inoyama Elementary School, Hakui-shi, Japan. He collected the holotype (1971) while cleaning the schoolyard drains. He has left teaching and devotes his time to helping prevent the salamander from becoming extinct.

Talada

Cuban Spotted Toad *Peltophryne taladai* **Schwartz**, 1960

James A. Talada '....greatly assisted the author in the field.' Beyond that dedication in Schwartz's description, we can find nothing about him.

Talamanca

Almirante Robber Frog *Craugastor talamancae* **Dunn**, 1931

Talamanca Glass Frog *Hyalinobatrachium talamancae* **Taylor**, 1952

Named for the Cordillera de Talamanca, a mountain range in Costa Rica and Panama where these frogs occur.

Talberg

Kaieteur Tepui Treefrog *Tepuihyla talbergae* **Duellman** & Yoshpa, 1996

Alisa Talberg is described as 'an enthusiastic young herpetologist and friend of the collector'. We know nothing more about her.

Tamsitt

San Adolfo Robber Frog *Pristimantis tamsitti* **Cochran** & **Goin**, 1970

Dr James R. Tamsitt is a herpetologist who was Professor of Biology, University of the Andes, Bogotá (1960). He lectured at the Institute of Natural Sciences, Colombia (1982). Earlier he was a member of the faculty at the Department of Zoology, University of Manitoba, Winnipeg, Canada (1958). He co-wrote *Records and observations on Colombian reptiles* (1963).

Tandy

Tandy's Smalltongue Toad *Werneria tandyi* **Amiet**, 1972

Tandy's Sand Frog *Tomopterna tandyi* **Channing** & Bogart, 1996

Dr Robert Mills Tandy is a biologist, herpetologist and wildlife photographer at the Department of Zoology, University of Texas, Austin. The University awarded his master's degree (1971) and doctorate (1972). His doctoral thesis was *The evolution of African Bufo*. Much of his work is about African amphibians. His wife Jocelyn is also a herpetologist.

Tanner, J & L

Tanners' Reed Frog *Hyperolius tanneri* **Schiøtz**, 1982

Tanner's Squeaker *Arthroleptis tanneri* **Grandison**, 1983 [Alt. Tanzania Screeching Frog]

John and Lucie Tanner owned a tea estate at Mazumbai, Tanzania. They donated part of it to the University of Dar es Salaam and it is now the Mazumbai Forest Reserve. The screeching frog is named after John alone, the reed frog after both of them – it has been suggested that the binomial be amended to the plural form *tannerorum*. The couple were noted for welcoming visiting zoologists and other collectors.

Tantant

Robber Frog sp. *Pristimantis tantanti* **Lehr**, Torres-Gastello & Suarez-Segovia, 2007

The binomial is derived from a name in the language of the Asháninka people of western Amazonia, and refers to the frog's call.

Tasan

Tasan Frog *Ingerana tasanae* **Smith**, 1921

Tasan (or Tha San) is the place in Thailand where the holotype was collected.

Tautz

Andes Frog sp. *Phrynopus tautzorum* **Lehr** & **Aguilar**, 2003

The Tautz family of Germany was honoured in recognition of their support of taxonomic research and nature conservation. They are patrons of BIOPAT.

Tatayo

Glass Frog sp. *Hyalinobatrachium tatayoi* **Castroviejo-Fisher**, **Ayarzagüena** & Vilà, 2007

Dr Santiago Castroviejo Bolibar (1946–2009) was a Spanish botanist who was Director of the Royal Botanic Gardens, Madrid (1984–1994). To his friends and relations he was known as 'Tatayo'. He was the senior describer's uncle and credited with part of the 'blame' for his dedication to taxonomy.

Taylor, EH

Oriental Tree Frog genus *Edwardtayloria* Marx, 1975 NCR
[In *Nyctixalus*]
Dicroglossid frog genus *Taylorana* **Dubois**, 1987 NCR
[Usually included in *Limnonectes*]
Taylor's Toad *Hypopachus cuneus* **Cope**, 1889 NCR
[Alt. Sheep Frog; JrSYN *Hypopachus variolosus*]
Taylor's Stream Frog *Craugastor vocalis* Taylor, 1940
Taylor's Barking Frog *Craugastor occidentalis* Taylor, 1941
Taylor's Pygmy Salamander *Thorius troglodytes* Taylor, 1941
Taylor's Yellow Treefrog *Dendropsophus sartori* **Smith**, 1951
Taylor's Worm Salamander *Oedipina taylori* **Stuart**, 1952
Peralta Frog *Lithobates taylori* Smith, 1959
Taylor's Tree Frog *Kurixalus bisacculus* Taylor, 1962
Taylor's Caecilian *Ichthyophis pseudangularis* Taylor, 1965
[Alt. Lesser Yellow-banded Caecilian]
Taylor's Robber Frog *Craugastor taylori* **Lynch**, 1966
Taylor's Ecuador Caecilian *Caecilia subterminalis* Taylor, 1968
Taylor's Glass Frog *Hyalinobatrachium taylori* **Goin**, 1968
Cerro Cituro Salamander *Bolitoglossa taylori* **Wake, Brame** & **Myers**, 1970
Tiny Taylor's Caecilian *Microcaecilia taylori* **Nussbaum** & **Hoogmoed**, 1979
Taylor's Salamander *Ambystoma taylori* Brandon, Maruhka & Rumph, 1982
Taylor's Forest Frog *Platymantis taylori* **Brown, Alcala** & **Diesmos**, 1999
Dicroglossid frog sp. *Limnonectes taylori* **Matsui** et al., 2010
Red-eyed Treefrog ssp. *Agalychnis callidryas taylori* Funkhouser, 1957

Dr Edward Harrison Taylor (1889–1978) was an American herpetologist. He went to teach in a Philippines village school (1912), returning briefly to his university, Kansas, to finish his master's. He was Chief, Division of Fisheries, Manila (1916–1920). He was Head of the Zoology Department, University of the Philippines (1923–1927) then worked at Kansas University (1927–1949), becoming full Professor (1934). He was an early broadcaster, giving a series of ten radio talks on herpetology (1932–1933). He took students on trips to Mexico (1937–1948), travelling in 'marginally reliable vehicles'; collecting took place at planned or unplanned stops. He collected after retirement in Costa Rica (1949), Thailand and Brazil, and wrote 19 papers on Philippines herpetology (1915–1928). *Platymantis taylori* was named 21 years after his death in recognition of the 11 years he devoted to the study of Philippine herpetofauna (1912–1923). Twenty-nine reptiles and a mammal are named after him. (See **Sartor**)

Taylor, RC

Richard's Salamander *Nototriton richardi* **Taylor**, 1949
Fringe-limbed Treefrog sp. *Hyla richardtaylori* Taylor, 1954 NCR
[JrSYN *Ecnomiohyla fimbrimembra*]

Richard Clark Taylor (1926–2002) was the son of the describer, Edward Harrison Taylor. He served in US Army Air Corps in the Second World War and graduated with a bachelor's degree (1949) from Haverford College. He worked as a labour relations manager in Massachusetts, New Jersey, Illinois and Nebraska. He was a talented carpenter, had a fine tenor voice, and was passionately devoted to fishing. He collected the holotype of the salamander (1947) during an expedition with his father.

Tecuan

Treefrog sp. *Charadrahyla tecuani* **Campbell,** Blancas-Hernández & Smith, 2009

This frog is not named after a person but after another animal: in the Náhuatl language 'tecuáni' refers to the jaguar (*Panthera onca*).

Tecun Uman

Cave Spikethumb Frog *Plectrohyla tecunumani* **Duellman** & **Campbell**, 1984

Tecun Uman (c.1500–1524) was the last king of the Quiché Maya people, in what is now Guatemala, where he is regarded as the national hero. He was killed in battle against the invading Spanish conquistadors.

Tellini

Tellini's Grass Frog *Ptychadena tellinii* **Peracca**, 1904
[Alt. Peracca's River Frog]

Professor Achille Tellini (1866–1938) was an Italian geologist, botanist and folklorist who collected in Eritrea (1900–1901). He was a speaker and promoter of Esperanto.

Tenchall

Tenchall's False Brook Salamander *Pseudoeurycea tenchalli* **Adler**, 1996
[Alt. Bearded Salamander]

We have no idea who first 'personalized' this vernacular name, but the etymology makes it clear that it is a word in the Nahuatl language and means 'chin' – referring to 'the enormous mental-gland cluster on the chin of mature males of this species'.

Terentjev

Terentjev's Frog *Pelophylax terentievi* Mezhzherin, 1992

Dr Pavel Victorovich Terentjev (1903–1970) was a herpetologist at the Russian Academy of Sciences (1934–1970), and was Professor of Zoology, University of Leningrad (St Petersburg). His initial studies were at Moscow State University where he acted as Director of the Zoo (1922–1926). He was in Leningrad throughout the Second World War and its siege (1941–1944) and played a major part in protecting collections. He co-wrote *Field Guide to the Reptiles and Amphibians of the USSR* (1936).

Ternetz

Chimbo Frog *Barycholos ternetzi* **Miranda-Ribeiro**, 1937
Ternetz's Swamp Frog *Pseudopaludicola ternetzi* Miranda-Ribeiro, 1937
[Alt. Goias Swamp Frog]

Dr Carl Ternetz (1870–1928) was a Swiss ichthyologist and naturalist who first collected near Basel (1892), and later in South America for more than 30 years from about 1895. Several fish species and a reptile are named after him.

Tetzlaff

Bomberai Rainforest Frog *Cophixalus tetzlaffi* **Günther**, 2003

Immo Tetzlaff of Zepernick, Germany is a friend of the describer and collected in New Guinea with him (2000).

Thampi

Thampi's Torrent Frog *Micrixalus thampii* **Pillai**, 1981
[Alt. Silent Valley Tropical Frog]

Shri P. H. Thampi led an expedition to the Silent Valley in Kerala, India, where the holotype was acquired.

Thelen

Thelen Madagascar Frog *Gephyromantis thelenae* **Glaw** & Vences, 1994

Martina Thelen, who is a biologist, and Wolfgang Thelen of Antananarivo, Madagascar showed great friendship and hospitality to the describers. Martina observed for them the development of embryo frogs from tadpoles they had taken back to the Thelens' house.

Theo

Climbing Frog sp. *Anodonthyla theoi* Vences et al., 2010

Theophilus 'Theo' Rajeriarison is a Malagasy experimental naturalist and nature guide working at Ranomafana National Park. His brother is Emile Rajeriarison (See **Emile**).

Theobald

Theobald's Ranid Frog *Hylarana tytleri* Theobald, 1868
[Alt. Dacca Frog]

William Theobald (1829–1908) was a malacologist and herpetologist who became Deputy Superintendent, Geological Survey of Bengal (1876). He worked in India (1847–1881), and published *Descriptive catalogue of the reptiles of British India* (1876). Despite his herpetological knowledge, he once picked up a king cobra by the tail, thinking it to be another species. He was also interested in Shakespeare and wrote *On the Authorship of the Sonnets Attributed to Shakespeare* (1896). A mammal and three reptiles are named after him.

Theuerkauf

Theuerkauf's Bush Frog *Raorchestes theuerkaufi* Zachariah et al., 2011

Wolfgang Theuerkauf is a German-born Indian botanist and naturalist who is Director, Gurukula Botanical Garden, Periya, Wayanad, Kerala. He first went to Kerala in 1970 and bought land there, subsequently founding the Garden (1981).

Thin Thin

Thin Thin's Stream Toad *Ansonia thinthinae* Wilkinson, Sellas & Vindum, 2012

Ms Daw Thin Thin was a dedicated member of the Myanmar Herpetological Survey Team, and greatly aided in the understanding of amphibian and reptile natural history in Myanmar. At the conclusion of the

Indawgyi Lake trip, Thin Thin with others designed and implemented a long-term amphibian and reptile-monitoring program for Hlawga Wildlife Park and at Alaungdaw Kathapa National Park. In 2003 she began the task of curating specimens collected by the survey at the newly established Herpetology Section of the Myanmar Biodiversity Museum.

Thoda

Thoda Bush Frog *Raorchestes thodai* Zachariah et al., 2011

Thoda is the name of an indigenous tribal community in the Nilgiris, Tamil Nadu, South India.

Thomas

Thomas' Robber Frog *Eleutherodactylus thomasi* **Schwartz**, 1959

Dr John Paul Richard Thomas (b.1938) is a zoologist and herpetologist who is Professor, Department of Biology, University of Puerto Rico. He gained his doctorate in zoology (1976) from the University of Louisiana. He collected amphibians and birds in Peru with Duellman for the Natural History Museum, University of Kansas (1971) and was a member of the Louisiana State University Museum of Zoology expedition to Peru (1974). He co-wrote *The Status of* Sphaerodactylus gilvitorques *Cope and of* Sphaerodactylus nigropunctatus *Gray (Sauria: Gekkonidae)*. Six reptiles are named after him.

Thomasset

Thomasset's Frog *Sooglossus thomasseti* **Boulenger**, 1909

H. P. Thomasset was a planter and naturalist who collected on Seychelles, Aldabra and other Indian Ocean islands in the first two decades of the twentieth century. He collected the frog holotype (1905) on the island of Mahé.

Thompson, FG

Thompson's Water Frog *Telmatobius thompsoni* **Wiens**, 1993

Andes Frog sp. *Phrynopus thompsoni* **Duellman**, 2000

Dr Fred Gilbert Thompson (b.1934) is Curator of Non-marine Malacology, Florida State Museum. He collected the holotypes of both these frog species in Peru (1972). Two reptiles are named after him.

Thompson, K

Thompson's Caecilian *Caecilia thompsoni* **Boulenger**, 1902

Mr Kay Thompson collected the holotype in Colombia, but Boulenger says nothing more about him.

Thomson

Thomson's Toothless Frog *Genyophryne thomsoni* **Boulenger**, 1890

Sir Basil Home Thomson KCB (1861–1939) was a British intelligence officer, police officer, prison governor, colonial administrator and writer. Educated at Hendon and Eton, he attended New College, Oxford for two terms but suffered bouts of depression and left, spending time in Iowa, USA as a farmer (1881–1882). Needing to be secure in order to marry he joined the Colonial Office (1883) and was posted to Fiji (1884). He was then posted to New Guinea but was invalided out owing to malaria and returned to London where he married (1890). He returned to Fiji with his wife, becoming Commissioner of Native Lands. He was given the job of establishing a British Protectorate there (1890) despite native opposition. He resigned from the Colonial Office (1893) and returned to England because of his wife's ill health. He took up writing, producing *South Sea Yarns* (1894), *The Diversions of a Prime Minister* (1894), and *The Indiscretions of Lady Asenath* (1898). He read for the Bar in the 1890s and was admitted (1896) but did not practise, instead taking a post as Deputy Governor of HM Prison Liverpool. He later served as Governor at Northampton, Cardiff, Dartmoor and Wormwood Scrubs, then becoming Secretary of the Prison Commission (1908–1913). He was appointed Assistant Commissioner of the Metropolitan Police (1913) as head of CID and head of the Secret Service Bureau when war broke out (1914). He became known as a 'spy catcher', which he documented in his memoirs *The Scene Changes*. Notably he interrogated Mata Hari. On a darker note he was foremost in the suppression of the Suffragettes, English Marxists and Irish republicans. He went on to become Director of Intelligence at the Home Office (1919–1921). In December 1925 Thomson was arrested in London's Hyde Park, and charged with 'committing an act in violation of public decency' with a young woman, Miss Thelma de Lava. Thomson rejected the charges, insisting that he was engaged in conversation with the woman for the purposes of research for a book he was writing on London vice. Nonetheless, he was found guilty of public indecency and fined £5. He obtained a single specimen of the frog on Sudest Island, Papua New Guinea. A bird is named after him.

Thornton

Thornton's Rocket Frog *Colostethus thorntoni* **Cochran** & **Goin**, 1970

Dr. Wilmot 'Bill' Arnold Thornton (1921–2008) was a zoologist whose bachelor's degree (1949) was from the University of Texas, followed by a master's (1950) and a doctorate (1954). His Colombian collections, made (early 1960s) when he was at the Universidad del Valle, greatly impressed the describers. He was (1963) Professor of Biology at Fort Lewis College, Durango, Colorado, and became (1964–1967) Assistant Professor, Zoonotic Ecology, Veterinary Pathology and Hygiene Department, University of Illinois. He was at the Biology Department, Angelo State University, Texas (1967–1983).

Thwaites

Thwaites' Shrub Frog *Pseudophilautus adspersus* **Günther**, 1872

George Henry Kendrick Thwaites (1811–1882) was an accountant who studied botany in his spare time. He became a lecturer in botany at the Schools of Pharmacy and the Bristol Medical School from 1847. He arrived in Ceylon (Sri Lanka) in 1849 and in the same year became the Superintendent, and then (1857) Director, of the Botanical Garden at Peradeniya, a post he held until his retirement (1880). He published the first flora of Ceylon in 1864 under the title *Enumeratio Plantarum Zeylaniae*. A reptile is named after him.

Tica

Tico Salamander *Bolitoglossa tica* Garcia-Paris, Parra-Olea & **Wake**, 2008

To quote the etymology: 'This species is named in honour of the Costa Rican people, who refer to themselves as 'ticos', in recognition of their leadership in habitat preservation and conservation.'

Tipton

Robber Frog sp. *Eleutherodactylus tiptoni* **Lynch**, 1964 NCR

[JrSYN *Pristimantis cerasinus*]

Vernon J. Tipton (1920–2001) was an entomologist who was Professor of Zoology, Brigham Young University, Utah. He was in the US Army during the Second World War. He collected the holotype in Panama (1961). At one time he was Chairman of the Education Committee of the Entomological Society of America. He was co-editor, with Wenzel (q.v.), of *Ectoparasites of Panama* (1966) and wrote *Introductory Entomology* (1973) among other books on parasites in the Americas.

Tlahcuiloh

Green-flecked Salamander *Pseudoeurycea tlahcuiloh* **Adler**, 1996

David M. Dennis is a photographer and artist, colleague and field companion of the describer, Kraig Adler, who chose the Nahuatl word 'tlahcuiloh'– meaning painter or artist – to symbolize his long-time friend.

Tlaloc

Treefrog genus *Tlalocohyla* **Faivovich** et al., 2005

Tlaloc's Leopard Frog *Lithobates tlaloci* Hillis & **Frost**, 1985

Tlaloc was an Olmec rain god.

Tlilicxitl

Black-footed Salamander *Pseudoeurycea tlilicxitl* Lara-Góngara, 2003

The original description contains no etymology and we have not been able to find any explanation of *tlilicxitl*. The name appears to be Aztec.

Toachi

Pichincha Rocket Frog *Hyloxalus toachi* **Coloma**, 1995

Toachi is a native name for the basin and river system in Ecuador where this frog is found.

Toft

Pachitea Robber Frog *Pristimantis toftae* **Duellman**, 1978

Dr Catherine A. Toft (b.1950) is Professor Emerita, Evolution and Ecology Department, University of California, Davis. Her bachelor's degree (1972) was awarded by Iowa State University and her doctorate (1977) by Princeton. She wrote *Feeding ecology of Panamanian litter anurans: Patterns in diet and foraging mode* (1981).

Tomamorum

Moss Salamander sp. *Nototriton tomamorum* Townsend et al., 2010

The binomial refers to the gods of the Tolupan people of Honduras, where the salamander originates.

Tomas

Tomas' Worm Salamander *Oedipina tomasi* **McCranie**, 2006

Tomás Manzanares Ruiz, one of the author's friends and field companions, collected the holotype. A reptile is also named after him.

Tonkawa

Jollyville Plateau Salamander *Eurycea tonkawae* Chippindale, Price, **Wiens** & Hillis, 2000

This salamander is named after the Tonkawa tribe of Native Americans, who inhabited central Texas until the 1850s, when they were relocated to Oklahoma. The Tonkawa made camps around the springs where the salamander occurs.

Tony

Cabo Cruz Frog *Eleutherodactylus tonyi* **Estrada** & **Hedges**, 1997

Dr Antonio 'Tony' R. Pérez Asso, an entomologist who specializes in millipedes, was on the Penn State Cuba herpetological expeditions of 1989, 1990 and 1994. He worked at the National Natural History Museum, Havana (1995). He wrote *Revisión del género Amphelictogon (Diplopoda: Polydesmida: Chelodesmidae) en Cuba* (1996).

Tornier

Ethiopian Frog genus *Tornierella* **Ahl**, 1924 NCR
[Usually included in *Paracassina*]
Tornier's Forest Toad *Nectophrynoides tornieri* **Roux**, 1906
[Alt. Usambara Viviparous Toad]
Digging Frog sp. *Kaloula tornieri* **Vogt**, 1913 NCR
[JrSYN *Kaloula borealis*]
Tornier's Frog *Litoria tornieri* **Nieden**, 1923
[Alt. Black-shinned Rocket Frog]
Madagascar Frog sp. *Guibemantis tornieri* Ahl, 1928
Tornier's Reed Frog *Hyperolius tornieri* Ahl, 1931

Dr Gustav Tornier (1858–1938) was a German zoologist, anatomist, palaeontologist and taxonomist. He received his doctorate from Heidelberg University (1892). He worked at the zoological museum of the Friedrich-Wilhelms-University, Berlin. He is best remembered (rather unfairly) for supporting the incorrect view that *Diplodocus* and other sauropod dinosaurs walked with a sprawling, lizard-like gait with widely splayed legs. Three reptiles are named after him.

Toulson

Reed Frog sp. *Hyperolius toulsonii* **Bocage**, 1868 NCR
[JrSYN *Hyperolius parallelus*]

A Toulson collected in Angola (1860s). Bocage described him as 'an amateur…former trader with Loanda'. A bird is also named after him.

Toumanoff

Toumanoff's Wart Frog *Rana toumanoffi* **Bourret**, 1941 NCR
[Jr SYN *Limnonectes dabanus*]

Dr Constantin Toumanoff (1903–1967) was a Russian-born biologist, bacteriologist and parasitologist who took French nationality (1933). His early education was at secondary school in Irkutsk, Siberia (1917–1919) and the University of Irkutsk (1919–1922). He was persecuted for not having a sufficiently proletarian background, left for Manchuria (1922), and moved to France (1923). He was awarded a bachelor's degree by the University of Paris (1926). He worked at the Institut Pasteur, Paris (1923–c.1950), first as Assistant, studying bees and their diseases (1926–1930). He was in Vietnam (1930–1946), becoming head of laboratory, Hanoi from 1934, and in overall charge of the Institut Pasteur, Indochina from 1936. He conducted a survey of malarial mosquitoes in Corsica (1947). His later career involved working with the Rockefeller Foundation and the WHO. He became senior French medical entomologist and insect pathologist (1953). He wrote *Les tiques (Ixodoidea) de l'Indochine* (1944).

Tournier

Tournier's Rocket Frog *Ptychadena tournieri* **Guibé** & **Lamotte,** 1955

Jean-Luc Tournier was Director of the Institut Français d'Afrique Noire, Abidjan, Ivory Coast (1944). He and Lamotte established Lamto Ecology Station, Ivory Coast (1961) – where 'Lam' stands for Lamotte and 'to' for Tournier.

Townsend

Townsend's Salamander *Parvimolge townsendi* **Dunn**, 1922

Prescott Townsend (1894–1973) was co-collector (with Dunn) of the holotype near Jalapa, Veracruz, Mexico (1921). He found fame as a gay activist long before it became safe, let alone fashionable. He was a Boston blue blood with conservative political views but was a cultural radical. He graduated from Harvard (1918) and began law school there, but left after one term. He was arrested on 29 January 1943 for participating in what was quaintly described as an 'unnatural and lascivious act' and was sentenced to eighteen months in the Massachusetts House of Corrections. Most of his family ostracized him because of his sexuality.

Tradouw

Tradouw's Toad *Capensibufo tradouwi* **Hewitt**, 1926
[Alt. Tradouw Mountain Toadlet]

The toad is named after (part of) the type locality, 'Swellendam Mountains and in Tradouw Pass at 3500–5500 feet altitude', Western Cape Province, South Africa.

Tran Kien

Trankien Frog *Odorrana trankieni* **Orlov**, Ngat & Ho, 2003

Tran Kien is a Vietnamese herpetologist who co-wrote *Species of Venomous Snakes in Vietnam* (1995).

Trapicheiro

Three-lined Snouted Treefrog *Scinax trapicheiroi* **Lutz** & **Lutz**, 1954

Trapicheiro is the name of a stream in Rio de Janeiro state, Brazil, where the holotype was collected.

Tremblay

Tremblay's Salamander *Ambystoma tremblayi* Comeua, 1943

[Originates from hybridization between *A. jeffersonianum* and *A. laterale*]

Dr Jean-Louis Tremblay (b.1906) was a Canadian biologist. He graduated in chemistry (1931) at Laval University, Canada and in biology at the University of Strasbourg, France (1934), later gaining a doctorate at Laval (1935). He was Professor of Biology, Faculty of Science and founded the Biology Department at Laval, retiring as Professor Emeritus (1974).

Troschel

Troschel's Treefrog *Hypsiboas calcaratus* Troschel, 1848

Dr Franz Hermann 'Fritz' Troschel (1810–1882) was a German malacologist, herpetologist and ichthyologist. The University of Berlin awarded his doctorate (1834). He became assistant to Martin Lichtenstein at the Humboldt Museum of Natural History, Berlin (1840–1849). He was later Professor of Zoology, University of Bonn (1849>). A number of other taxa are named after him including a reptile.

True

Coastal Tailed Frog *Ascaphus truei* **Stejneger**, 1899

Frederick William True (1858–1914) was Head Curator, Department of Biology, US National Museum. He graduated (1878) and worked for the US Government Fish Commission for a couple of years before returning to New York University to do his master's degree. He worked at the Smithsonian (1881–1914), first as Librarian, then (1883–1909) as Curator, Mammals and Assistant Secretary of the Smithsonian (1909–1914). He was an acknowledged expert on whales. He wrote *The useful aquatic reptiles and batrachians* (1884). Six mammals are named after him.

Trueb

Toad genus *Truebella* Graybeal & **Canatella** 1995

Trueb's Madagascar Treefrog *Madecassophryne truebae* **Guibé**, 1974

Trueb's Cochran Frog *Nymphargus truebae* **Duellman**, 1976

Trueb's Beaked Toad *Rhinella truebae* **Lynch** & Renifo, 1990

Trueb's Water Frog *Telmatobius truebae* **Wiens,** 1993

Robber Frog sp. *Pristimantis truebae* Lynch & Duellman, 1997

Dr Linda Trueb (b.1942) is a herpetologist at the Department of Ecology and Evolutionary Biology, University of Kansas, where she gained her doctorate (1968), and is Professor and Curator-in-Charge, herpetology collection. She specializes in the study of amphibians. She co-wrote *Biology of the Amphibians* (1985). She is married to W. E. Duellman (q.v.). (See also **Linda**)

Tschenk

Tschenk's Madagascar Frog *Gephyromantis tschenki* **Glaw** & Vences, 2001

[Syn. *Mantidactylus tschenki*]

Michael Tschenk was honoured 'in recognition of his generous support to the biosystematic research at the ZSM [Zoologische Staatssammlung München].' He is a supporter of BIOPAT.

Tschudi

Tschudi's African Bullfrog *Pyxicephalus adspersus* Tschudi, 1838

Tschudi's Button Frog *Cycloramphus fuliginosus* Tschudi, 1838

Tschudi's Caribbean Toad *Peltophryne peltocephala* Tschudi, 1838

[Alt. Cuban Toad]

Tschudi's Frog *Leptobrachium hasseltii* Tschudi, 1838

[Alt. Hasselt's Litter Frog]

Tschudi's Froglet *Crinia georgiana* Tschudi, 1838

[Alt. Quacking Frog]

Baron Dr Johann Jakob von Tschudi (1818–1889) was a Swiss explorer, physician, diplomat, naturalist, hunter, anthropologist, cultural historian, language researcher and statesman. In 1838 he travelled to Peru, where he spent five years exploring and collecting. He was appointed Swiss ambassador to Brazil (1860–1868). He wrote *Untersuchungen uber die Fauna Peruana Ornithologie* (1844). Ten birds, six reptiles and five mammals are named after him.

Tunku

Tunku's Flying Frog *Rhacophorus tunkui* Kiew, 1987 NCR

[JrSYN *Rhacophorus prominanus*]

Tunku Abdul Rahman Putra Al-Haj ibni Almarhum Sultan Abdul Hamid Halim Shah (1903–1990) was generally known simply as 'Tunku'. He was the first Prime Minister of the Federation of Malaya from its independence (1957) and of Malaysia (1963–1970). He was also known as *Bapa*

Kemerdekaan (Father of Independence). He became the first Secretary-General of the Organisation of Islamic Co-operation (1971–1973).

Tunn

Masoala Bright-eyed Frog *Boophis ulftunni* Wollenberg et al., 2008

Dr Ulf Walter Tunn has given significant support to BIOPAT.

Turpin

Robber Frog sp. *Pristimantis turpinorum* **Hardy**, 2001

The Turpin family have, for over 200 years, been prominent in Tobago, where they had sugar plantations. Today they own and operate tourist businesses, including cottages at Man-O-War Bay, Charlotteville, where David Hardy (who described this frog) stayed during his almost-annual visits since 1962. Charles Turpin gave (1968) the St Giles Islands to the Government of Trinidad and Tobago as a wildlife sanctuary, and today Mrs Pat Turpin is President, Environment Tobago, established (1996) to safeguard Tobago's environmental future.

Tweedie

Tweedie's Wart Frog *Limnonectes tweediei* **Smith**, 1935

Michael Wilmer Forbes Tweedie (1907–1993) was a herpetologist, ichthyologist and malacologist who worked (1932–1971) at Raffles Museum (now called the Singapore National Museum), becoming its Director (1946). A reptile is named after him.

Twitty

American newt subgenus *Twittya* **Dubois** & Raffaëlli, 2009

[Subgenus of *Taricha*]

Twitty's Newt *Taricha granulosa twittyi* **Bishop**, 1941

[Alt. Southern Rough-skinned Newt; validity questionable]

Dr Victor Chandler Twitty (1901–1967) was an American biologist, embryologist and herpetologist, specializing in amphibians. His bachelor's degree in chemistry was awarded by Butler College, Princeton University (1925), and his doctorate in zoology by Yale (1929). He stayed at Yale as Instructor until becoming Research Fellow, Kaiser Wilhelm Institute, Berlin (1931). He returned to the USA (1932) as Assistant Professor, Stanford, Palo Alto, becoming Professor (1936) and Executive Head, Biological Sciences (1948–1963). The University honoured him (1964) by naming him Herzstein Professor of Biology. He wrote *Migration and speciation in newts* (1959). He died of self-inflicted poisoning.

Tyler, M

Ranid frog genus *Tylerana* **Dubois**, 1992 NCR

[Included in *Hylarana*]

Tyler's Tree Frog *Litoria tyleri* **Martin** et al., 1979

Tyler's Toadlet *Uperoleia tyleri* **Davies** & **Littlejohn**, 1986

Sepik Mehely Frog *Copiula tyleri* **Burton**, 1990

Tyler's Big-eyed Treefrog *Litoria michaeltyleri* **Frost** et al., 2006

Professor Michael James Tyler (b.1937) is an Australian herpetologist at the Department of Environmental Biology, University of Adelaide, South Australia. He is an expert on Australian frogs and wrote *Australian Frogs: A Natural History* (1976).

Tytler

Dacca Frog *Hylarana tytleri* **Theobald**, 1868

[Alt. Theobald's Ranid Frog]

Colonel Robert Christopher Tytler (1818–1872) was a British soldier, naturalist, photographer and collector. He served throughout India (1835–1864) and in Kabul and other parts of Afghanistan, only interrupted by two periods of sick leave in England. He and his wife Harriet (1827–1907) were keen photographers and took c.500 large-format calotype negatives of scenes associated with the Indian Mutiny (1857). He was the third Superintendent of the convict settlement at Port Blair, part of the Andaman Islands Administration. He spent the last six months of his life in charge of the museum in Simla. He was a keen amateur naturalist, being particularly expert on birds. Six birds and a reptile are named after him.

U

Uakari

Loreto Poison Frog *Ranitomeya uakarii* Brown, **Schulte** & **Summers**, 2006
[Syn. *Dendrobates uakarii*]

This frog is named after a monkey, the Peruvian Red Uakari *Cacajao calvus ucayalii*. 'The epithet was chosen because both the red uakari and *D. uakarii*, **sp. nov.** shared similar historical distributions and have bright red dermal pigmentation.'

Uis

Robber Frog sp. *Pristimantis uisae* **Lynch**, 2003

This species is named after a university: UIS stands for Universidad Industrial de Santander, Bucaramanga, Colombia.

Underwood

Robber Frog sp. *Craugastor underwoodi* **Boulenger**, 1896
Underwood's Yellow Treefrog *Dendropsophus microcephalus underwoodi* Boulenger, 1899

Cecil F. Underwood (1867–1943) left London for Costa Rica (1889) to collect natural history specimens for a living, and stayed for the rest of his life. He was an all-round naturalist who collected for a number of foreign museums, combining this with his job as a taxidermist at Costa Rica's National Museum. He described many new taxa from Central America, often in association with George Goodwin. He is remembered in the names of four mammals and two birds.

Unterstein

Unterstein's Newt *Pachytriton labiatus* Unterstein, 1930
[Alt. Spotless Stout Newt]

Walter Unterstein, who worked at the Berlin Museum (late 1920s to early 1930s), was the describer.

Urich

Lesser Antilles Robber Frog *Pristimantis urichi* **Böttger**, 1894

Frederick William Urich (1872–1936) was a Trinidadian naturalist and one of the co-founders of the Trinidad Field Naturalists' Club (now the Trinidad and Tobago Field Naturalists' Club). Two mammals and a bird are named after him.

Uthaman

Uthaman's Bush Frog *Raorchestes uthamani* Zachariah et al., 2011

Although described in the singular, this species is actually named after two Indian naturalists in Kerala: P. K. Uthaman is an ornithologist and active environmentalist, whilst his namesake K. V. Uthaman is a forest officer with the Kerala Forests & Wildlife Department.

Utsunomiya

Utsunomiyas' Tip-nosed Frog *Odorrana utsunomiyaorum* **Matsui**, 1994

Dr Taeko Utsunomiya (1926–2008) was a Japanese biologist who specialized in amphibians. She worked at the Faculty of Applied Biological Science, Hiroshima University, with her husband, Dr Yasuaki Utsunomiya. They are both honoured in the (plural form) binomial *utsunomiyaorum*. They co-wrote *Karyotypes and banding patterns of* Tylototriton andersoni *Boulenger, a newt endemic to the Ryukyu islands* (1982). Matsui wrote an obituary of her.

V

Vaillant

Vaillant's Frog *Lithobates vaillanti* **Brocchi**, 1877
White-lined Leaf Frog *Phyllomedusa vaillantii* **Boulenger**, 1882

Professor Léon Louis Vaillant (1834–1914) was a French herpetologist, ichthyologist and malacologist at the Muséum National d'Histoire Naturelle, Paris. He was on four French naval expeditions (1880–1883). Three reptiles are named after him.

Val

Gruta Button Frog *Cycloramphus valae* **Heyer**, 1983

Dr Francisca Carolina do Val is a Brazilian entomologist and artist. She was Professor of Entomology, Universidade de São Paulo and is now retired. She is Heyer's friend and field companion. He wrote that she 'has consistently aided our night-collecting after she collected drosophilid flies by day.'

Valerio

La Palma Glass Frog *Hyalinobatrachium valerioi* **Dunn**, 1931

Professor Manuel Valerio (b.1887) was a Costa Rican botanist, entomologist and naturalist who collected the frog holotype (1928). He was Director, National Museum of Costa Rica, San José (1932–1934). He taught medical botany at the Faculty of Pharmacy and made a malacological collection which is now lost.

Valhalla

Pulo Weh Toad *Duttaphrynus valhallae* Meade-Waldo, 1909

The describer, Geoffrey Meade-Waldo, was travelling near Sumatra on a yacht named 'Valhalla' when the toad holotype was taken. The yacht belonged to the Earl of Crawford.

Valkyries

Hidden Valley Rainforest Frog *Albericus valkuriarum* **Menzies**, 1999

In Norse legend, and in Wagner's opera cycle *Der Ring des Nibelungen*, there were nine Valkyries, the chief of whom was Brunhilda. The name means 'chooser of the slain' and their role was to conduct those slain in battle to Valhalla, the great hall of the god Odin.

Vallan

Climbing Frog sp. *Anodonthyla vallani* Vences et al., 2010

Dr Denis Vallan (b.1965) is a Swiss herpetologist at Natur-Museum Luzern. He has studied the effect of forest fragmentation on amphibians at the frog's type locality, Ambohitantely Special Reserve, Madagascar.

Van Dijk

Toad genus *Vandijkophrynus* **Frost** et al., 2006
Van Dijk's River Frog *Amietia vandijki* Visser & **Channing**, 1997

Eddie Van Dijk (b.1925) is a noted South African herpetologist and indefatigable tadpole specialist. He studied at the University of Witwatersrand and Johannesburg Teachers' Training College, then at Stellenbosch University where he obtained his master's (1953) and doctorate (1959). Many years later he undertook a second master's in palaeontology at Stellenbosch University (2000). While studying he worked as a technician in the Bacteriology Laboratory of the South African Institute for Medical Research, and thereafter lectured in the Zoology Department, University of Natal for about 30 years. He has compiled bibliographies on African anurans, African tadpoles, African fossil frogs, and African vertebrate ichnology (tracks and other trace fossils). He has written a number of textbooks including *Introduction to Modern Biology* (1970) and *Senior Secondary Biology 9* (1988). He is also commemorated in the names of a fossil plant bug, *Australoprosoboloides vandijki*.

Van Dyke

Van Dyke's Salamander *Plethodon vandykei* Van Denburgh, 1906

Dr Edwin Cooper Van Dyke (1869–1952) took a bachelor's degree (1893) at the University of California and qualified as a physician (1895) at Cooper Medical School, Stanford. He practised medicine in San Francisco (1895–1913), but then abandoned general practice and became an instructor in entomology at the University of California. He stayed as a member of the faculty, becoming Professor of Entomology (1927–1939), retiring as Professor Emeritus.

Vannutelli

Vannutelli's Tree Frog *Leptopelis vannutellii* **Boulenger**, 1898
[Alt. Dime Forest Treefrog]

Lamberto Vannutelli (1871–1966) was an officer in the Italian navy and was on the Italian Geographical Society's second Bottego Expedition to East Africa (1895). Bottego was killed fighting Ethiopian forces and his men, including Vannutelli, were captured and held prisoner for two years before the Italian authorities became aware of the fate of the expedition (1897). Vannutelli co-authored an account of the expedition, *L'Omo. Viaggio d'esplorazione nell'Africa Orientale* (1899).

Van Vliet

Van Vliet's Frog *Hyla vanvlietii* **Baird**, 1854 NCR
[JrSYN *Smilisca baudinii*]

Brigadier General Stewart Leonard Van Vliet (1815–1901) was a US Army officer who fought on the side of the Union during the American Civil War. He was commissioned in 1840 and attached to the US Third Artillery during the Seminole Wars, seeing action in several engagements (in one of which he killed an Indian chief in hand-to-hand fighting). He also served in the Mexican-American War. He collected natural history specimens, notably fishes on the Texas-Mexico border in 1856. Baird described this species without mentioning who he was naming it after. As Baird's type specimen came from Brownsville, Texas – near the Mexican border – we think we have identified the right man.

Vanzolini

Leptodactylid frog genus *Vanzolinius* **Heyer**, 1974 NCR
[Now usually included in *Leptodactylus*]
Vanzolini's Amazon Frog **Leptodactylus discodactylus* **Boulenger,** 1884
Vanzolini's Spiny-chest Frog *Alsodes vanzolinii* **Donoso-Barros**, 1974
Brazilian Poison Frog *Ranitomeya vanzolinii* **Myers**, 1982
Vanzolini's Tree Toad *Hylodes vanzolinii* Heyer, 1982
Vanzolini's Leaf Frog *Phrynomedusa vanzolinii* **Cruz**, 1991
Poison Frog sp. *Allobates vanzolinius* Morales, 2002

Professor Dr Paulo Emilio Vanzolini (b.1924) is a Brazilian herpetologist, who worked at the AMNH, New York (late 1970s–1980s). He was Director (1962–1993), Zoological Museum, University of São Paulo, Brazil, and was responsible for assembling one of the largest collections of herpetofauna in the Americas. He (compulsorily) retired at 70, but still has a research affiliation with the Museum. He is also famous in Brazil as a composer of samba music. He wrote *Elementary Statistical Methods in Zoological Systematics* (1993). Two mammals and twelve reptiles are named after him.

* William Heyer proposed a new genus, *Vanzolinius,* for *Leptodactylus discodactylus* but this has not gained wide acceptance – other than inspiring Frank & Ramus (1995) to invent the common name 'Vanzolini's Amazon Frog'.

Varaldi

Varaldi's Spadefoot Toad *Pelobates varaldii* Pasteur & Bons, 1959
[Alt. Moroccan Spadefoot Toad]

Marcel Varaldi collected reptiles and amphibians in Morocco (1950s). He wrote an article, *Quelques lézards marocains* (1958).

Varela

Carrizo's Treefrog *Hypsiboas varelae* **Carrizo**, 1992

Esperanza Varela de Olmedo worked with Carrizo, and they were two of the authors of *Catalogue of the voices of Argentine Amphibians Part 1* (1993).

Varley

Varley's Robber Frog *Eleutherodactylus varleyi* **Dunn**, 1925

James B. Varley of Guabairo, Cuba was, according to Dunn, '…a hospitable friend', who helped the author to collect the frog holotype (1924).

Vasanth

Kalakad Wrinkled Frog *Nyctibatrachus vasanthi* Ravichandran, 1997

M. Vasanth of the Zoological Survey of India collected the holotype. He is an entomologist and wrote *Studies on Insecta: Orthoptera: Gryllidae from NE India* (1993).

Vásquez

Bush Toad sp. *Oreophrynella vasquezi* **Señaris**, **Ayarzagüena** & **Gorzula**, 1994

Dr Enrique Vásquez was a young biologist who died in an accident; he fractured his skull when he slipped on a rock whilst trying to help his wife, who had fallen into a stream in the Venezuelan Amazon.

Veith

Reed Frog sp. *Hyperolius veithi* Schick et al., 2010

Dr Michael Veith is a biologist and biogeographer. Johannes Gutenberg-Universität Mainz awarded his bachelor's degree (1986) and doctorate (1991). He worked at the Zoology Department, Johannes Gutenberg-Universität Mainz (1991–1997), and was Assistant Professor at Universität Hohenheim, Stuttgart (1998–2000). He returned to Mainz

(2000–2005), becoming Professor of Zoology (2004). He was Professor of Animal Systematics and Zoogeography at the Institute for Biodiversity and Ecosystem Dynamics, Universiteit van Amsterdam (2005–2007), and since 2007 has been Professor of Biogeography, Geography Department, Universität Trier, Germany. He co-wrote *Naturschutzgenetik bei Amphibien: Von der Theorie zur Praxis.* (2009).

Velasco

Plateau Tiger Salamander *Ambystoma velasci* **Dugès**, 1888

José Maria Velasco Gómez (1840–1912) originally described this animal as *Siredon Tigrina* in *Descripción, metamorfosis y costumbres de una especie nueva del género Siredo* (1879). Because the binomial *tigrina* had already been used for another salamander species, it was necessary to change the name, so Dugès adopted Velasco's name as the new binomial. Velasco was a Mexican landscape and nature painter of extraordinary talent who became Professor of Landscape Perspective at the Academia de San Carlos (1868). He lived and worked at La Villa de Guadelupe Hidalgo, Valle de México, which is where the holotype was acquired, and is now within the urban sprawl that is Mexico City.

Vellard

Vellard's Water Frog *Telmatobius vellardi* Munsterman & Leviton, 1959

Alto Maranon Toad *Rhinella vellardi* Leviton & **Duellman**, 1978

Marsupial Frog ssp. *Gastrotheca bifasciata vellardi* **Laurent**, 1969 NCR

[JrSYN *Gastrotheca marsupiata*]

Dr Jehan Albert Vellard (1901–1996) was a Tunisian-born French physician, ethnologist, herpetologist and arachnologist. He qualified as a physician (1924), then went to Brazil and worked at the Butantan Institute, São Paulo (1925–1929). He ran the Paraguayan Natural History Museum (1930–1933). He travelled widely, always accompanied by his mother, and became caught up in the Chaco War between Paraguay and Bolivia (1932–1935). They found an abandoned encampment in which there was a coati, a pot of honey and a one-year-old girl. The fate of the coati is not reported, but they certainly ate the honey and adopted the little girl whom they called Maryvonne; she was brought up under the regime of a strict French classical education. He was Director, National Museum, La Paz, Bolivia (1940–1943) and Head, Zoology Department, University of Tucumán, Argentina. He founded (1946) and was Professor and Director (1947–1956) at the Herpetology Department, Natural History Museum, San Marcos University, Lima, Peru. He created, and was Director (until 1962) of, the French Institute of Andean Studies, Lima. He was also Director (until 1966), Bolivian Institute of the Biology of Altitude, being very interested in the physiological problems of adjusting to limited air at high altitudes. He retired (1966) to Buenos Aires. He wrote *Civilisations des Andes* (1963). A reptile is also named after him.

Venancio

Venancio's Robber Frog *Ischnocnema venancioi* **Lutz**, 1958

Joaquim Venancio (1895–1957) was the faithful factotum of the Lutz family. He was born in very humble circumstances in Minas Gerais province, Brazil, and made his way to Rio de Janeiro (1916). He lived in a house next to the building occupied by the Institute Oswaldo Cruz. He met Adolpho Lutz and became his laboratory assistant (1920), went on collecting expeditions with him and accompanied him on his visits to Venezuela and the USA (1925). Lutz lost mobility with old age, so Venancio physically carried him on his back. After Lutz's death (1940), Venacio worked for Bertha Lutz, who named this frog after him.

Vera

Vera's Tree Frog *Litoria verae* **Günther**, 2004

Vera Heinrich works at the Museum für Naturkunde, Humboldt University Berlin. She and the author have been friends for about 30 years; he expressed his appreciation for the high-quality graphics she has produced over that period for papers he has written.

Verbeek/Snyder

Rocket Frog sp. *Anomaloglossus verbeeksnyderorum* Barrio-Amoros, Santos & Jovanovic, 2010

The Verbeek and Snyder families from Dallas, Texas, appear to have pooled resources as they won the 2008 'Year of the Frog' auction run by Amphibian Ark. The funds raised by the auction went to the 'Atelopus Project' to protect Andean amphibians.

Vercammen

Mwana Screeching Frog *Arthroleptis vercammeni* **Laurent**, 1954

Dr Paul-Henry Vercammen-Grandjean (1915–1995) was an entomologist and virologist. The Université de Paris awarded his doctorate. He worked at the

Musée Royal de l'Afrique Centrale, Belgium. He wrote *Révision du genre Schoutedenichia Jad. et Verc* (1958).

Vergel

Hellmich's Rocket Frog *Hyloxalus vergeli* **Hellmich**, 1940

This frog is named after the type locality, Finca El Vergel, Departamento Cundinamarca, Colombia.

Verreaux

Verreaux's Tree Frog *Litoria verreauxii* **Duméril**, 1853 [Alt. Whistling Tree Frog]

Jules Pierre Verreaux (1807–1873) and his brother Jean Baptiste Edouard Verreaux (1810–1868) were French naturalists, collectors and dealers. Jules was an ornithologist and plant collector for the Muséum National d'Histoire Naturelle, Paris, which sent him (1842) to Australia. He and his servant Emile collected in Tasmania, New South Wales and Queensland (1842–1850). He returned to France (1851) with a huge natural history collection. Earlier he helped Andrew Smith (q.v.) in founding the South African National Museum, Cape Town. Between them, the Verreaux brothers also have two reptiles, three mammals and twenty-five birds named after them.

Vestergaard

Vestergaard's Forest Toad *Nectophrynoides vestergaardi* Menegon, Salvidio & Loader, 2004

Martin Vestergaard is a Danish biologist who works for COWI, a consulting organization based in Denmark, and specializing in economics, engineering and environmental science. Vestergaard collected (1992) the holotype of this toad in Tanzania.

Vicente

Vicente's Poison Frog *Oophaga vicentei* Jungfer, **Weygoldt** & Juraske, 1996 [Syn. *Dendrobates vicentei*]

Vicente Augusto Fernández helped to find and collect the frogs in Panama. He died in a tragic accident.

Vieillard

Smooth Horned Frog sp. *Proceratophrys vielliardi* **Martins** & **Giaretta**, 2011

Dr Jacques Marie Edme Vielliard (1944–2010) was a French ornithologist, zoologist and ecologist. The Université de Paris awarded his bachelor's degree (1967), master's (1968) and doctorate (1971). Interested in birds from an early age, he made his first expedition at 18 to Coto Doñana, Spain. He continued to travel widely from Romania to the Himalayas (1967–1971). He went on to study birds in Africa before being invited by the Brazilian Academy of Science to help develop ecological study in the country (1976). After four years he took a post at the Universidade Estadual de Campinas, Brazil, where he created the Laboratory of Bioacoustics of Brazilian Fauna. He was a professor who had the junior author as one of his pupils. His particular interest was in the migratory ecology of aquatic birds. He died following kidney surgery. A bird is named after him.

Vilars

Rio Uaupes Robber Frog *Pristimantis vilarsi* Melin, 1941

Arthur Vilars was a Brazilian engineer on the Swedish Amazonas Expedition (1923–1925). Melin, with whom he collected the holotype (1924), was the expedition leader. They co-wrote *Amazon Basin, Rio Negro drainage: Manaus area, small white water river* (1923).

Vilcabamba

Robber Frog sp. *Pristimantis vilcabambae* **Lehr**, 2007

This species is named after the type locality, Cordillera de Vilcabamba, Peru.

Vilhena

Vilhena's Reed Frog *Hyperolius vilhenai* **Laurent**, 1964 [Alt. Luita River Reed Frog]

Comandante Ernesto Jardim de Vilhena (1876–1967) was a Portuguese naval officer (1898–1920), during which time he was also a colonial administrator in Africa (1902–1912) and a politician, including being Minister for the Colonies (1917–1920). He combined his public career with being a businessman and was Director and Chairman of the Angola Diamond Company (1919–1955) and Chairman of the Portuguese Diamond Society from 1958. The original description was published in the diamond company's own cultural journal.

Villiers

Villiers' Toad *Amietophrynus villiersi* **Angel**, 1940

Yapo River Frog *Phrynobatrachus villiersi* **Guibé**, 1959

Dr André Villiers (1915–1983) was a French entomologist and herpetologist. He joined the Muséum National d'Histoire Naturelle in Paris (1937) as a technical assistant in entomology. He gained his doctorate (1943), and in 1945 moved to the Institut Français d'Afrique Noire, Senegal. He and his assistants made large collections of insects and herpetofauna on their expeditions in French West Africa. Returning to France (1956), he worked in the entomology laboratory at the Paris Museum. Among his publications is *Les serpents de L'Ouest africain* (1950). A reptile is named after him.

Vilma

Slender-legged Treefrog sp. *Osteocephalus vilmae* Ron et al., 2012

Vilma Duran is a Peruvian herpetologist at the Centro de Ornitología y Biodiversidad, where she has done much to improve the herpetological collection. The etymology says that she collected 'the holotype and tissues of this new species.'

Vinha

Vinha's Robber Frog *Ischnocnema vinhai* **Bokermann**, 1975

Dr Sergio G. da Vinha is a Brazilian botanist engaged in research for the cocoa industry. He was the first curator (1966) of the herbarium of the Centro de Pesquisas do Cacau, Ileus, Bahia, which is where the frog holotype was collected. He co-wrote *Ocorrência do Jacarandá no sul da Bahia* (1975).

Viosca

Viosca's Red Salamander *Pseudotriton ruber vioscai* **Bishop**, 1928

[Alt. Southern Red Salamander]

Paul Percy Viosca Jr (1892–1961) was a professional biologist who graduated with a bachelor's degree from Tulane University, New Orleans (1913), and was Instructor of Biology there (1913–1917). He was Curator, Reptiles and Amphibians and Curator, Fishes, Louisiana State Museum (1915–1926), combining these duties with also being (1917–1926) Biologist, Louisiana State Department of Conservation. He became Director of Fisheries, Louisiana (1926). After 1929 he held a number of other posts, including being Senior Biologist, Fisheries, Louisiana (1936–1942), after which he became an independent consultant biologist. Among Viosca's claims to fame was discovering the importance of fire in the ecology of the Louisiana marshes. He wrote *Louisiana Out-of-Doors* (1933).

Vizcacheras

Vizcacheras' White-lipped Frog *Leptodactylus bufonius* **Boulenger**, 1894

[Alt. Viscacha Frog, Oven Frog]

The English vernacular name seems to have arisen as a pseudo-translation of the Spanish name, 'Rana de las vizcacheras'; a reference to the Viscacha, a burrowing rodent. The apostrophe in the common name is completely redundant!

Vizotto

Robber Frog sp. *Ischnocnema vizottoi* **Martins** & **Haddad**, 2010

Dr Luiz Dino Vizotto (b.1931) has been, since 2007, Professor Emeritus, Department of Zoology and Botany, Universidade de São Paulo, which awarded his original bachelor's degree in natural history (1955). He spent his entire career on the faculty of the Universidade de São Paulo, filling many roles including managing the Department of Zoology and Botany (1957–1985), and being Deputy Director of the Institute (1975–1979). A species of ant is also named after him.

Vogl

Vogl's Harlequin Toad *Atelopus vogli* **Müller**, 1934

Father Cornelius Vogl (1884–1959) was a German Benedictine priest. He was a missionary (1910–1919) in Dar Es Salaam (Tanzania), and then in Venezuela (1925–1959). He collected (1933–1948) botanical, entomological and herpetological specimens, which he sent to the Natural History Museum, Munich. A reptile is named after him.

Vogt

Vogt's River Frog *Phrynobatrachus vogti* **Ahl**, 1924

Admiralty Island Webbed Frog *Discodeles vogti* Hediger, 1934

Theodor Vogt (1881–>1932) was a German naturalist. He wrote *Reptilien und Amphibien aus Neu-Guinea* (1911) and *Beitrag zur Reptilien und Amphibienfauna der Südseeinseln* (1912). A reptile is named after him.

Volz

Sumatra Squat Frog *Calluella volzi* **Van Kampen**, 1905

Dr Walter Volz (1875–1907) was a Swiss zoologist, traveler and ichthyologist. He was in Indonesia and Thailand (1901–1902), also visiting Japan and Hawaii en route to Bern. He published on the fishes of Sumatra (1903). He travelled in Sierra Leone and Liberia (1906–1907). He appears to have been killed ('collateral damage') by French troops conducting a punitive raid. A bird is named after him.

Vozmediano

Glass Frog sp. *Celsiella vozmedianoi* **Ayarzagüena** & **Señaris**, 1997

[Syn. *Cochranella vozmedianoi*]

Jesus Vozmediano y Gomez-Feu is a Spanish lawyer who qualified at the Universidad de Sevilla, Spain (1978). He specializes in environmental cases, including criminal proceedings for the mortality of birds occurring in the Doñana National Park (1986).

He was honoured in the frog's binomial 'for his enormous and selfless efforts on behalf of the Spanish nature, most notably the PN de Doñana.'

Vulcan

Volcan San Martin Rainfrog *Craugastor vulcani* Shannon & **Werler**, 1955

In Roman mythology, Vulcan was the god of fire and of volcanoes. The frog holotype was collected on the slopes of a Mexican volcano.

Vulpia

African Water Frog sp. *Petropedetes vulpiae* Barej et al., 2010

Dr Christine Fuchs works at the Institut für Mikroskopische Anatomie und Neurobiologie der Universitätsmedizin der Johannes Gutenberg-Universität Mainz, Germany. This is one of those joke eponyms that some zoologists indulge in. Fuchs is German for 'fox', which in Latin is 'vulpes'; hence *vulpiae* is a feminine form of Fuchs.

W

Wade

Wade's Grass Frog *Ptychadena wadei* **Largen**, 2000
[Alt. Tisisat Grassland Frog]

Professor Edward O. Z. Wade (b.1940) is an artistic illustrator who is also interested in herpetology and taxonomy. He was a lecturer at Hornsey College of Art (1965–2005). He developed a method of reconstructing severely mutilated road casualties to a lifelike appearance, thereby accessing valuable morphological data which would have otherwise been overlooked.

Wager

Wager's Stream Frog *Strongylopus wageri* Wager, 1961

Dr Vincent A. Wager (1904–1989) was a South African pathologist, herpetologist and an expert on plant diseases. He wrote *The Frogs of South Africa* (1965). We have read that this frog 'was accidently named after Vincent Wager' - and don't know what to make of it, as he appears to have named it after himself.

Wagler

Australian Frog genus *Wagleria* **Girard**, 1853 NCR
[In *Limnodynastes*]
Wagler's Spadefoot Toad *Pelobates cultripes* **Cuvier**, 1829
[Alt. Western Spadefoot, Spanish Spadefoot Toad]

Johann Georg Wagler (1800–1832) was a German herpetologist. He was Spix's assistant and upon his death (1826) became Director, Zoological Museum, University of Munich, continuing to work on the extensive Brazilian collections. In addition he worked on systematics of amphibians and reptiles in the Museum. Wagler died accidentally (1832) from a self-inflicted gunshot wound while out collecting. He wrote the highly regarded *Monographia Psittacorum*. Eleven birds and eight reptiles are named after him.

Wagner

Wagner's White-lipped Frog *Leptodactylus wagneri* **Peters**, 1862

Johann Andreas Wagner (1797–1861) was a German palaeontologist, zoologist and archaeologist. He became Professor of Zoology and Assistant Curator, University of Munich, Zoological Museum. Among his publications is *Diagnosen neuer Arten brasilischer Säugethiere* (1842). A mammal and a reptile are named after him.

Wagter

Wagter's Robber Frog *Pristimantis wagteri* Venegas, 2007

Ron Wagter is a biologist and archaeologist who graduated from the University of Amsterdam. Since 1997 he has been organizer of scientific expeditions into the Peruvian cloud forests. He works for the conservation of the area around Lake Los Cóndores.

Wahlberg

Wahlberg's Frog *Arthroleptis wahlbergii* **Smith**, 1849
[Alt. Bush Squeaker, Wahlberg's Forest Frog]

Johan August Wahlberg (1810–1856) was a naturalist and collector. He studied chemistry and pharmacy at Uppsala (1829) and worked in a Stockholm chemist's shop while studying at the Forestry Institute. He travelled and collected in southern Africa (1838–1856), returned briefly to Sweden (1853) but was soon back in Africa at Walvis Bay (Namibia) until 1854. He was exploring the headwaters of the Limpopo when a wounded elephant killed him. Four birds, four reptiles and a mammal are named after him.

Wake

Mantellid frog genus *Wakea* **Glaw** & Vences, 2006
Wake's Moss Salamander *Cryptotriton wakei* **Campbell** & Smith, 1998

Dr David Burton Wake (b.1936) is an expert on salamanders. The Pacific Lutheran College, Tacoma awarded his bachelor's degree (1958); the University of Southern California awarded both his master's (1960) and doctorate (1964). He spent much of his career at the Museum of Vertebrate Zoology, University of California, Berkeley – including being Director of the Museum (1971–1998), Professor of Integrative Biology (1989), and Curator of Herpetology. He is presently Emeritus Professor.

The frog genus is named after David and his wife, Dr Marvalee Hendricks Wake, who is an expert on caecilians. She joined (1969) the faculty of the University of California, Berkeley, after her doctorate was awarded (1968) by the University of Southern California. She became Chairman, Department of Zoology and its successor, Department of Integrative Biology. She was President of the American Society of Ichthyologists and Herpetologists (1983) and is now Professor of the Graduate School. A reptile is named after them.

Walford

Treefrog sp. *Dendropsophus walfordi* **Bokermann**, 1962

Dr Roy Lee Walford (1924–2004) qualified as a physician at the University of Chicago (1948). After working at the Gorgas Medical Center, Panama, and at a medical centre in Los Angeles, he served two years in the US Air Force during the Korean War. He joined the faculty at the University of California at Los Angeles (1954) and became Professor of Pathology, UCLA School of Medicine (1966). He became Professor of Pathology and Laboratory Medicine, Emeritus (1991), leaving to become resident physician and one of the eight crew members of Biosphere 2, where they lived, sealed in, for two years. With a fellow graduate student, a mathematician called Albert Hibbs, who eventually became an astronaut in NASA's Apollo programme, he went to Nevada to 'break the bank'. They used statistical analysis to determine which wheels were biased and then betted heavily on them. The casino owners caught on and threw Walford and Hibbs out; but not before they had made a lot of money. With the winnings Walford paid for most of his medical school education and a house and, together, they bought a yacht and spent more than a year sailing the Caribbean.

Walker, B

Walker's Sierra Frog *Geobatrachus walkeri* **Ruthven**, 1915

Dr Bryant Walker (1856–1936) was a leading malacologist of the late nineteenth and early twentieth centuries. He was a wealthy lawyer who retired early, devoting the rest of his life to the study of molluscs, collecting many himself and financing expeditions. He published many scholarly works on molluscs. He had a very fine library and mollusc collection (containing about two million specimens), which were moved to the University of Michigan and became the backbone of their collection. He started the series of Occasional Papers at the University; Ruthven, who was also on the Walker-Newcomb Expedition collecting amphibians in northeastern Nevada, wrote many of them. Walker organised a number of such expeditions and Ruthven took part in some, and wrote descriptions of the herpetological specimens collected by many more. Most pertinent is the paper by Ruthven, *The Bryant Walker Expedition, of the University of Michigan, to the Santa Marta Mountains, Colombia, in the Summer of 1913.*

Walker, CF

Walker's Treefrog *Hyla walkeri* **Stuart**, 1954
Walker's Stubfoot Toad *Atelopus walkeri* **Rivero**, 1963
Walker's Mushroomtongue Salamander *Bolitoglossa walkeri* **Brame** & **Wake**, 1972
Walker's Robber Frog *Pristimantis walkeri* **Lynch**, 1974
Walker's Marsupial Frog *Gastrotheca walkeri* **Duellman**, 1980

Professor Charles Frederick Walker (1904–1979) was Curator, Division of Reptiles and Amphibians, University of Michigan Museum of Zoology (1947–1975), and obtained the frog holotype in Ecuador. He graduated from Ohio State University (1930) and went on to take his doctorate at the University of Michigan (1935). His first position was with the Soil Conservation Service, but he then (1938) joined the staff of the Stone Laboratory, Ohio State University, where he remained until rejoining the University of Michigan (1947). He was considered an authority on New World amphibians, and his publications were considered the acme of scientific accuracy. He began publishing herpetological papers very early (1928), including describing new taxa, and continued to publish until retirement (1974). He was also President of the American Society of Ichthyologists and Herpetologists (1968).

Wallace, AR

Wallace's Flying Frog *Rhacophorus nigropalmatus* **Boulenger**, 1895

Alfred Russel Wallace (1823–1913) was an English naturalist, evolutionary scientist, geographer, anthropologist, social critic and theorist and a follower of the utopian socialist Robert Owen. He was one of the giants of Victorian science, with claims to be the father of zoogeography. He discovered and described the faunal discontinuity that now bears his name – Wallace's Line – a natural boundary separating the Oriental and Australasian faunal regions. He started out as an apprentice surveyor, but his interest in natural history took over and he went to Brazil on a self-sustaining natural history collecting expedition (1948–1852). On his way home, his ship caught fire and sank with all his specimens. He went to the Indonesian archipelago, where he covered 14,000 miles over about 70 expeditions (1862–1869), visiting every important island at least once and collecting 125,660 specimens. He was thinking along the same lines as Darwin and sent him his essay *On the law which has regulated the introduction of new species,* which encapsulated his most profound theories on evolution, and later another essay *On the tendency of varieties to depart indefinitely from the original type,* presenting the theory of 'survival of the fittest'. Darwin and Lyell presented this essay, together with Darwin's own work, to the Linnean Society. Wallace's thinking spurred Darwin to encapsulate these ideas in *The Origin of Species;* the rest is history, but Wallace never has been as well known to the general public as Darwin. He wrote many books and papers, including *The Malay Archipelago* (1869), the most celebrated of all writings on Indonesia and one of the nineteenth century's best scientific travel books. Many taxa are named after him, including a reptile, two mammals and twelve birds.

Wallace, HK

Georgia Blind Salamander *Eurycea wallacei* **Carr**, 1939
[Syn. *Haideotriton wallacei*]

Dr Howard Keefer Wallace (b.1907) was an arachnologist at the Department of Biology, University of Florida, where he was awarded his bachelor's degree (1929). His doctorate was gained (1938) at the University of Florida, where he was employed (1932–1972), first as Instructor, Biology and Geology, then as Assistant Professor. His career was interrupted by service in the US Army in Europe (1942–1945). He returned to the University of Florida (1945) as Professor and became (1949) Professor of Zoology there until his retirement, having also been Head, Zoology Department (1955–1969).

Walpole

Walpole's Frog *Geocrinia lutea* **Main**, 1963
[Alt. Main's Ground Froglet, Nornalup Frog]

This amphibian is named after Walpole, Western Australia, near where the holotype was collected.

Walter

Walter's Narrow-mouthed Frog *Kaloula walteri* **Diesmos, Brown** & **Alcala**, 2002
Walter's Limestone Forest Frog *Platymantis bayani* Siler et al., 2009

Dr Walter Creighton Brown (see **Brown, WC**). The binomial *bayani* is a Filipino term meaning 'hero' or 'highly respected' and refers to Walter Brown 'in recognition of his numerous contributions to Philippine herpetology'.

Waltl

Sharp-ribbed Salamander *Pleurodeles waltl* Michahelles, 1830
[Alt. Spanish Ribbed Newt]

Dr Joseph Waltl (1805–1888) was a German entomologist who first qualified as a physician in Munich in 1819 (you could qualify very young in those days!), then travelled in Austria, France and Spain. He became a teacher (1833) at Passau (Germany) and Professor of Natural History, Universität Passau (1835), where he stayed for the rest of his life. The describer, Michahelles, was one of his friends and colleagues.

Wanda

Villavicencio Snouted Treefrog *Scinax wandae* **Pyburn** & **Fouquette**, 1971

Wanda Pyburn is the senior author's widow.

Wandolleck

Wandolleck's Button Frog *Cycloramphus ohausi* Wandolleck, 1907
Wandolleck's Land Frog *Liophryne schlaginhaufeni* Wandolleck, 1911
Wandolleck's White-lipped Treefrog *Litoria albolabris* Wandolleck, 1911

Dr Bernhard 'Benno' Edward Max Julius Wandolleck (b.1864) was an entomologist and herpetologist. He was also a member of the German pistol shooting team at the Stockholm Olympics (1912). His weapon was the dueling pistol!

Waorani

Robber Frog sp. *Pristimantis waoranii* McCracken, Forstner & **Dixon**, 2007

The Waorani are an ancient indigenous tribe in Ecuador.

Warren, AN

Warren's Treefrog *'Hyla' warreni* **Duellman** & **Hoogmoed**, 1992
[Taxonomic (genus) allocation unclear]

Adrian Neil Warren (1949–2011) was a British zoologist, filmmaker and expert on Mount Roraima in Guyana, having led an expedition to its North Ridge (1971) and collected on the previously unexplored summit. He was a member of the advance party of four who dropped in by parachute. Film of this expedition was shown on the BBC *World About Us* series, and was entitled 'To Catch an Orchid'. They discovered several new species of frog, including the above eponymous one. He wrote *Report of the 1971 British Expedition to Mt Roraima in South America* (1973). He made dozens of films about wildlife and the environment, including 'Secrets of Wild Pandas' (1992–1993). His first degree was from London University and he also had a pilot's licence. His adventurous life included being a storm chaser, a judge at a film festival, a BBC producer, a skydiver and an officer in the RAF. He died as the result of aggressive melanoma.

Warren, CR

Warren's Robber Frog *Eleutherodactylus warreni* **Schwartz**, 1976

C. Rhea Warren is a member of the University of Miami's Iron Arrow Society (based on Seminole Indian ritual and spirituality). He collected specimens on the Île de la Tortue (off Haiti) (1968–1970). Two reptiles are named after him.

Warszewiscz

Warszewitsch's Frog *Lithobates warszewitschii* **Schmidt**, 1857

[Syn. *Rana warszewitschii*]

Joseph Ritter von Rawicz Warszewiscz (1812–1866) was a Lithuanian-Polish botanist who collected in Peru (1860). He is regarded as one of the greatest orchid collectors of all time. Six birds are named after him.

Watson

Ngezi Reed Frog *Hyperolius watsonae* **Pickersgill**, 2007

Christine Ann Watson has been on expeditions with the describer and is his partner.

Wavrin

Upper Orinoco Treefrog *Hypsiboas wavrini* **Parker**, 1936

Marquis Robert de Wavrin (1888–1971) was a Belgian traveller, explorer and film-maker who studied South American fauna, flora and ethnography. He visited the most inaccessible areas of the continent (1913–1930). He wrote *Les bêtes sauvages de l'Amazonie* (1939).

Weale

Weale's Running Frog *Semnodactylus wealii* **Boulenger**, 1882

[Alt. Long-toed Running Frog, Rattling Frog]

James Philip Mansel Weale (1838–1911) was an amateur entomologist who went from England to South Africa to farm. He lived in Bedford, Kaffraria (now Eastern Cape Province) and sent specimens back to England.

Webb

Webb's Madagascar Frog *Gephyromantis webbi* **Grandison**, 1953

[Syn. *Mantidactylus webbi*]

Cecil Stanley Webb (b.1895), known as 'Webbie', was an animal collector who was Curator of Mammals and Birds at London Zoo. He was responsible for restocking both London and Whipsnade Zoos after the Second World War. He was later Superintendent of Dublin Zoo (1952), but resigned the same year. He wrote *The Odyssey of an Animal Collector* (1954). A mammal is named after him.

Weber

Malatgan River Caecilian *Caudacaecilia weberi* **Taylor**, 1920

Weber's Wrinkled Ground Frog *Platymantis weberi* **Schmidt**, 1932

Dr Max Wilhelm Carl Weber van Bosse (1852–1937) was a German-Dutch physician and zoologist who became Director, Zoological Museum, Amsterdam (1883) and a Dutch citizen. He did German military service, half the time as a doctor and half as a hussar. He combined the roles of watch-keeping officer, ship's doctor and naturalist on a voyage (1881) in the 'Willem Barents', appropriately to the Barents Sea. His wife Anna was a botanist and the Webers spent summers in Norway, where he dissected whales and she collected algae. They went (1888) to Sumatra, Java, Sulawesi and Flores, and to South Africa (1894). He led (1899–1900) the Siboga expedition. 'Weber's Line', an important zoogeographical line between Sulawesi and the Moluccas, is sometimes preferred over Wallace's Line (between Sulawesi and Borneo) as the dividing line between the Oriental and Australasian faunas. He co-wrote *The Fishes of the Indo-Australian Archipelago* (1911). Two mammals, three reptiles and a bird are named after him.

Webster

Webster's Salamander *Plethodon websteri* Highton, 1975

Thomas Preston Webster III (1947–1975) was a Harvard student. E. E. Williams wrote of him: 'He published extensively while he was a graduate student, although he never finished his PhD. He died in an auto accident in MT just after he had gone there as an Assistant Professor.' He had taken up the post at the University of Montana. A reptile is named after him.

Wehrle

Wehrle's Salamander *Plethodon wehrlei* **Fowler** & **Dunn**, 1917

Richard White Wehrle (c.1853–1937) was a jeweler and naturalist. He donated hundreds of specimens to Philadelphia Academy of Natural Sciences and the Carnegie Museum, including over 500 specimens of salamanders. He collected the holotype (1911) and was described as '...an indefatigable collector of cold-blooded vertebrates...'

Weidholz

Weidholz's Banana Frog *Afrixalus wiedholzi* **Mertens**, 1938
[Alt. Weidholz's Leaf-folding Frog]

Alfred Weidholz was an Austrian wildlife dealer, explorer and traveller. He collected in North and West Africa, including fishes from the River Niger and rodents from Tunisia (1912) for Vienna's Schönbrunn Zoo. He made a film about his travels, *Für Tierpark Schönbrunn nach Afrika* (1928), and wrote *Bei den bergheiden in Nordkamerun* (1943). A reptile is named after him.

Weigel

Weigel's Toad *Notaden weigeli* Shea & **Johnston**, 1988
[Alt. Grey-limbed Spadefoot Toad]

John Randall Weigel (b.1955) is Director of the Australian Reptile Park in New South Wales, where he has worked for 28 years. He dropped out of university and went to work at the California Alligator Farm as a reptile keeper and showman. He left the USA and settled in Australia (1981), taking up a job at the reptile centre. When the business failed in 1984 he and his wife teamed up with a local businessman, bought the attraction and revitalized it. He co-founded the Reptile Keepers Association (1985) to encourage the keeping and breeding of native species. He wrote *Care of Australian reptiles in captivity* (1988). He has made more than 20 expeditions to the Kimberley District, during one of which he discovered the eponymous toad as well as the Pygmy King Brown Snake *Pseudechis weigeli*. He is also a talented photographer. When the centre burnt down and the collection destroyed he went about collecting replacements and established a captive breeding programme that in turn led to his becoming the sole supplier of venom used in the production of antivenins in Australia – he continues to milk 200 snakes every two weeks! In 2008 he was awarded Australia's highest civilian accolade, the 'Order of Australia', in recognition of his contributions to Australian tourism, herpetology and snake venom production.

Weigold

Plateau Brown Frog *Rana weigoldi* **Vogt**, 1924 NCR
[JrSYN *Rana kukunoris*]
Weigold's Toothed Toad *Oreolalax weigoldi* Vogt, 1924

Dr Max Hugo Weigold (1886–1973) was a German zoologist and pioneer bird bander. He worked for the Scientific Commission for Marine Research in Heligoland, setting up a bird observatory (1910) to catch and band migratory birds. He may have been the first European to see a living giant panda. He bought a cub (1916), which did not survive long, whilst he was in China as a member of the Walter Stotzner Expedition (1914–1916) for the Staatlichen Museum für Volkerkunde Dresden. He later captured a wild giant panda in Szechwan. He was back in China in Chengtu (1931) with the First Dolan Expedition from the Academy of Natural Sciences, Philadelphia. He became Director of Natural Sciences at the Niedersachsisches Landesmuseum, Hanover. Twelve birds are named after him.

Weinland

Weinland's Marsupial Frog *Gastrotheca weinlandii* **Steindachner**, 1892
Weinland's Robber Frog *Eleutherodactylus weinlandi* **Barbour**, 1914

David Friedrich Weinland (1829–1915) was a German zoologist, theologian and best-selling author, whose book *Rulaman* (1878) – set in Stone Age times – is still in print. He originally studied theology at the University of Berlin but graduated as a zoologist. He acted as Assistant to Louis Agassiz at Harvard (1855–1858). He made research trips to Canada and the Caribbean, especially Haiti (1858), and then returned to Germany where he founded and edited *Der Zoologische Garten*, lectured on zoology at the Senckenberg Museum, Frankfurt, and was Director, Senckenberg Naturalist Society (1859 to 1863).

Weller

Weller's Salamander *Plethodon welleri* **Walker**, 1931

Worth Hamilton Weller (c.1913–1931) collected the holotype (1930) and in the following year, after collecting more specimens of this amphibian, fell and died on Grandfather Mountain in North Carolina, just a week after graduating from high school.

Welter

Black Mountain Salamander *Desmognathus welteri* **Barbour**, 1950

Dr Wilfred August Welter (1906–1939) was a zoologist and teacher who was Professor of Biology and Head, Science Department, Morehead State Teacher's College, Kentucky (1932–1939). The University of Minnesota awarded his bachelor's degree (1926) and his master's was awarded (1927) by Indiana State College; Cornell awarded his doctorate (1932). He was killed in an automobile accident. Barbour stated that he named this salamander in memory of his old friend.

Wendy

Wendy's Forest Toad *Nectophrynoides wendyae* **Clarke**, 1988
[Alt. Uzungwe Scarp Tree Toad]

Wendy Clarke is the describer's wife.

Werler

Werler's False Brook Salamander *Pseudoeurycea werleri* Darling & **Smith**, 1954
Robber Frog sp. *Eleutherodactylus werleri* **Lynch** & Fritts, 1965 NCR
[JrSYN *Craugastor laticeps*]

John E. Werler (1922–2004) was born in Germany from where he emigrated to the USA (1926). He was appointed Reptile Keeper, Staten Island Zoo (1940). He became Curator of Reptiles, San Antonio Zoo (1946), then moved to Houston Zoo (1954) as General Curator and later became the Zoo's Director (1963–1992). He wrote *Miscellaneous notes on the eggs and young of Texan and Mexican reptiles* (1951).

Wermuth

Wermuth's Reed Frog *Hyperolius wermuthi* **Laurent**, 1961

Professor Dr Heinz Wermuth (1918–2002) was a German herpetologist. He graduated from Humboldt University, Berlin, in zoology, botany, palaeontology and chemistry. He became (1952) Head, Herpetological Section, Zoological Museum, Berlin. The construction of the Berlin Wall (1961) put him out of work, as he lived in West Berlin and the Museum was in East Berlin. He was at the Staatlisches Museum für Naturkunde, Stuttgart (1961–1983), first as Curator and retiring as Head, Division of Zoology. He lived in a small flat but kept a variety of reptiles as pets, including an American alligator – quite small at about 80 centimetres long compared with his caiman, which was over 2.5 metres long! Among his publications is, with Robert Mertens, *Liste der Rezenten Amphibien und Reptilien, Testudines, Crocodylia, Rhynchocephalia* (1977). A reptile is named after him.

Werner

Smalltongue Toad genus *Werneria* Poche, 1903
Werner's Button Frog *Cycloramphus asper* Werner, 1899
Werner's Snouted Treefrog *Scinax dolloi* Werner, 1903
Werner's Toad *Rhaebo nasicus* Werner, 1903
Werner's River Frog *Phrynobatrachus werneri* **Nieden,** 1910
Treefrog sp. **Dendropsophus werneri* **Cochran**, 1952

Professor Dr Franz Werner (1867–1939) was an Austrian explorer, zoologist and herpetologist who taught at the Natural History Museum, Vienna. Here the Director, Steindachner (q.v.), disliked him and forbade him access to the herpetological collection. He collected in North and East Africa, being in Egypt (1904) and the Sudan (1905), and made regular visits south to Uganda and west to Morocco until the outbreak of the First World War. His publications include *Amphibien und Reptilien* (1910). Twenty-eight reptiles are named after him.

* Cochran coined the name *Dendropsophus werneri* as a replacement name for *Hyla pygmaea* (Werner, 1894), which proved to be unavailable under taxonomic diktat.

Wetmore

Wetmore's Robber Frog *Eleutherodactylus wetmorei* **Cochran**, 1932

(Frank) Alexander Wetmore (1886–1978) was an American ornithologist and avian palaeontologist. He was a bird taxidermist at the Denver Museum of Natural History, Colorado (1909). He was in Puerto Rico (1911) and later travelled throughout South America for two years, investigating bird migration between continents, while working for the US Bureau of Biological Survey. He worked at the Smithsonian as Assistant Secretary (1925–1945) then Secretary (1945–1952). He conducted (1946–1966) an annual research programme in Panama. He was President of the American Ornithologists' Union (1926–1929). He wrote *A Systematic Classification for the Birds of the World* (1930), wherein he devised the Wetmore Order, a sequence of bird classification, which had widespread acceptance until very recently, but is still in use. A glacier and a canopy bridge in Panama are named after him as are numerous taxa, including eight birds, five reptiles and a mammal.

Wettstein

Wettstein's Andes Frog *Psychrophrynella wettsteini* **Parker,** 1932

Dr Otto von Wettstein Ritter von Westersheim (1892–1967) was an Austrian zoologist who was Professor of Natural History, Natural History Museum, Vienna. Originally interested in birds and mammals, he found that herpetology was the only department with a vacancy (1915). During the Second World War, he successfully kept the collection of tens of thousands of specimens, preserved in alcohol, safe in bunkers or in mines. He took over many other duties, even succeeding in publishing the *Annals of the Museum* (1941–1944). The Allies (1945) officially barred him from the

Museum, so he retired to private life and worked for the Department of Forest Protection, studying insects and their parasites. He published 60 scientific papers on herpetology; *Herpetologia Aegaea* was the result of field trips to Greece. A reptile and a bird are named after him.

Weygoldt

Weygoldt's Treefrog *Aplastodiscus weygoldti* **Cruz** & **Peixoto**, 1987

Dr Peter Johann Henning Weygoldt (b.1933) is an arachnologist and herpetologist who has worked with Peixoto, with whom he described *Scinax heyeri* (1986). He read zoology at Kiel and Freiburg Universities, being awarded his doctorate (1958) by Christian Albrechts University, Kiel for his thesis on the embryology of crustaceans. He joined the Free University of Berlin (1960) as an assistant professor. He was a research associate and Visiting Professor at Duke Marine Laboratory, North Carolina (1965–1967). It was during this period that his interest changed from crustaceans to spiders; he later (1970s) developed an interest in neo-tropical frogs. He became full Professor, Freiburg's Albert-Ludwigs University (1967), retiring as Professor Emeritus (1995).

White

White's Tree Frog *Litoria caerulea* White, 1790

[Alt. Australian Green Treefrog]

John White (c.1756–1832) served as a surgeon in the Royal Navy (1781–1820), visiting India and the West Indies before 1788, when he was appointed Chief Naval Surgeon of the 'First Fleet' – the 11 ships which took the first 1,500 colonists to Australia, among whom were 778 convicts. Despite scurvy and dysentery only 34 people died on that voyage. He stayed in Australia and, as a keen naturalist, he accompanied Governor Phillip on two journeys of exploration. A colorful character, he fought a duel with his Third Assistant, William Balmain, which left them both slightly wounded. He had three legitimate children and also a son by a convict. The boy was brought up as part of his legitimate household. He kept a journal, which he sent to a friend in London, Thomas Wilson. This was published as *Journal of a Voyage to New South Wales* (1790). White returned to London in July 1795. He was reluctant to return to New South Wales and in August 1796, faced with the alternatives of doing so immediately or of resigning his appointment, he chose to resign. A reptile is named after him.

Whitehead

Whitehead's Borneo Frog *Meristogenys whiteheadi* **Boulenger**, 1887

John Whitehead (1860–1899) was a British explorer who collected in Borneo (1885–1888), the Philippines (1893–1896) and Hainan (1899). He wrote *Explorations of Mount Kina Balu, North Borneo* (1893); he may have been the first European to reach its summit. Thirteen birds and five mammals are named after him.

Whymper

Tanti Rocket Frog *Hyloxalus whymperi* **Boulenger**, 1882

Edward Whymper (1840–1911) was an English mountaineer who was also an illustrator and an explorer. He made the first ascent of the Matterhorn (1865). He led an expedition to Greenland (1867) during which he made an important collection of fossil plants. He returned to Greenland (1872) to survey its coastline. He was on an expedition to Ecuador (1880) to study the effects of altitude sickness and reduced pressure on the human body, and made the first ascent of Chimborazo (6,267 metres / 20,561 feet). During this trip he collected natural history specimens including the eponymous frog. He went to the Canadian Rockies (1901) and made the first ascent of Stanley Peak and of another mountain, now named Mount Whymper after him. He wrote *Travels amongst the Great Andes of the Equator* (1892).

Wiens

Wiens' Robber Frog *Pristimantis wiensi* **Duellman** & **Wild**, 1993

Wiens' Water Frog *Telmatobius necopinus* Wiens, 1993

Dr John Joseph Wiens is a herpetologist and Associate Professor, Department of Ecology and Evolution, Stony Brook University, New York (2003). The University of Kansas awarded his bachelor's degree (1991); his doctorate (1995) is from the University of Texas. He was Curator, Section of Amphibians and Reptiles, Carnegie Museum of Natural History, Pittsburgh (1995–2002). He wrote *Systematics and Herpetology in the Age of Genomics* (2008).

Wightman

Wightman's Robber Frog *Eleutherodactylus wightmanae* **Schmidt**, 1920

[Alt. Wrinkled Coqui]

Mrs Margaret Wightman Schmidt was the describer's wife. He wrote that her '...loyal assistance contributed largely to the success of the work in Porto Rico in 1919.'

Wilcox

Wilcox's Frog *Litoria wilcoxii* **Günther**, 1864
[Alt. Stony-Creek Frog]

James Fowler Wilcox (1823–1881) was a collector born in England whose family moved to Sydney the year he was born. He was a naturalist, most interested in botany and taxidermy. He accompanied Captain Owen Stanley in HMS Blazer, which conveyed Sir John Franklin on his last and ill-fated expedition (1845). He was later on board HMS Rattlesnake, under Captain Stanley, as collector in the interests of the Norwich Museum (1846–1850). This expedition visited the north and northeast coasts of Australia, southern New Guinea and part of the Arafura Sea. The voyage was terminated by Stanley's death in Sydney. Wilcox was then engaged in natural history pursuits in New South Wales, conducted business in Sydney for five years, and settled in South Grafton (1857) on the Clarence River where the holotype of the frog was taken. He was Commissioner at the Melbourne Exhibition of 1866, and exhibited many specimens illustrative of the natural history of the Clarence, Richmond and Tweed Rivers.

Wild

Wild's Egg Frog *Leptodactylodon wildi* **Amiet** & Dowsett-Lemaire, 2000

Christopher Wild managed San Diego Zoo's Center for Reproduction of Endangered Species (CRES) project at Nyasoso, Cameroon. He has helped a number of expeditions collecting in southwest Cameroon. He wrote *Notes on the rediscovery and the congeneric associations of the Pfeffer's Chameleon* Chamaeleo pfefferi *(Tornier 1900) (Sauria: Chamaeleonidae) with a brief description of the hitherto unknown female of the species* (1993). A reptile is named after him.

Wilder

Blue Ridge Two-lined Salamander *Eurycea wilderae* **Dunn**, 1920
Wilder's Treefrog* *Osteopilus wilderi* Dunn, 1925

Professor Dr Inez Luanne Wilder née Whipple (1871–1929) went from being a high school teacher to Professor of Zoology, Smith College, Massachusetts (1914). She was one of the first graduate students (1901) in the Zoology Department, Smith College, where Harris Hawthorne Wilder (1864–1928) taught her and whom she married (1906). They published a paper that is considered a landmark in the field of genetics: *The Ventral Surface of the Mammalian Chiridium: With Special Reference to the Conditions Found in Man* (1904). This suggests that the development of the surfaces of the hands and feet (*chiridia*) of all mammals are to some extent similar. Her work is regarded as seminal in the development of the practice of identifying criminals through fingerprints.

* The treefrog should have been named *wilderorum* (plural) as Dunn mentions Dr and Mrs H. H. Wilder in the description.

Wiley

Glass Frog sp. *Nymphargus wileyi* Guayasamin et al., 2006

Dr Edward Orlando Wiley (b.1944) is a biologist who is Curator, Ichthyology, Natural History Museum, University of Kansas. Southwest Texas State University awarded his bachelor's degree (1966), Sam Houston State University his master's (1972), and City University of New York his doctorate (1976). He wrote *Phylogenetics. The Theory and Practice of Phylogenetic Systematics* (1981), which ties in with the words used in the etymology of the frog's description '....one of the most influential persons in the development of phylogenetic systematics and the use of the evolutionary species concept.'

Wilhelm

Wilhelm Callulops Frog *Callulops wilhelmanus* **Loveridge**, 1948

This species is named after the type locality, Mount Wilhelm, Papua New Guinea.

Wilhelmina

Wilhelmina Cochran Frog *Cochranella geijskesi* **Goin**, 1966

This frog's vernacular name comes from the type locality, the Wilhelmina Mountains in Suriname.

Williams, EE

Williams' Bright-eyed Frog *Boophis williamsi* **Guibé**, 1974

Dr Ernest Edward Williams (1914–1998) was an American herpetologist at the Museum of Comparative Zoology, Harvard, retiring as Professor Emeritus of both Biology and Zoology. He greatly expanded the holdings of the Museum's Herpetology Department. He was an expert on *Anolis* lizards, and named many new taxa. Among his publications is *The ecology of colonization as seen in the zoogeography of anoline lizards on small islands* (1969). Eleven reptiles are named after him.

Williams, FX

Williams' Bubble-nest Frog *Philautus williamsi* **Taylor**, 1922 NCR
[JrSYN *Philautus surdus*]

Dr Francis Xavier Williams (1882–1967) was an American entomologist who worked for the Hawaiian Sugar Planters' Association (1917–1949), after which he retired back to California and donated his library and collections to the California Academy of Sciences. St Ignatius College (now the University of San Francisco) awarded him a bachelor's degree (1903), as did Stanford University. The University of Kansas awarded his master's (1912) and Harvard his doctorate (1915). He was in the Galápagos Islands for 17 months (1905–1906) and collected about 4,000 insects. He wrote *Handbook of the Insects and Other Invertebrates of Hawaiian Sugar Cane Fields* (1931).

Williamson

Williamson's Marsupial Frog *Gastrotheca williamsoni* **Gaige**, 1922

Edward Bruce Williamson (1878–1933) was an entomologist and botanist. He graduated from Ohio State University (1898) and became an assistant curator at the Carnegie Museum (1898–1899). He was President of the Wells County Bank at Bluffton, Indiana (1905–1930) and was a research associate at the Museum of Zoology, University of Michigan (1930–1933). He wrote *The Dragonflies of Indiana* (1900). Twenty-one dragonflies are named after him.

Wilsmore

Wilsmore's Frog *Neobatrachus wilsmorei* **Parker**, 1940
[Alt. Goldfields Bull Frog, Plonking Frog]

The original description contains no etymology, but the frog may be named after Australian scientist Dr Norman Thomas Mortimer Wilsmore (1868–1940). He took his bachelor's degree (1890), master's (1893) and doctorate (1907) at the University of Melbourne. He did research in London (1894–1897), at Göttingen (1897–1901) and at Zurich (1901–1902). He taught at University College, London (1903–1913) before taking up the foundation chair of chemistry at the new University of Western Australia. During the First World War he was concerned with the supply of explosives for the armed forces. He returned to Perth and the University (1919–1937), becoming Vice-Chancellor (1924–1925). On his retirement he refused the title emeritus because its dictionary definition included 'having become unfit for service, worn out'.

Witte

Molo Frog *Amietia wittei* **Angel**, 1924
[Alt. De Witte's River Frog; Syn. *Afrana wittei*]
Witte's Madagascar Frog *Blommersia wittei* **Guibé**, 1974

(See **de Witte**)

Wolf

Wolf's Sticky-toed Frog *Batrachylodes wolfi* **Sternfeld**, 1920

Dr Eugen Wolf was a member of the 1909 Hanseatische Südsee-Expedition, the year in which he became Director, Senckenberg Museum, Frankfurt. He wrote the expedition's travel report (published 1909–1911). A reptile is also named after him.

Wollaston

Highland Treefrog *Litoria wollastoni* **Boulenger**, 1914

Dr Alexander 'Sandy' Frederick Richmond Wollaston (1875–1930) was a physician, naturalist and explorer. He led an expedition to Dutch New Guinea (1912–1913), having been on an earlier expedition (1910–1911) which had been deliberately misdirected by Dutch authorities. During the First World War he was a naval surgeon in East Africa. He was on an Everest expedition (1922) as physician and botanist. He wrote *An Expedition to Dutch New Guinea* (1914). While teaching, he was shot dead in his Cambridge rooms by a deranged student, who then shot and killed the policeman who had come to arrest him and finally committed suicide (1930). Two mammals and a bird are named after him.

Wolterstorff

Toad genus *Wolterstorffina* **Mertens**, 1939
Wolterstorff's Cross Frog *Oreophryne wolterstorffi* **Werner**, 1901
Wolterstorff's Newt *Hypselotriton wolterstorffi* **Boulenger**, 1905
[Alt. Yunnan Lake Newt; Syn. *Cynops wolterstorffi*]

Dr Willy Georg Wolterstorff (1864–1943) was a geologist and herpetologist. An illness deprived him of his hearing and power of speech (1871) but he learned to lip-read. He was also very myopic and so had a lonely childhood, compensating for this by collecting and keeping amphibians; they remained his major lifelong interest. He was spared seeing the total destruction of the Magdeburg Museum and all his work, including 12,000 specimens in glass jars, by the RAF (1945). A reptile and a bird are named after him.

Woodhouse

Woodhouse's Toad *Anaxyrus woodhousii* **Girard**, 1854
[Syn. *Bufo woodhousii*]

Dr Samuel Washington Woodhouse (1821–1904) was an American army surgeon, explorer and naturalist. He qualified as a physician (1847). He was on the Creek Indian Boundary Survey commanded by Abert (1849–1850). He was doctor and naturalist on the Sitgreaves Expedition to the Colorado and Zuni rivers (1851). This proved to be quite an eventful and hazardous affair, involving fighting hostile 'Indians' and Woodhouse being wounded by an arrow in his leg. He was on an expedition to Nicaragua (1853–1854) to investigate the possibility of building an Atlantic-to-Pacific railway. When he returned from Central America he practised medicine again. He resigned his commission (1856) and until 1859 was employed as a surgeon in a military prison. He became a ship's doctor (1859) on a line trading between Philadelphia and Liverpool, and after the American Civil War he practised medicine in Philadelphia. He wrote *A Naturalist in Indian Territory: The Journal of S. W. Woodhouse, 1849–50*. Two birds and a mammal are named after him.

Woodley

Woodley's Treefrog *Stefania woodleyi* **Rivero**, 1968

Dr Jeremy David Woodley was Director, Centre for Marine Sciences, University of West Indies, Jamaica. Both his bachelor's degree and doctorate were awarded by Oxford. He collected the treefrog holotype in British Guiana (Guyana) (1959). He co-wrote *Field guidebook to the modern and ancient reefs of Jamaica* (1977).

Woodworth

Woodworth's Wart Frog *Limnonectes woodworthi* **Taylor**, 1923
[Syn. *Rana woodworthi*]

Professor Harold Evans Woodworth (1894–1971) was Professor of Entomology, College of Agriculture, Los Baños, Philippines. He wrote a number of papers, mostly on crop pests, such as *A host index of insects injurious to Philippine crops III* (1922).

Worcester

Mindanao Bubble-nest Frog *Philautus worcesteri* **Stejneger**, 1905

Dean Conant Worcester (1866–1924) was an American zoologist, collector and public official. He was a member of the United States Philippine Commission (1899–1901), and thereafter served as Secretary of the Interior for the Insular Government of the Philippine Islands. He wrote *The Philippine Islands and their People* (1898). Four birds are named after him.

Wright, AH & AMA & MR

Oregon Slender Salamander *Batrachoseps wrighti* **Bishop**, 1937
Wright's Mountain Treefrog *Hyla wrightorum* **Taylor,** 1939
[Alt. Arizona Treefrog]
Toad sp. *Bufo wrighti* **Schmidt** & **Liu**, 1940 NCR
[JrSYN *Bufo gargarizans*]
Wrights' Leopard Frog *Lithobates chiricahuensis* Platz & Mecham, 1979
[Alt. Chiricahua Leopard Frog; Syn. *Rana chiricahuensis*]

Dr Albert Hazen Wright (1879–1970) was an American herpetologist. He attended Cornell, where he was awarded his doctorate in vertebrate zoology (1908) and met Anna Maria Allen (1882–1964), whom he married (1910). He worked at Cornell (1908–1946), first as an instructor then assistant professor from 1915, and professor from 1925. Dr and Mrs Wright's main interest was the study of amphibians and they co-authored *Handbook of Frogs and Toads of the United States and Canada* (1933). Two reptiles are named after him and Anna has a share in one of them.

The salamander is named after Dr. Albert Wright and 'Miss Margaret R. Wright', who was his daughter. The binomial is sometimes 'corrected' to the plural *wrightorum*. The treefrog is named after Albert and Anna.

Wright, C

Cuban Treefrog sp. *Trachycephalus wrightii* **Cope**, 1863 NCR
[JrSYN *Osteopilus septentrionalis*]

Charles Wright (1811–1885) was a botanist, teacher and collector who explored the western USA for the Pacific Railroad Company. He sent a collection of plants to Professor Asa Gray at Harvard (1844). Gray helped him get places on various surveys and expeditions, including the Boundary Survey Commission, Texas. He was on the United States North Pacific Exploring and Surveying Expedition (1853–1856) but left at San Francisco and went alone to Nicaragua. He explored in Cuba (1856–1857). He went with a United States Commission to Santo Domingo (1871). Between trips, he spent time at the Gray Herbarium, Cambridge, Massachusetts. Many plants, a bird and a reptile are named after him.

Wright, GM

Pygmy Salamander *Desmognathus wrighti* King, 1936

George Melendez Wright (1904–1936) was an American naturalist who graduated (1927) from the College of Forestry, University of California. He travelled to the Mount McKinley area of Alaska (1926) with Joseph Scattergood Dixon. After graduating he became a ranger in Yosemite National Park, and began (1929) conducting a 4-year wildlife survey program for national parks that he funded himself. The results were published (1932 & 1933) in the series *Fauna of the National Parks of the United States*. He was appointed (1934) Chief, Wildlife Division, National Park Service. He was travelling with the Superintendent of Yellowstone National Park, Roger Toll, when they were killed in an automobile accident.

Wright, JW

Robber Frog sp. *Eleutherodactylus johnwrighti* **Lynch**, 1996 NCR

[Invalid replacement name for *Pristimantis petersi*]

Dr John William Wright (b.1936) is a biologist who specializes in whiptail lizards. He studied under Professor Charles Herbert Lowe at the University of Arizona, from where he received his doctorate (1965). He was Curator of Herpetology, Los Angeles Museum (1992). A reptile is named after him.

Wucherer

Heart-tongued Frog sp. *Phyllodytes wuchereri* **Peters**, 1873

Dr Otto Edward Henry Wucherer (1820–1874) was a Portuguese-born German physician and herpetologist. He qualified at Eberhard Karls Universität Tübingen, and practised at St Bartholemew's Hospital, London and in Lisbon. He discovered the cause of the tropical disease elephantiasis. He left Europe and settled in Salvador Bahia, Brazil (1843). He wrote *Sobre a mordedura das cobras venenosas e seu tratamento* (1867). Three reptiles and a bird are named after him.

Wuensche

Wuensche's Wrinkled Ground Frog *Platymantis wuenscheorum* **Günther**, 2006

Rosi and Jochen Wuensche of Blumberg Village, near Berlin, are described by Günther as 'his good friends'. He writes that he also intends to commemorate other friends with whom he and his wife have collected and celebrated many birthdays and New Year's Eves – Sybille & Claus Pohle, Uschi & Heinz Streiffeler, Heidi & Ulli Thieke, and Ulla & Conny Wiegank. This rather seems to be stretching the etymology!

Wurstner

African Water Frog sp. *Petropedetes juliawurstnerae* Barej et al., 2010

Julia Wurstner is a photographer who collected in Cameroon with Barej (2007).

X

Xucaneb

Xucaneb Robber Frog *Craugastor xucanebi* **Stuart**, 1941

Stuart wrote: 'The second species I offer to Xucaneb, the mountain king of the Alta Verapaz, on whose domain I trespassed to collect.' It appears that Stuart 'personalized' Cerro Xucaneb, the highest mountain peak in that area of Guatemala.

Y

Yakusin

Southern Torrent Frog *Petropedetes yakusini* **Channing**, Moyer & **Howell**, 2002

[Syn. *Arthroleptides yakusini*]

This is not an eponym despite appearances that the frog is named after someone called Yakusin. As Allan Channing explained: 'The species name is derived from 'ya kusini', which is Swahili for 'from the south'. This species is found south of the range of *Petropedetes martiensseni*, with which it was previously confused. There is also a slight pun as the discovery that this was a new species was made by a scientist 'from the south' (me).'

Yalden

Yalden's Tree Frog *Leptopelis yaldeni* **Largen**, 1977

[Alt. Grassland Forest Treefrog]

Dr Derek William Yalden (b.1940) is a zoologist who received his doctorate (1966) in London. He worked (1965–2005) at the School of Biological Sciences, University of Manchester. He was on the Great Abbai Expedition (1968) and has written many scientific papers and longer works, including *The History of British Mammals* (1999). He co-authored *The Lives of Bats* (1975) and *Mammals of Ethiopia* (7 parts 1976–1997), which involved five collecting trips to Ethiopia. A mammal is also named after him.

Yang

Yang's Salamander *Hynobius yangi* Kim, Min & **Matsui**, 2003

[Alt. Kori Salamander]

Dr Suh-yung Yang is a biologist and herpetologist who was (1972) at the Biology Department, University of California, San Diego. He is now a professor at the Department of Biology, Inha University, Incheon, South Korea. He co-wrote *Genetic Relationships among Six Korean* Rana *Species (Amphibia; Ranidae) Based on the Mitochondrial Cytochrome b Gene* (2008).

Yasun

Yasuni Broad-headed Treefrog *Osteocephalus yasuni* Ron & Pramuk, 1999

Yasuni Rocket Frog *Hyloxalus yasuni* Páez-Vacas, **Coloma** & Santos, 2010

These frogs are named after the Yasuní National Park, Ecuador, where the holotypes were collected.

Yatsu

Yatsu's Salamander *Hynobius yatsui* Oyama, 1947

Dr Naohide Yatsu (1877–1947) was Professor of Zoology, Tokyo Imperial University. He had studied at Columbia University, New York, and had worked at the Biological Station in Naples. He wrote *An Experimental Study on the Cleavage of the Ctenophore Egg* (1910). He was the describer's supervisor.

Ye

Ye's Spiny-vented Frog *Quasipaa yei* Chen, Qu & Jiang, 2002

[Alt. Henan Spiny Frog; Syn. *Nanorana yei*]

Professor Ye Chang-yuan (b. 1938) is a biologist and herpetologist at the Herpetology Department, Chengdu Institute of Biology, Chinese Academy of Sciences, Chengdu. She wrote *Rare and Economic Amphibians of China* (1993).

Yehudah

Lycian Salamander sp. *Lyciasalamandra yehudahi* Göçmen & Akman, 2012

Dr Yehuda Leopold Werner (b.1931) was born in Germany and emigrated as a child to the Palestine Protectorate (Israel). He graduated in zoology (1956) from the Hebrew University, which awarded his doctorate (1961). He was an assistant in the Department of Zoology (1953), becoming Curator of Amphibians and Reptiles (1973) and Director, Life Sciences Collections (1990). He was Professor, Alexander Silberman Institute of Life Sciences, Hebrew University, Jerusalem, and is now Emeritus. He wrote in English, Hebrew, German, Dutch, Hungarian and Czech. His works include *A Guide to our Reptiles and Amphibians* (1995). Two reptiles are named after him.

Yong

Yong's Grainy Frog *Kalophrynus yongi* **Matsui**, 2009

Dr Yong Hoi-Sen is Professor Emeritus at the University of Malaya, which awarded all his degrees – a bachelor's in botany and zoology (1964), a master's in zoology (1965) and a doctorate in genetics and zoology (1968). His entire career was spent at the University of Malaya, as student demonstrator (1964–1965), Tutor, Department of Zoology (1965–1968), Lecturer in Biological Science (1968–1974), Associate Professor of Genetics and Cellular Biology (1974–1986), Professor of Zoology (1986–1998), Senior Research Fellow (1999) and Research Associate (2000–2007). He wrote *Malaysian butterflies: An introduction* (1983). Various other species including a beetle are named after him.

Young

Chang Mai Caecilian *Ichthyophis youngorum* **Taylor**, 1960

Gordon's Warty Frog *Theloderma gordoni* Taylor, 1962 [Alt. Gordon's Bug-eyed Frog]

Oliver Gordon Young (b.1927) and his family were associated with Chiang Mai, Thailand. Young was normally known by his second name and was a third generation American Baptist missionary. He was born in Burma (currently Myanmar) and was evacuated (1942) upon the Japanese invasion of Burma, only returning to Asia (1953) to settle in Chiang Mai. He knew the Lahu tribe from his youth, contacted them again and became recognized by them as an efficient hunter, and wrote a book about those experiences. Taylor collected with him and other members of his family and reported Harold Young (Oliver's father) as having collected lizards and snakes for him; indeed he named a lizard *Lygosoma haroldyoungi* after him. Gordon combined missionary work with helping his father work for the CIA in providing translation services of intercepted Chinese messages (1950s–1960s). He wrote *The hill tribes of Northern Thailand*. Oliver collected the holotype of the frog. The caecilian's etymology reads: 'This species is named for the family of Oliver Gordon Young of Chiang Mai, Thailand. Mr. Young has been untiring in his efforts to assist my work.'

Yuracare

Sehuencas Water Frog *Telmatobius yuracare* **De la Riva**, 1994

The Yuracaré people are an indigenous group in Bolivia.

Yustiz

Yacambu Poison Frog *Mannophryne yustizi* **La Marca**, 1989

Yustiz's Rain Frog *Pristimantis yustizi* Barrio-Amorós & Chacón-Ortiz, 2004

Professor Enrique Elias Yustiz (d.2003) was an entomologist and herpetologist at the Department of Entomology and Zoology, Faculty of Agronomy, University Centro-Occidental 'Lisandro Alvarado', Barquisimeto, Venezuela. He was interested in spiders and scorpions. He wrote *Aspectos biogeográficos de la herpetofauna de la cuenca hidrográfica del Río Turbio (Estado Lara, Venezuela)* (1996).

Z

Zaparo

Zaparo's Poison Frog *Allobates zaparo* **Silverstone**, 1976
[Alt. Sanguine Poison Frog]

The Zaparo are a small tribe of indigenous people in Ecuador, where the holotype was collected.

Zetek

Zetek's Treefrog *Isthmohyla zeteki* **Gaige**, 1929
Zetek's Golden Frog *Atelopus zeteki* **Dunn**, 1933
[Alt. Panamanian Golden Frog, Cerro Campano Stubfoot Toad]

Professor James Zetek (1886–1959) was an American entomologist. He graduated from the University of Illinois and was employed by the US Government in the Panama Canal Zone (1911–1953). He became Professor of Biology and Hygiene, National Institute of Panama (1916) and the first Director of the Smithsonian Tropical Research Institute on Barro Colorado Island, Gatun Lake, Panama Canal (1923). He wrote *Report on reptiles from Barro Colorado Island* (1950). A reptile is also named after him.

Zeus

Zeus' Robber Frog *Eleutherodactylus zeus* **Schwartz**, 1958

In Greek mythology, Zeus was supreme ruler of the gods on Mount Olympus. The name was perhaps inspired by its large size (for a member of this genus), but Schwartz does not give a definitive etymology.

Zhang

Zhang's Horned Toad *Xenophrys zhangi* **Ye** & **Fei**, 1992
[Syn. *Megophrys zhangi*]
Jilin Clawed Salamander *Onychodactylus zhangyapingi* Che et al., 2012

Ya-ping Zhang (b.1965) was Director of Kunming Institute of Zoology, Chinese Academy of Sciences (2005-2012), and is currently Vice-President of the Chinese Academy of Sciences.

Zhao

Zhao's Odorous Frog *Odorrana zhaoi* Li, Lu & **Rao**, 2008

Dr Zhao Da-Yu is President of, and Professor at, Shenyang Normal University, China. He was a strong supporter of the survey in Xizang which collected the holotype.

Zhao Ermi

Liaoning Clawed Salamander *Onychodactylus zhaoermii* Che, Poyarkov & Yan, 2012

Dr Zhao Er-mi (see under **Ermi Zhao**).

Zheng

Zhangcun Frog *Rana zhengi* **Zhao**, 1999

Zheng Mingquan collected the holotype.

Zimmer

Zimmer's Screeching Frog *Arthroleptis zimmeri* **Ahl,** 1925
Rummassala Shrub Frog *Pseudophilautus zimmeri* Ahl, 1927
Zimmer's Cross Frog *Oreophryne zimmeri* Ahl, 1933

Dr Carl Wilhelm Erich Zimmer (1873–1950) was a German zoologist. He was Director of the Zoological Museum and Professor of Zoology, University of Berlin (1926–1943). Ahl worked under Zimmer. He wrote *Anleitung zur Beobachtung der vogelwelt* (1917). Several crustaceans, his specialty, are named after him as is a reptile.

Zimmerman

Zimmerman's Robber Frog *Pristimantis zimmermanae* **Heyer** & Hardy, 1991

Dr Barbara Lewis Zimmerman (b.1954) is a herpetologist and ecologist at the Faculty of Forestry, University of Toronto, Canada. She is Director of the Kayapo Project in Brazil.

Zimmermann, H & E

Zimmermann's Poison Frog *Ranitomeya variabilis* Zimmermann & Zimmermann, 1988

Helmut and Dr Elke Zimmermann are German herpetologists who work closely together. He is President, Foundation for Tropical Nature and Species Conservation Team, Stuttgart, Germany and she is Director and Professor at Deutsches Primatenzentrum Göttingen. In addition to writing together, they have published independently: Helmut (1978) *Tropical Frogs – Care and Breeding* (1978), and Elke (1986) *Breeding Terrarium Animals: Amphibians and Reptiles Care – Behavior – Reproduction*. Despite the position of the apostrophe we believe both are referred to in the vernacular name.

Zippel

Rocket Frog sp. *Aromobates zippeli* Barrio-Amorós & Santos, 2012

Dr Kevin C Zippel is a biologist who is programme director at *Amphibian Ark* which, in the words of the etymology, does '...endless work in spreading the alarming message of the global decline of frogs....' In recognition of this, he was honoured in the binomial. Cornell University awarded his bachelor's degree (1994) and the University of Florida his doctorate (2000). He worked at the Department of Herpetology, Wildlife Conservation Society / Bronx Zoo (1999) and in the National Amphibian Conservation Center, Detroit Zoo (1995–2000). He wrote *Climate change and amphibians* (2010).

Zipper

Zipper's Madagascar Frog *Mantidactylus zipperi* Vences & **Glaw**, 2004

Zipper is the nickname of Claus Zimmermann, a supporter of BIOPAT. Viola Zimmermann arranged to have this species named after him.

Zograf

Frog sp. *Rana zografi* **Terentjev**, 1922 NCR
[JrSYN *Rana dybowskii*]

G N Zograf was the describer's first zoology teacher.

Zoila

Robber Frog sp. *Pristimantis zoilae* Mueses-Cisneros, 2007

Zoila Rosa Cisneros is the author's mother.

Zolitschka

Zolitschka Madagascar Frog *Mantidactylus zolitschka* **Glaw** & Vences, 2004

The family of Joachim and Lissi Zolitschka of Wesendorf, Germany are honoured for their contribution to conservation through the BIOPAT progamme.

Zorro

Zorro Bubble-nest Frog *Pseudophilautus zorro* Manamendra-Arachchi & Pethiyagoda, 2005
[Alt. Gannoruwa Shrub Frog]

Zorro, the masked hero and subject of a number of Hollywood films, first appeared in Johnston McCulley's (1919) book *The Curse of Capistrano*. The frog was so named because it has a mask-like eye band.

Zug

Zug's Robber Frog *Eleutherodactylus zugi* **Schwartz**, 1958

Dr George Robert Zug (b.1938) joined the Smithsonian (1968) as Assistant Curator, becoming Curator of Amphibians and Reptiles (1975) then Curator Emeritus Director, Department of Systematic Biology – Vertebrate Zoology. His master's degree was awarded by the University of Florida (1963) and his doctorate (1968) by the University of Michigan. Among his publications is *Data resources for Tropical Asian Dry Forest Amphibians and Reptiles* (2012). Two reptiles are named after him and a third after him and his wife, Patricia.

Zugmayer

Baloch Green Toad *Pseudepidalea zugmayeri* **Eiselt** & Schmidtler, 1973

Professor Dr Erich Johann Georg Zugmayer (1879–1938) was an Austrian explorer, zoologist, ichthyologist and herpetologist at the Bavarian State Zoological Collection, Munich. He visited Iceland (1902). He explored the area around Lake Urmia, Persia (Iran), and collected in Tibet, Ladakh and Baluchistan (then part of India, now of Pakistan). He published *Bericht über eine Reise in Westtibet* (1909). Two reptiles and two birds are named after him.

Zweifel

Ranid frog genus *Zweifelia* **Dubois**, 1992 NCR
[In *Lithobates*]
Zweifel's Rainforest Frog *Cophixalus cryptotympanum* Zweifel, 1956
Zweifel's Snouted Frog *Xenorhina similis* Zweifel, 1956
Zweifel's Papua Frog *Barygenys exsul* Zweifel, 1963
Zweifel's Big-eyed Treefrog *Litoria zweifeli* **Tyler**, 1967
Tributary Caecilian *Oscaecilia zweifeli* **Taylor**, 1968
Zweifel's Frog *Lithobates zweifeli* Hillis, **Frost** & **Webb**, 1984
[Syn. *Rana zweifeli*]
Zweifel's Frog *Cophixalus zweifeli* **Davies** & **McDonald**, 1998
Zweifel's Snouted Frog *Xenorhina zweifeli* Kraus & **Allison**, 2002
Zweifel's Barking Frog *Craugastor augusti fuscofemora* Zweifel, 1956

Dr Richard George Zweifel (b.1926) is an American herpetologist and leading expert on Australasian frogs. He joined the AMNH, New York (1954), served as Chairman of the Department of Herpetology (1968–1980), retired as Curator Emeritus (1989) and now lives in Arizona continuing his studies at the Museum's research station at Portal. Among his many publications he co-wrote *Encyclopedia of Reptiles & Amphibians* (1998). Seven reptiles are named after him.

Confusingly, two unrelated frogs have been given the common name 'Zweifel's Frog', and two members of the genus *Xenorhina* have been called 'Zweifel's Snouted Frog'.

BIBLIOGRAPHY

American Museum Novitates. New York, 1921–.

Amphibia-Reptilia Leiden. 1980–.

Annales des Sciences Naturelles, Zoologie et Paléontologie. Paris, 1864–1915.

Annals and Magazine of Natural History. London, 1841–1966.

Annals of the Transvaal Museum. Pretoria, 1911–.

Annuaire du Musée Zoologique de l'Académie Imperiale des Sciences. St. Petersburg, 1896–1932.

Beolens, B. & M. Watkins. (2003) Whose Bird? London, Christopher Helm.

Beolens, B., M. Watkins & M. Grayson. (2009) The Eponym Dictionary of Mammals. Baltimore, Johns Hopkins University Press.

Beolens, B., M. Watkins & M. Grayson. (2011) The Eponym Dictionary of Reptiles. Baltimore, Johns Hopkins University Press.

Bericht über die Senckenbergischer Naturforschende Gesellschaft. Frankfurt, 1869–1896.

Bocourt, M. F. (1870) Description de quelques sauriens nouveaux originaires de l'Amérique méridionale. *Nouvelles archives du Muséum d'Histoire Naturelle,* Paris, 2nd ser., 6.

Bonner zoologische Beiträge. Bonn, 1950–.

Bulletin du Museum National d'Histoire Naturelle. Paris, 1895–1970.

Bulletin of the American Museum of Natural History. New York, 1881–.

Bulletin of the Museum of Comparative Zoology. Harvard University, Cambridge, Mass., 1863–.

Channing, A. (2001) Amphibians of Central and Southern Africa. Ithaca, Cornell University Press.

Copeia. New York, 1913–.

Daudin, F. M. (1802–1803) Histoire naturelle, générale et particulière des reptiles. Vols. 4–6. Paris, Dufart.

Dictionary of National Biography. (1992) Oxford, Oxford University Press.

Duméril, A. M. C. & G. Bibron (1839) Erpétologie générale ou Histoire naturelle complète des reptiles. Vol. 5. Paris, Roret Fain et Thunot.

Duméril, A. M. C., G. Bibron & A. H. A. Duméril. (1854) Erpétologie générale ou Histoire naturelle complète des reptiles. Vol. 7, pt. 1. Paris.

Dumerilia. Paris, 1994–.

Frank, N. & E. Ramus. (1995) A Complete Guide to Scientific and Common Names of Reptiles and Amphibians of the World. Pottsville, Pa., Ramus NG Publishing.

Glaw, F. & M. Vences. (1994) A Fieldguide to the Amphibians and Reptiles of Madagascar. Cologne, Vences and Glaw.

Gotch, A. F. (1995) Latin Names Explained—A Guide to the Classification of Reptiles, Birds and Mammals. London, Blandford Press.

Hamadryad. Madras, 1976–2000.

Herpetologica. Lawrence, Kans., 1936–.

The Herpetological Journal. London, 1985–.

Herpetological Monographs. Austin, 1982–.

Herpetological Review. Lawrence, Kans., 1967–.

Jornal de Sciencias Mathematicas, Physicas e Naturaes. Lisbon, 1866–1910.

Journal of Herpetology. Lawrence, Kans., 1968–.

Journal of the Bombay Natural History Society. Bombay, 1886–.

Mémoires de l'Academie Impériale des Sciences. St. Petersburg, 1809–1830.

Memoirs of the Queensland Museum. Brisbane, 1912–.

Ménétries, E. (1832) Catalogue raisonné des objets de zoologie recueillis dans un voyage au caucase et jusqu'aux frontières actuelles de la Perse. St. Petersburg, L'Académie Imperiale des Sciences.

Mittellungen aus dem Naturhistorisches Museum. Hamburg, 1882–1908.

Mocquard, F. (1894) Reptiles nouveaux ou insuffisamment connus de Madagascar. Compte rendu sommaire de séances de la Société Philomathique de Paris, 17.

Monatsberichte der Königlichen. Preussischen. Akademie der Wissenschaften zu Berlin. Berlin, 1856–1881.

Murray, J. A. (1892) The Zoology of Beloochistan and Southern Afghanistan (Reptiles and Batrachia). Bombay, Bombay Education Society Press.

Nouvelles archives du Muséum de l'Histoire Naturelle. Paris, 1865–1914.

Petranka, J.W. (1998) Salamanders of the United States and Canada. Washington, Smithsonian Institution Press.

Philippine Journal of Science. Manila, 1906–1994.

Proceedings of the Zoological Society of London. London, 1831–1965.

Publicaciones ocasionales del Museo de Ciencias Naturales. Caracas, Venezuela.

Ramírez Leyton, G. & D. Pincheira Donoso. (2005) Fauna del altiplano y desierto de Atacama. Phrynosaura Ediciones.
Records of the Albany Museum. Grahamstown, 1903–1935.
Records of the Australian Museum. Sydney, 1890–.
Records of the South Australian Museum. Adelaide, 1918–2002.
Records of the Western Australian Museum. Perth, 1974–.
Russian Journal of Herpetology. Moscow, 1994–.
Salamandra: Zeitschrift für Herpetologie und Terrarienkunde. Frankfurt am Main, 1965–.
Sitzungsberichte der Kaiserlichen Akademie der Wissensschaften. Vienna, 1850–1947.
Sterling, Keir B., Richard P. Harmond, George A. Cevasco & Lorne F. Harmond. (1997) Biographical Dictionary of American and Canadian Naturalists and Environmentalists. Westport, Conn., Greenwood Press.
The Fauna of British India, Ceylon and Burma, Reptilia and Amphibia, Volume III, Serpentes. (1943) London, Taylor and Francis.
Thomson, Keith S. (1995) HMS Beagle—The Ship That Changed the Course of History. New York, W. W. Norton and Co.
Titschack, E., ed. (1942) Beiträge zur Fauna Perus. Vol. 2. Hamburg.
Tropical Zoology. Florence, 1988–.
Tyler, M.J. & Knight, F. (2011) (Revised edition) Field Guide to the Frogs of Australia (revised edition). Melbourne, CSIRO Publishing.
Vestnik zoologii. Kiev, 1967–.
Videnskabelige Meddelelser fra den Naturhistoriske Forening i Kjöbenhavn. Copenhagen, 1849–1912.
Wynne, Owen E. (1969) Biographical Key—Names of Birds of the World—to Authors and Those Commemorated. Privately published.
Zoologischer Anzeiger. Jena, 1878–.
Zootaxa. Auckland, 2001–.

CPSIA information can be obtained at www.ICGtesting.com
Printed in the USA
BVOW09*0549071015

421235BV00003B/18/P